Operator Theory
Advances and Applications
Vol. 104

Nonselfadjoint Operator Algebras, Operator Theory, and Related Topics

The Carl M. Pearcy Anniversary Volume

Hari Bercovici
Ciprian Foias
Editors

Springer Basel AG

Editors:
Hari Bercovici and Ciprian I. Foias
Department of Mathematics
Indiana University
Bloomington, IN 47405-4301
USA

1991 Mathematics Subject Classification 47-xx, 46L05, 46L30

A CIP catalogue record for this book is available from the
Library of Congress, Washington D.C., USA

Deutsche Bibliothek Cataloging-in-Publication Data
Nonselfadjoint operator algebras, operator theory, and related topics : the
Carl M. Pearcy anniversary volume / Hari Bercovici ; Ciprian Foias ed. –
Basel ; Boston ; Berlin : Birkhäuser, 1998
(Operator theory ; Vol. 104)
ISBN 978-3-0348-9771-6 ISBN 978-3-0348-8779-3 (eBook)
DOI 10.1007/978-3-0348-8779-3

Originally published by Birkhäuser Verlag Basel Switzerland in 1998
Softcover reprint of the hardcover 1st edition 1998
Printed on acid-free paper produced from chlorine-free pulp. TCF ∞
Cover design: Heinz Hiltbrunner, Basel

ISBN 978-3-0348-9771-6

9 8 7 6 5 4 3 2 1

CONTENTS

EDITORIAL PREFACE vii

PORTRAIT OF CARL PEARCY viii

CARL M. PEARCY: A BIOGRAPHICAL SKETCH ix

PUBLICATIONS OF CARL M. PEARCY xiii

E. A. AZOFF & L. DING,
A good side to non–reflexive transformations 1

H. BERCOVICI, C. FOIAS, & A. TANNENBAUM,
On skew Toeplitz operators, II 23

H. BERCOVICI & D. VOICULESCU,
Regularity questions for free convolution 37

S. W. BROWN & E. KO,
Operators of Putinar type 49

R. E. CURTO & L. A. FIALKOW, *Flat extensions of positive moment matrices: Relations in analytic or conjugate terms* 59

R. G. DOUGLAS & G. MISRA,
Geometric invariants for resolutions of Hilbert modules 83

G. R. EXNER & I. B. JUNG,
Some multplicities for contractions with Hilbert-Schmidt defect 113

D W. HADWIN & D. R. LARSON, *Strong limits of similarities* 139

A. LAMBERT, *L^p multipliers and nested Sigma-algebras* 147

W. S. LI & D. TIMOTIN, *On isometric intertwining liftings* 155

M. MARSALLI, *The predual of a type I Von Neumann algebra* 169

M. MARTIN & N. SALINAS, *The canonical complex structure of flag manifolds in a C^*-algebra* 173

P. S. MUHLY & B. SOLEL,
An algebraic characterization of boundary representations 189

A. OCTAVIO & S. PETROVIC,
Joint spectrum and nonisometric functional calculus 197

CONTENTS

[illegible]

Editorial Preface

This volume is dedicated to Carl Pearcy on his 60th birthday. It collects recent contributions to operator theory, nonselfadjoint operator algebras, measure theory, and the theory of moments by several of the leading specialists in those areas. Many of the contributors are collaborators or former students of Carl Pearcy, and the variety of the topics bears witness to the wide range of his work and interests.

The editors were helped by many in the compilation of this volume. Srdjan Petrovic helped compile Carl's list of publications, while Arlen Brown and George Exner helped in writing the biographical and mathematical sketch. The work of many referees, who must remain anonymous, was very valuable. Israel Gohberg suggested that we publish this volume in the distinguished series *Operator Theory: Advances and Applications*.

The whole volume was expertly typeset by Elena Fraboschi. We wish to extend to all of these people our heartfelt thanks.

Carl M. Pearcy

Carl M. Pearcy: A Biographical Sketch

H. Bercovici & C. Foias

Carl Mark Pearcy, Jr. was born on August 23, 1935 in Beaumont, Texas. He was the eldest of two sons of Carl Mark Pearcy, Sr., and Carrie Edith (Tilbury) Pearcy. His family moved to Galveston in 1940, and that is where Carl resided until he left home to attend the university. Carl entered Texas A.&M. University in 1951 at age sixteen. He graduated (B.S.) in 1954 and, again (M.S.) in 1956. Carl entered Rice Institute (later Rice University) as a candidate for the Ph.D. in Mathematics in 1956. The degree was conferred in 1960 under the direction of Arlen Brown. In 1957–58 Carl was a fellow in the mathematics department at the University of Chicago, and in 1959–60 he held an appointment as Assistant Professor at Texas A.&M. He was then appointed a post-doctoral fellow at Rice for the year 1960–61. From 1961 to 1963 Carl was employed at the Houston research center of Humble Oil Co. In 1963 he was appointed Assistant Professor of Mathematics at the University of Michigan, where he remained until his "retirement" in the rank of Full Professor in 1990, at which time he accepted reappointment in the mathematics department of Texas A.&M. Carl was an A.P. Sloan foundation fellow from 1966 to 1968, and was the main speaker at two conferences sposored by the Conference Board of Mathematical Sciences of the National Science Foundation. These conferences were at Bucknell University in 1975, and Arizona State University in 1984.

While Carl appreciates good mathematics regardless of the field, most of his mathematical career was closely intertwined with the development of modern operator theory. There is practically no area of operator theory where Carl did not contribute. Many of these contributions were fundamental or trendsetting. We will highlight just some of the most influential of these contributions. The numbers in brackets refer to Carl's publication list.

The characterization of commutators, i.e., operators of the form $T = AB - BA$ on a Hilbert space was posed as a problem by P. Halmos. It was known that operators of the form $zI + C$, with z a scalar and C a compact operator, are not commutators. Carl Pearcy and Arlen Brown proved in [15] that all other operators are in fact commutators. This is one of the earliest deep results pertaining to arbitrary operators on a Hilbert space. The methods used in the proof contain the germs of the development of the general approximation theory of Hilbert space operators. In a different direction, this work led to developments in the theory of operator algebras. Indeed, it became possible to characterize commutators in various kinds of C^* and von Neumann algebras, and this also yielded results about the radical structure of such algebras.

Another circle of ideas introduced by Halmos is the study of quasitrangular operators. These operators were introduced in relation with the invariant subspace problem. Quasitriangularity is a geometric condition related with the behavior of an operator on a chain of finite dimensional subspaces. Quite surprizingly, Carl Pearcy and Ronald Douglas found in [35] a necessary condition for quasitriangularity expressed purely in spectral (and Fredholm index) terms. This was again a very general result, applying to a wide class of Hilbert space operators subject only to mild conditions, and it raised the possibility that quasitriangularity might be entirely characterized in spectral terms. This characterization was indeed realized (by Apostol, Foias, and Voiculescu) when the condition discovered by Pearcy and Douglas was shown to be sufficient as well. These developments led to the search for other relations between geometric and spectral properties of operators, and there is now a vast body of work on this subject. It should be mentioned that techniques from Carl's work on commutators also turned out to be inspiring in the study of quasitriangularity and related questions.

The invariant subspace problem, already present in quasitriangularity, was one of Carl's long lasting preoccupations. One line of research is illustrated by his work on the Lomonosov technique (see [50], [55], and [56]). Another line was inaugurated by his joint work with S. Brown and B. Chevreau [65] in which the existence of invariant subspaces is deduced from a richness condition on the spectrum. Carl recognized very early that the methodology of this work can yield much more. In fact it was seen through the work of Carl (and his collaborators and students) that a great variety of results about the structure of the invariant subspaces of an operator can be derived from factorization methods. To give an interesting example, it is well known that the invariant subspaces of the usual shift operator S (multiplication by the variable on the Hardy space H^2) are classified by the inner functions defined in the unit disk. Thus, for instance, if M is one of these invariant subspaces then $S|M$ is unitarily equivalent to S. It was known that the corresponding multiplication operator T on the Bergman space of the unit disk is not as tractable, and very few facts were known about the invariant subspaces of T. As an application of the factorization techniques, it was shown in [82] and [83] that $T \oplus T \oplus T \oplus \cdots$ can be realized as the compression of T to a semiinvariant subspace. This result has amazing consequences for the structure of invariant subspaces of T. The result has redirected the work of researchers in function theory, who wanted to approach it from a more classical point of view, and were in fact able to derive independently some of its consequences. The number of results in this general area was such that it required a separate entry in the 1991 Mathematics Subject Classification: 47D27 Dual Algebras. We cannot conclude the discussion of dual algebras without mentioning the definitive result of [95]: every contraction whose spectrum contains the unit circle has nontrivial invariant subspaces.

Besides these areas, to which Carl Pearcy contributed in a major way, there are many areas which he kept alive by getting his students and colleagues interested. One such area is the similarity problem for polynomially bounded operators on a Hilbert space (which was eventually solved in the negative by Gilles Pisier).

Carl has influenced research in Operator Theory not only directly through his work, but also indirectly through his numerous students, many of whom became leaders in the field. Carl's courses introduced his students to all significant aspects of modern operator theory. Much of the material in these courses is contained in the — still unpublished — Part II of his book on operator theory written jointly with A. Brown.

Carl's contributions to mathematics were not limited to his research and teaching. For many years Carl organized sessions at the annual meetings of the American Mathematical Society. He helped launch successfully the *Journal of Operator Theory*, and he supervised for several years the operations of *Mathematical Reviews*. He maintained close relations with his colleagues in Eastern Europe at a time when they had little access to current mathematical publications, and only sporadic occasions to travel outside their countries. Carl ultimately encouraged or helped many of his Romanian colleagues to relocate in the United States; some came as senior mathematicians (C. Foias, C. Apostol, D. Voiculescu), some as students (H. Bercovici, G. Popescu, R. Gadidov, A. Ionescu). All of them owe a debt of gratitude to Carl.

For more than thirty years, Carl Pearcy was one of the most influential personalities in operator theory. His strength as a mathematician, his dedication as a teacher, and his warmth as a friend helped him achieve this elevated status. Besides that, he was one of the few who had the vision of what the most important feasible problems at each stage in pure operator theory were. He not only worked on those problems, but he always succeded in making other people work on them as well. For that reason, the whole operator theory community wishes him the best for many more fruitful years in the profession.

Publications of Carl M. Pearcy

Research Papers

1. C. Pearcy, *A complete set of unitary invariants for* 3×3 *complex matrices*, Trans. Amer. Math. Soc. **104** (1962), 425–429.
2. C. Pearcy, *A complete set of unitary invariants for operators generating finite* W^**-algebras of type I*, Pacific J. Math. **12** (1962), 1405–1416.
3. C. Pearcy, W^**-algebras with a single generator*, Proc. Amer. Math. Soc. **13** (1962), 831–832.
4. C. Pearcy, *On convergence of alternating direction procedures*, Numer. Math. **4** (1962), 172–176.
5. D. Deckard and C. Pearcy, *On matrices over the ring of continuous complex valued functions on a Stonian space*, Proc. Amer. Math. Soc. **14** (1963), 322–328.
6. D. Deckard and C. Pearcy, *Another class of invertible operators without square roots*, Proc. Amer. Math. Soc. **14** (1963), 445–449.
7. C. Pearcy, *On unitary equivalence of matrices over the ring of continuous complex-valued functions on a Stonian space*, Canad. J. Math. **15** (1963), 323–331.
8. J. Douglas, Jr. and C. Pearcy, *On convergence of alternating direction procedures in the presence of singular operators*, Numer. Math. **5** (1963), 175–184.
9. D. Deckard and C. Pearcy, *On algebraic closure in function algebras*, Proc. Amer. Math. Soc. **15** (1964), 259–263.
10. C. Pearcy, *On certain von Neumann algebras which are generated by partial isometries*, Proc. Amer. Math. Soc. **15** (1964), 393–395.
11. C. Pearcy, *Entire functions on infinite von Newmann algebras of type I*, Michigan Math. J. **11** (1964), 1–7.
12. D. Deckard and C. Pearcy, *On continuous matrix-valued functions on a Stonian space*, Pacific J. Math. **14** (1964), 857–869.
13. A. Brown and C. Pearcy, *Structure theorem for commutators of operators*, Bull. Amer. Math. Soc. **70** (1964), 779–780.

14. C. Pearcy, *On commutators of operators on Hilbert space*, Proc. Amer. Math. Soc. **16** (1965), 53–59.
15. A. Brown and C. Pearcy, *Structure of commutators of operators*, Ann. of Math. **82** (1965), no. 2, 112–127.
16. A. Brown, P. R. Halmos and C. Pearcy, *Commutators of operators on Hilbert space*, Canad. J. Math. **17** (1965), 695–708.
17. R. G. Douglas and C. Pearcy, *Spectral theory of generalized Toeplitz operators*, Trans. Amer. Math. Soc. **115** (1965), 433–444.
18. D. Deckard and C. Pearcy, *On unitary equivalence of Hilbert-Schmidt operators*, Proc. Amer. Math. Soc. **16** (1965), 671–675.
19. A. Brown and C. Pearcy, *Multiplicative commutators of operators*, Canad. J. Math. **18** (1966), 737–749.
20. A. Brown and C. Pearcy, *Commutators in factors of type III*, Canad. J. Math. **18** (1966), 1152–1160.
21. A. Brown and C. Pearcy, *Spectra of tensor products of operators*, Proc. Amer. Math. Soc. **17** (1966), 162–166.
22. J. Douglas, Jr., A. O. Garder and C. Pearcy, *Multistage alternating direction methods*, SIAM J. Numer. Anal. **3** (1966), 570–581.
23. C. Pearcy, *An elementary proof of the power inequality for the numerical radius*, Michigan Math. J. **13** (1966), 289–291.
24. D. Deckard and C. Pearcy, *On rootless operators and operators without logarithms*, Acta Sci. Math. (Szeged) **28** (1967), 1–7.
25. C. Pearcy and D. Topping, *Sums of small numbers of idempotents*, Michigan Math. J. **14** (1967), 453–465.
26. C. Pearcy and J. R. Ringrose, *Trace-preserving isomorphisms in finite operator algebras*, Amer. J. Math. **90** (1968), 444–455.
27. A. Brown, C. Pearcy and D. Topping, *Commutators and the strong radical*, Duke Math. J. **35** (1968), 853–859.
28. A. Brown and C. Pearcy, *Operators of the form $PAQ - QAP$*, Canad. J. Math. **20** (1968), 1353–1361.
29. R. G. Douglas and C. Pearcy, *On a topology for invariant subspaces*, J. Funct. Anal. **2** (1968), 323–341.
30. R. G. Douglas, P. S. Muhly and C. Pearcy, *Lifting commuting operators*, Michigan Math. J. **15** (1968), 385–395.
31. R. G. Douglas and C. Pearcy, *A characterization of thin operators*, Acta Sci. Math. (Szeged) **29** (1968), 295–297.
32. C. Pearcy and D. Topping, *Commutators and certain II_1-factors*, J. Funct. Anal. **3** (1969), 69–78.
33. R. G. Douglas and C. Pearcy, *Von Neumann algebras with a single generator*, Michigan Math. J. **16** (1969), 21–26.

34. D. Deckard, R. G. Douglas and C. Pearcy, *On invariant subspaces of quasitriangular operators*, Amer. J. Math. **91** (1969), 637–647.
35. R. G. Douglas and C. Pearcy, *A note on quasitriangular operators*, Duke Math. J. **37** (1970), 177–188.
36. R. G. Douglas and C. Pearcy, *On the spectral theorem for normal operators*, Math. Proc. Cambridge Philos. Soc. **68** (1970), 393–400.
37. C. Pearcy and D. Topping, *On commutators in ideals of compact operators*, Michigan Math. J. **18** (1971), 247–252.
38. A. Brown and C. Pearcy, *Compact restrictions of operators*, Acta Sci. Math. (Szeged) **32** (1971), 271–282.
39. A. Brown, C. Pearcy and N. Salinas, *Ideals of compact operators on Hilbert space*, Michigan Math. J. **18** (1971), 373–384.
40. R. G. Douglas and C. Pearcy, *Hyperinvariant subspaces and transitive algebras*, Michigan Math. J. **19** (1972), 1–12.
41. A. Brown and C. Pearcy, *Compact restrictions of operators. II*, Acta Sci. Math. (Szeged) **33** (1972), 161–164.
42. C. Pearcy and N. Salinas, *An invariant-subspace theorem*, Michigan Math. J. **20** (1973), 21–31.
43. C. Pearcy and N. Salinas, *Compact perturbations of seminormal operators*, Indiana Univ. Math. J. **22** (1972; 1973), 789–793.
44. R. G. Douglas, C. Pearcy and N. Salinas, *Hyperinvariant subspaces via topological properties of lattices*, Michigan Math. J. **20** (1973), 109–113.
45. A. Brown, C. Pearcy and N. Salinas, *Perturbations by nilpotent operators on Hilbert space*, Proc. Amer. Math. Soc. **41** (1973), 530–534.
46. R. G. Douglas and C. Pearcy, *Invariant subspaces of non-quasitriangular operators*, Proc. Conf. Operator Theory, (Dalhousie Univ., Halifax, N.S.), Lecture Notes in Math., Vol. 345, Springer, Berlin, 1973, pp. 13–57.
47. C. Pearcy and N. Salinas, *Operators with compact self-commutator*, Canad. J. Math. **26** (1974), 115–120.
48. C. Pearcy and N. Salinas, *Finite-dimensional representations of separable C^*-algebras*, Bull. Amer. Math. Soc. **80** (1974), 970–972.
49. C. Pearcy, J. R. Ringrose, and N. Salinas, *Remarks on the invariant-subspace problem*, Michigan Math. J. **21** (1974), 163–166.
50. C. Pearcy and A. L. Shields, *A survey of the Lomonosov technique in the theory of invariant subspaces*, Topics in operator theory, Math. Surveys, No. 13, Amer. Math. Soc., Providence, R.I., 1974, pp. 219–229.
51. C. Foias and C. Pearcy, *A model for quasinilpotent operators*, Michigan Math. J. **21** (1974), 399–404.
52. C. Pearcy and N. Salinas, *The reducing essential matricial spectra of an operator*, Duke Math. J. **42** (1975), no. 3, 423–434.

53. C. Pearcy and N. Salinas, *Finite dimensional representations of C^*-algebras and the reducing matricial spectra of an operator*, Rev. Roumaine Math. Pures Appl. **20** (1975), no. 5, 567–598.

54. C. Foias, C. Pearcy and D. Voiculescu, *The staircase representation of biquasitriangular operators*, Michigan Math. J. **22** (1975), no. 4, 343–352.

55. H. W. Kim, C. Pearcy and A. L. Shields, *Rank-one commutators and hyperinvariant subspaces*, Michigan Math. J. **22** (1975), no. 3, 193–194.

56. H. W. Kim, C. Pearcy and A. L. Shields, *Sufficient conditions for rank-one commutators and hyperinvariant subspaces*, Michigan Math. J. **23** (1976), no. 3, 235–243.

57. C. Apostol, C. Foias and C. Pearcy, *Quasiaffine transforms of compact perturbations of normal operators*, INCREST preprint series in Mathematics, No. 13 (1976).

58. C. Pearcy and N. Salinas, *Extensions of C^*-algebras and the reducing essential matricial spectra of an operator*, K-theory and operator algebras (Proc. Conf., Univ. Georgia, Athens, Ga., 1975), Lecture Notes in Math., vol. 575, Springer, Berlin, 1977, pp. 96–112.

59. C. Apostol, C. Pearcy and N. Salinas, *Spectra of compact perturbations of operators*, Indiana Univ. Math. J. **26** (1977), no. 2, 345–350.

60. A. Brown and C. Pearcy, *Jordan loops and decompositions of operators*, Canad. J. Math. **29** (1977), no. 5, 1112–1119.

61. C. Foias, C. Pearcy and D. Voiculescu, *Biquasitriangular operators and quasisimilarity*, Linear spaces and approximation (Proc. Conf., Math. Res. Inst., Oberwolfach, 1977), Lecture Notes in Biomath., vol. **21**, Springer, Berlin New York, 1978, pp. 47–52.

62. D. Deckard, C. Foias and C. Pearcy, *Compact operators with root vectors that span*, Proc. Amer. Math. Soc. **76** (1979), no. 1, 101–106.

63. C. Apostol, C. Foias and C. Pearcy, *That quasinilpotent operators are norm-limits of nilpotent operators revisited*, Proc. Amer. Math. Soc. **73** (1979), no. 1, 61–64.

64. H. W. Kim and C. Pearcy, *Subnormal operators and hyperinvariant subspaces*, Illinois J. Math. **23** (1979), no. 3, 459–463.

65. S. Brown, B. Chevreau and C. Pearcy, *Contractions with rich spectrum have invariant subspaces*, J. Operator Theory **1** (1979), no. 1, 123–136.

66. C. Pearcy and A. Shields, *Almost commuting matrices*, J. Funct. Anal. **33** (1979), no. 3, 332–338.

67. H. W. Kim, R. Moore and C. Pearcy, *A variation of Lomonosov's theorem*, J. Operator Theory **2** (1979), no. 1, 131–140.

68. H. W. Kim and C. M. Pearcy, *Extensions of normal operators and hyperinvariant subspaces*, J. Operator Theory **3** (1980), no. 2, 203–211.

69. C. Foias, C. Pearcy and B. Sz.-Nagy, *The functional model of a contraction and the space L^1*, Acta Sci. Math. (Szeged) **42** (1980), no. 1–2, 201–204.

70. H. W. Kim, R. L. Moore and C. Pearcy, *A variation of Lomonosov's theorem. II*, J. Operator Theory **5** (1981), no. 2, 283–287.

71. B. Chevreau, C. Pearcy and A. L. Shields, *Finitely connected domains G, representations of $H^\infty(G)$, and invariant subspaces*, J. Operator Theory **6** (1981), no. 2, 375–405.

72. H. Bercovici, C. Foias, C. Pearcy and B. Sz.-Nagy, *Functional models and extended spectral dominance*, Acta Sci. Math. (Szeged) **43** (1981), no. 3–4, 243–254.

73. C. Foias, C. Pearcy and B. Sz.-Nagy, *Contractions with spectral radius one and invariant subspaces*, Acta Sci. Math. (Szeged) **43** (1981), no. 3–4, 273–280.

74. C. Bosch, C. Hernandez, E. de Oteyza and C. Pearcy, *Spectral pictures of functions of operators*, J. Operator Theory **8** (1982), no. 2, 391–400.

75. C. Apostol, H. Bercovici, C. Foias and C. Pearcy, *Quasiaffine transforms of operators*, Michigan Math. J. **29** (1982), no. 2, 243–255.

76. H. Bercovici, C. Foias, J. Langsam and C. Pearcy, *(BCP)-operators are reflexive*, Michigan Math. J. **29** (1982), no. 3, 371–379.

77. C. Foias and C. Pearcy, *(BCP)-operators and enrichment of invariant subspace lattices*, J. Operator Theory **9** (1983), no. 1, 187–202.

78. H. Bercovici, C. Foias and C. Pearcy, *Dilation theory and systems of simultaneous equations in the predual of an operator algebra. I*, Michigan Math. J. **30** (1983), no. 3, 335–354.

79. H. Bercovici, B. Chevreau, C. Foias and C. Pearcy, *Dilation theory and systems of simultaneous equations in the predual of an operator algebra. II*, Math. Z. **187** (1984), no. 1, 97–103.

80. H. Bercovici, C. Foias, C. Pearcy and B. Sz.-Nagy, *Factoring compact operator-valued functions*, Acta Sci. Math. (Szeged) **48** (1985), no. 1–4, 25–36.

81. H. Bercovici, C. Foias and C. Pearcy, *Factoring trace-class operator-valued functions with applications to the class $\mathbb{A}_{\aleph_0}$*, J. Operator Theory **14** (1985), no. 2, 351–389.

82. C. Apostol, H. Bercovici, C. Foias and C. Pearcy, *Invariant subspaces, dilation theory, and the structure of the predual of a dual algebra. I*, J. Funct. Anal. **63** (1985), no. 3, 369–404.

83. C. Apostol, H. Bercovici, C. Foias and C. Pearcy, *Invariant subspaces, dilation theory, and the structure of the predual of a dual algebra. II*, Indiana Univ. Math. J. **34** (1985), no. 4, 845–855.

84. B. Chevreau and C. Pearcy M., *Sur le probleme du sous-espace invariant pour les contractions [On the invariant subspace problem for contractions]*, C. R. Acad. Sci. Paris Ser. I Math. **301** (1985), no. 15, 735–738.

85. C. Apostol, H. Bercovici, C. Foias and C. Pearcy, *On the theory of the class $\mathbb{A}_{\aleph_0}$ with applications to invariant subspaces and the Bergman shift operator*, Advances in invariant subspaces and other results of operator theory (Timişoara and Herculane, 1984), Oper. Theory: Adv. Appl., vol. **17**, Birkhäuser, Basel Boston, Mass., 1986, pp. 43–49.

86. H. Bercovici, C. Foias and C. Pearcy, *On the reflexivity of algebras and linear spaces of operators*, Michigan Math. J. **33** (1986), no. 1, 119–126.

87. B. Chevreau and C. Pearcy, *On the structure of contraction operators with applications to invariant subspaces*, J. Funct. Anal. **67** (1986), no. 3, 360–379.

88. B. Chevreau and C. Pearcy, *Growth conditions on the resolvent and membership in the classes $\mathbb{A}$ and $\mathbb{A}_{\aleph_0}$*, J. Operator Theory **16** (1986), no. 2, 375–385.

89. H. Bercovici, C. Foias and C. Pearcy, *A spectral mapping theorem for functions with finite Dirichlet integral*, J. Reine Angew. Math. **366** (1986), 1–17.

90. B. Chevreau and C. Pearcy, *Membership in $\mathbb{A}_{\aleph_0}$ and invariant subspaces*, Operators in indefinite metric spaces, scattering theory and other topics (Bucharest, 1985), Oper. Theory: Adv. Appl., vol. **24**, Birkhäuser, Basel Boston, MA, 1987, pp. 41–49.

91. J. Barria, H. W. Kim and C. Pearcy, *On reflexivity of operators*, J. Math. Anal. Appl. **126** (1987), no. 2, 316–323.

92. S. Brown, B. Chevreau and C. Pearcy, *Sur le probleme du sous-espace invariant pour les contractions [On the invariant subspace problem for contractions]*, C. R. Acad. Sci. Paris Ser. I Math. **304** (1987), no. 1, 9–12.

93. B. Chevreau, G. Exner and C. Pearcy, *Sur la reflexivité des contractions de l'espace hilbertien [On the reflexivity of contraction operators in Hilbert space]*, C. R. Acad. Sci. Paris Ser. I Math. **305** (1987), no. 4, 117–120.

94. B. Chevreau and C. Pearcy, *On the structure of contraction operators. I*, J. Funct. Anal. **76** (1988), no. 1, 1–29.

95. S. Brown, B. Chevreau and C. Pearcy, *On the structure of contraction operators. II*, J. Funct. Anal. **76** (1988), no. 1, 30–55.

96. H. Bercovici, C. Foias and C. Pearcy, *Two Banach space methods and dual operator algebras*, J. Funct. Anal. **78** (1988), no. 2, 306–345.

97. B. Chevreau, G. Exner and C. Pearcy, *Structure theory and reflexivity of contraction operators*, Bull. Amer. Math. Soc. (N.S.) **19** (1988), no. 1, 299–301.

98. B. Chevreau and C. Pearcy, *On Sheung's theorem in the theory of dual operator algebras*, Special classes of linear operators and other topics (Bucharest, 1986), Oper. Theory: Adv. Appl., vol. **28**, Birkhäuser, Basel-Boston, MA, 1988, pp. 43–49,.
99. B. Chevreau, G. Exner and C. Pearcy, *On the structure of contraction operators. III*, Michigan Math. J. **36** (1989), no. 1, 29–62.
100. B. Chevreau and C. Pearcy, *On common noncyclic vectors for families of operators*, Houston J. Math. **17** (1991), no. 4, 637–650.
101. B. Chevreau and C. Pearcy, *The isolated Fredholm spectrum in the theory of dual algebras*, Houston J. Math. **17** (1991), no. 3, 395–403.
102. C. Pearcy and S. Petrovic, *On polynomially bounded weighted shifts*, Houston J. Math. **20** (1994), no. 1, 27–45.
103. V. Paulsen, C. Pearcy and S. Petrovic, *On centered and weakly centered operators*, J. Funct. Anal. **128** (1995), no. 1, 87–101.
104. B. Chevreau, G. Exner, and C. Pearcy, *Boundary sets for a contraction*, J. Operator Theory **34** (1995), no. 2, 347–380.
105. W. S. Li and C. Pearcy, *On polynomially bounded operators. II*, Houston J. Math. **21** (1995), no. 4, 719–733.

Books

[1] C. Pearcy (Ed.), *Topics in Operator Theory,*, Mathematical Surveys, No.13, Amer. Math. Soc., 1974.

[2] A. Brown and C. Pearcy, *Introduction to operator theory. I. Elements of functional analysis*, Graduate Texts in Mathematics, No. 55, Springer Verlag, New York Heidelberg, 1977, pp. xiv+474.

[3] C. Pearcy, *Some recent developments in operator theory*, CBMS, Regional Conference Series in Mathematics, No. 36, American Mathematical Society, Providence, R.I., 1978, pp. v+73.

[4] H. Bercovici, C. Foias and C. Pearcy, *Dual algebras with applications to invariant subspaces and dilation theory*, CBMS Regional Conference Series in Mathematics, 56, American Mathematical Society, Providence, R.I., 1985, pp. xi+108.

[5] A. Brown and C. Pearcy, *An introduction to analysis*, Graduate Texts in Mathematics, vol. **154**, Springer Verlag, New York, 1995, pp. viii+297.

Operator Theory:
Advances and Applications, Vol. 104

A Good Side to Non–Reflexive Transformations

EDWARD A. AZOFF AND LIFENG DING

To Carl Pearcy in honor of his 60th birthday

ABSTRACT. Let V be a vector space over an infinite scalar field and suppose that $a \in L(V)$. We show that, as a strictly closed algebra of transformations, $\operatorname{alg}\operatorname{lat} a$ is generated by its own rank one members and the original transformation a. To do so, we obtain a concrete description of the rank one members of $\operatorname{alg}\operatorname{lat} a$ when a is locally algebraic.

Applications include a unified approach to earlier reflexivity results and an explanation of the phenomenon that among locally nilpotent transformations, it is the *non–reflexive* ones which always admit reasonable Jordan canonical forms.

1. INTRODUCTION

Early on in our linear algebra experience, we are taught to regard invariant subspaces as "good"—operators with invariant subspaces admit block triangular representations and this is thought of as a first step in developing a structure theory for such operators. The Jordan canonical form theorem is a notable success of this point of view; basic blocks come from complemented invariant subspaces and the internal structure of these blocks reflects chains of further invariant subspaces.

An operator a is said to be reflexive if each operator leaving invariant all a–invariant subspaces of the underlying domain space must belong to the (suitably closed) operator algebra generated by a. From the perspective of the preceding paragraph, such operators should be particularly "simple". Indeed, many of the bounded Hilbert space operators we understand best— normal and Toeplitz operators—are reflexive. More recently, in Theorem 10.6 of [3], H. Bercovici, C. Foias, and C. Pearcy have shown that every weighted shift whose norm coincides with its spectral radius must be reflexive.

We are, of course, a long way from any global structure theory for bounded Hilbert space operators—reflexive or otherwise. The situation is different in the purely algebraic setting when the underlying vector space V is not equipped with any topology. In fact, if $\dim V = \aleph_0$, a theorem of Ulm featured in Section 11 of I. Kaplansky's monograph [12] provides a complete set of similarity invariants for the locally algebraic operators on V. Since recent work of M. B. Delaï [6] characterizes the reflexive transformations on such spaces, we are in the position of deciding whether the reflexive transformations are indeed simpler than the non–reflexive ones. The answer turns out to be no in a surprisingly strong sense.

Let V be a vector space over an infinite field. Given $a \in L(V)$, we write $\operatorname{alg}\operatorname{lat} a$ for the algebra of all transformations on V which leave invariant each a–invariant subspace of V. Following Delaï, we call $a \in L(V)$ *reflexive* if each member of $\operatorname{alg}\operatorname{lat} a$ must belong to the *strictly closed* algebra generated by a. A surprising result due to D. Hadwin [10] states that every transformation which is *not* locally algebraic must be reflexive. This is a *qualitative* blow against the simplicity of reflexive transformations.

The coup de grâce comes from Delaï's analysis of locally nilpotent transformations. Indeed, Theorem 6 of [6] shows that if such a transformation fails to be reflexive, some quotient space $\ker a^{n+1}/\ker a^n$ must be one–dimensional. In the last section of the present paper, we apply techniques from Kaplansky's monograph to show that this condition forces a to be a direct sum of (finite and infinite) backward shifts. In particular, the Ulm invariants of these transformations stabilize at the first infinite ordinal. By contrast, there are (necessarily reflexive) locally nilpotent transformations whose Ulm invariants take arbitrarily long to stabilize: this is *quantitative* evidence for the superiority of *non*–reflexive transformations.

The body of this paper attempts to explain this mystery. In operator theory, associating rank one operators with an object of study is often a good way to understand the structure of that object. To cite a few sample references, the analysis of nest–like algebras in [9], [13] is facilitated by the fact that they *contain* many rank one members, while the study of dual algebras in [3] makes extensive use of those members of their preduals which have rank one representatives. More to the point of the present paper, an algebra of bounded Hilbert space operators is reflexive if and only if rank one operators are weak*–total in its preannihilator.

Returning to the setting of a locally nilpotent transformation a, the algebra it generates contains few if any rank one members. $\operatorname{alg}\operatorname{lat} a$ can, however, be richer in rank one members. In fact, it is a corollary of our main result that this happens precisely when a is non–reflexive. Evidently, $\operatorname{alg}\operatorname{lat} a$ is more "closely related" to a than any putative "preannihilator", whence rank one members of the former yield more information about the structure of a than rank one members of the latter.

The structure of the balance of the paper is as follows. In Section 2, we review the basic definitions in the setting of linear subspaces of $L(V)$; this highlights the contrast between general and singly generated subalgebras of $L(V)$—a contrast which does not exist in the Hilbert space setting. Section 3 includes a self–contained exposition of earlier work from [5], [10], and [6], and a brief comparison with the situation for bounded Hilbert space operators.

In Sections 4 and 5, we concentrate on primary transformations: $a \in L(V)$ having the property that $p(a)$ is locally nilpotent for some irreducible polynomial p. A simple set of "block invariants" for such transformations is introduced and studied in Section 4. These are used to concretely characterize the rank one members of $\operatorname{alg}\operatorname{lat} a$ in Theorem 5.4. In Theorem 5.8 we obtain our main result that $\operatorname{alg}\operatorname{lat} a$ is always generated by its rank one members together with a. In particular, this approach explains the anomaly that the characterization of when a is reflexive is most complicated when p is linear and a is algebraic.

The final section of the paper fills in the details of our earlier discussion concerning the relative simplicity of non-reflexive local nilpotents as compared with their reflexive counterparts.

2. Reflexive closures

Throughout this paper, V will denote a vector space over an infinite field F; we write $L(V)$ for the algebra of all linear transformations on V. As usual, V^* denotes the dual space of V. Given $v \in V$ and $\phi \in V^*$, we write $v \otimes \phi$ for the transformation defined by

$$(v \otimes \phi)(y) = \phi(y)v \ , \quad y \in V.$$

Every rank one transformation takes this form and when the underlying space V is finite–dimensional (so that trace makes sense on $L(V)$), we have $\operatorname{tr}(v \otimes \phi) = \phi(v)$.

Definition 2.1. Let S be a linear subspace of $L(V)$ and suppose $b \in L(V)$.

(1) Given a positive integer k, we say that b belongs to the *k–reflexive closure* of S if for each sequence $v_1, \ldots, v_k$ in V, there is a transformation $a \in S$ satisfying $av_i = bv_i$ for $i = 1, \ldots, k$.
(2) b is said to belong to the *strict closure* of S if it belongs to the k–reflexive closure of S for each $k \in \mathbb{N}$.
(3) The k–reflexive and strict closures of S are denoted by $\operatorname{ref}_k S$ and $\operatorname{str} S$ respectively. S is said to be *k–reflexive* if $S = \operatorname{ref}_k S$.

Reference to k is suppressed when it is 1.

When $A \subset L(V)$ is an identity containing algebra, $b \in \operatorname{ref} A$ if and only if b leaves invariant each A–invariant subspace of V, i.e., $\operatorname{ref} A = \operatorname{alg}\operatorname{lat} A$ in the usual notation. A useful consequence of this observation is that all "closure operations" of Definition 2.1 respect direct products (see the proof of Proposition 3.4 below).

For any S, we have the chain of inclusions

$$S \subset \operatorname{str}(S) \subset \ldots \subset \underset{3}{\operatorname{ref}}(S) \subset \underset{2}{\operatorname{ref}}(S) \subset \underset{1}{\operatorname{ref}}(S). \tag{2.1}$$

Example 2.2. Subject to the obvious restriction, it is possible to specify the positions of proper inclusion in Display 2.1 arbitrarily.

Proof. The obvious restriction is that if $\operatorname{ref}_{k+1} S = \operatorname{ref}_k S$ for all $k \geq k_0$, then $\operatorname{str} S = \operatorname{ref}_k S$ for all such k as well.

Taking S to be the space of all finite rank transformations on an infinite–dimensional V, we see that it is possible to arrange proper inclusion at the leftmost position of (2.1), with equality elsewhere.

Given $k \in \mathbb{N}$, take $V = F^{k+1}$ and fix an invertible transformation $b \in L(V)$ with non–zero trace. (The identity transformation will do unless $k+1$ is a multiple of the characteristic of F.) Take

$$S \equiv \{a \in L(V) : \operatorname{tr}(ba) = 0\}.$$

We check that the identity transformation, e, belongs to $\operatorname{ref}_k S$. Indeed, given any set of k vectors in V, we can find a non–zero $\phi \in V^*$ which annihilates them. Next, apply invertibility of b to choose $v \in V$ with $\phi(bv) = \operatorname{tr} b$. Then $a \equiv e - v \otimes \phi$ belongs to S and agrees with e on the given set of k vectors.

Since S has codimension one in $L(V)$, we conclude that $\operatorname{ref}_k S = L(V)$. Since $\operatorname{ref}_{k+1} S = S$, we get proper inclusion at the kth position from the right in (2.1), with equality elsewhere.

The proof is completed by taking various direct products of the concrete spaces we have constructed so far. □

The time–honored trick of embedding linear spaces of transformations in algebras of upper–triangular block matrices (e.g. Proposition 3.9 of [1]) shows that arbitrariness in Display 2.1 persists even if we restrict attention to commutative algebras of transformations. The situation is quite different for the singly generated algebras which we will study in the balance of the paper.

3. Singly generated algebras

Definition 3.1. *Fix $a \in L(V)$ and regard the underlying vector space V as a module over the polynomial ring $F[x]$.*

(1) *We write $\operatorname{pol} a$ for the subalgebra of $L(V)$ consisting of polynomials in a; we also write A for this algebra.*
(2) *The strict and reflexive closures of $\operatorname{pol} a$ are denoted by $\operatorname{str} a$ and $\operatorname{ref} a$, respectively.*
(3) *We write $\langle E \rangle$ for the submodule of V generated by the subset E of V. As usual, $\langle \{v\} \rangle$ is abbreviated to $\langle v \rangle$.*
(4) *The minimal polynomial of the restriction of a to $\langle E \rangle$ is denoted by p_E.*

In other words, $\langle E \rangle$ is the smallest subspace of V which contains E and is invariant under a; in particular, $\langle v \rangle = Av$.

Part (1) of Theorem 3.3 below is due to L. Brickman and P. A. Fillmore [4]; part (3) is the result of Hadwin [10] mentioned in the Introduction. The proofs given here are essentially those of the original authors. The simple fact recorded in Proposition 3.2 is implicit in all proofs concerning reflexivity; fancier versions of this principle can be found in [7] and [8].

Proposition 3.2. *Let $a \in L(V)$ and suppose that $b \in \operatorname{ref} a$ satisfies $bv = 0$ for some $v \in V$. Then $bw = 0$ whenever $\langle v \rangle \cap \langle w \rangle = \{0\}$ and p_w divides p_v.*

PROOF. Choose a polynomial q satisfying $b(v+w) = q(a)(v+w)$. Since $\langle v \rangle$ and $\langle w \rangle$ are invariant under a, we must have $0 = bv = q(a)v$ and $bw = q(a)w$. Thus p_v divides q, so p_w must also divide q, whence $bw = 0$. □

Theorem 3.3. *Let $a \in L(V)$ and write A for the subalgebra* $\operatorname{pol} a$ *of $L(V)$ consisting of polynomials in a.*

(1) $\operatorname{str} A = A' \cap \operatorname{ref} A$.

(2) *In particular, Display 2.1 collapses to*

$$A \subset \operatorname{str} A = \operatorname{ref}_2 A \subset \operatorname{ref} A. \tag{3.1}$$

(3) *If a is not locally algebraic, then* $A = \operatorname{ref} A$.

PROOF.

(1) The opposite inclusion being obvious, suppose that $b \in A' \cap \operatorname{ref} A$, and let E be a finite subset of V. As a finitely generated module over a principal ideal domain, we can express $\langle E \rangle = \langle v_1 \rangle \oplus \cdots \oplus \langle v_n \rangle$ with the minimal polynomial p_{v_1} of the restriction of a to $\langle v_1 \rangle$ coinciding with the minimal polynomial p_E of $a|_{\langle E \rangle}$. Choose a polynomial q with $bv_1 = q(a)v_1$ and write $c = b - q(a)$. Then c annihilates v_1 by construction, whence c vanishes on $\langle v_1 \rangle$ because it commutes with a. On the other hand, Proposition 3.2 implies that c vanishes on $\langle v_2 \rangle, \ldots, \langle v_n \rangle$. We conclude that $b = q(a)$ on $\langle E \rangle$. The arbitrariness of E shows that $b \in \operatorname{str} A$, as desired.

(2) This follows from (1) and the obvious inclusion $\operatorname{ref}_2 A \subset A' \cap \operatorname{ref} A$.

(3) Choose a vector $v \in V$ which separates polynomials in the sense that $p(a)v = 0$ only when $p = 0$. Suppose that $b \in \operatorname{ref} A$ and E is a finite subset of V containing v. Then we can express $\langle E \rangle = \langle v_1 \rangle \oplus \ldots \oplus \langle v_n \rangle$ with v_1 separating polynomials. Choose a polynomial q with $bv_1 = q(a)v_1$ and write $c = b - q(a)$. We have $cv_1 = 0$ by construction, so Proposition 3.2 tells us that c vanishes on $\langle v_2, \ldots, v_n \rangle$.
Necessarily $cav_1 = p(a)v_1$ for some polynomial p. Given a scalar λ, we have $p(a)v_1 = c(a - \lambda e)v_1$ whence $p(a)v_1 \in (a - \lambda e)Av_1$. Since v_1 separates polynomials, this implies that $x - \lambda$ divides p. Because the underlying field is infinite, we conclude $cav_1 = 0$, whence an induction argument implies that c vanishes on $\langle v_1 \rangle$. Thus b agrees with $q(a)$ on $\langle E \rangle$. Recalling that the vector $v \in E$ separates polynomials, we see that q is independent of E, so $b \in A$ as desired. □

The following result, basically Theorem 3 of [6], adapts the primary decomposition of [5] to vector spaces of arbitrary dimension.

Proposition 3.4. *Suppose $a \in L(V)$ is locally algebraic and write A for the subalgebra of $L(V)$ it generates. For each irreducible polynomial $p \in F[x]$, set*

$$V_p = \bigcup_{n \in \mathbb{N}} \ker p^n(a).$$

Write a_p for the restriction of a to V_p, and A_p for the algebra generated by a_p.

(1) $V = \bigoplus_{p \text{ irr}} V_p$.
(2) $\operatorname{ref} A = \prod_{p \text{ irr}} \operatorname{ref} A_p$.
(3) $\operatorname{str} A = \prod_{p \text{ irr}} \operatorname{str} A_p$.
(4) *In particular,* $\operatorname{str} A$ *is reflexive if and only if* $\operatorname{str} A_p$ *is reflexive for each irreducible p.*

PROOF. We make the usual distinction between direct sums and direct products. Thus (1) means that the $\{V_p : p \text{ irreducible}\}$ are mutually independent and span V; these are consequences of the Chinese Remainder Theorem for polynomials. Since each V_p is invariant under A, each *individual* member $b \in \operatorname{ref} A$ is decomposable in the sense that $bv = \sum b_p v_p$ for each vector $v \in V$. Whether one thinks of this as a direct sum (reflecting the structure of V) or a direct product (reflecting the fact that b_p may fail to vanish for any p) is a matter of taste; we take the former course below. On the other hand, since A is a linear space, it would be wrong to think of A as a subset of the direct *sum* $\oplus A_p$. Thus, the most that can be said in general is $A \subset \prod A_p$ and of course this inclusion can be quite proper. In any case, the $\subset$ inclusions of (2) and (3) follow from Definition 2.1.

To establish the reverse inclusion for (2), suppose $\bigoplus b_p \in \prod \operatorname{ref} A_p$. Given a vector $v \in V$, choose a non-zero polynomial q satisfying $q(a)v = 0$ and write $q = \prod_{i=1}^{n} p_i^{k_i}$ for its prime factorization. Then we can write $v = \sum_{i=1}^{n} v_i$ where v_i belongs to $\ker p_i^{k_i} \subset V_{p_i}$. By definition of reflexive closure, for each i, there is a polynomial q_i such that $q_i(a)$ agrees with b_{p_i} on v_i. Apply the Chinese Remainder Theorem to get a polynomial r simultaneously satisfying $r \equiv q_i \mod p_i^{k_i}$ for $i = 1, \ldots, n$. Thus $r(a)$ agrees with $\bigoplus b_p$ on v, and we have shown that $\prod \operatorname{ref} A_p \subset \operatorname{ref} A$.

Replacing ref by ref_2 and v by a pair of vectors v, w in the preceding paragraph, we conclude that $\operatorname{ref}_2 A = \prod \operatorname{ref}_2 A_p$, whence (3) follows by Theorem 3.3(2).

Part (4) is a direct consequence of (2) and (3). □

The next result (from [10]) settles the question of when the left containment in Display 3.1 is proper. Following Delaï, we therefore focus on the right containment in the basic Definition 3.6.

Corollary 3.5. *If a is locally algebraic but not algebraic, then A is properly contained in* $\operatorname{str} A$*; otherwise they are equal.*

PROOF. If a is not locally algebraic, then $A = \operatorname{str} A$ by Theorem 3.3(3); since every finite–dimensional subspace of $L(V)$ is strictly closed, the same conclusion holds if a is algebraic.

Conversely, suppose that a is locally algebraic and $A = \operatorname{str} A$. For each irreducible p, the operator $p(a_p)$ is locally nilpotent on V_p, so the infinite series $\sum_{i \in \mathbb{N}} p^i(a_p)$ belongs to $\operatorname{str} A_p$. Since $A_p = \operatorname{str} A_p$, this series must in fact be finite, and we see that each a_p is algebraic. Since A has countable dimension, the equation $A = \operatorname{str} A = \prod \operatorname{str} A_p$ implies that V_p vanishes for all but finitely many p. Thus $a = \bigoplus a_p$ is algebraic. □

Definition 3.6. (*Delaï*) *An individual linear transformation $a \in L(V)$ is reflexive* if $\operatorname{str} a = \operatorname{ref} a$.

We conclude this section with a brief comparison of our purely algebraic setting with its topological analogue. For definiteness, let H be a Hilbert space. Attention is restricted to *closed* subspaces of H and *bounded* operators on H. Thus the definition of reflexive closure becomes

$$\operatorname{ref} S = \{b \in B(H) : \ bx \in \overline{Sx} \quad \text{for each } x \in H\}$$

where the bar indicates closure in the norm topology. The intersection $\bigcap_{k \in \mathbb{N}} \operatorname{ref}_k S$ is the closure of S in the *strong operator topology* on $B(H)$; this is denoted by $\operatorname{sot} S$. Display 2.1 remains valid in the Hilbert space setting; actually it can be lengthened by considering various other topologies on $B(H)$.

The Baire Category Theorem shows that if $a \in B(H)$ is non–algebraic, then $\operatorname{pol} a$ can not even be uniformly closed. In particular, there is no chance of having $\operatorname{pol} a$ reflexive for such operators. The universal convention is thus to call an individual operator $a \in B(H)$ *reflexive* if the **sot**–closed algebra generated by a is reflexive. Similar reasoning applies to the framing of Definition 3.6.

Theorem 3.3 says that singly generated subalgebras of $L(V)$ come close to being reflexive. The analoguous statement fails rather spectacularly in the Hilbert space setting. In [14], W. Wogen showed how to embed arbitrary subspaces of $B(H)$ in singly generated operator algebras, and he applied this technique to provide counterexamples to Parts (1) and (2). In fact [2], points of proper inclusion in Display 2.1 can still be arbitrarily specified for singly generated subalgebras of $B(H)$.

As for Theorem 3.3(3), a category argument (Theorem 15 of [12]) shows that every locally algebraic operator in $B(H)$ must be algebraic. In particular, the complementary subset of $B(H)$ has many non–reflexive members—we are thrilled to find new classes of them with any non-trivial invariant subspaces at all.

4. PRIMARY TRANSFORMATIONS

We return to the algebraic setting where the underlying vector space V is not equipped with any topology. Following Kaplansky [12], a transformation $a \in L(V)$

is called *primary* if $p(a)$ is locally nilpotent for some irreducible polynomial p. The transformation $a = 0$ is allowed, for which we take $p(x) = x$. Proposition 3.4 reduces the study of locally algebraic transformations to the primary ones and we concentrate on them in this section.

Answers to concrete questions should be phrased in terms of "simple" quantities. In this section, we discuss a sequence of "computable" numbers associated with primary transformations. Corollary 4.7 provides the bridge between these numbers and the concrete description of the rank one members of ref a of Theorem 5.4. The preparatory material in 4.1–4.6 adapts various classical results to our purposes.

Definition 4.1. *The block invariants* of a primary transformation $a \in L(V)$ are defined by

$$\mathcal{B}_k(a) = \dim\,[\ker p(a) \cap \operatorname{ran} p^k(a)], \quad k \in \mathbb{N}.$$

The sequence $\{\mathcal{B}_k(a)\}$ is clearly non–increasing, and invariant under similarity.

Example 4.2. All operators in this example act on finite–dimensional spaces.

Write s_n for the (backward) shift acting on F^{n+1}. Then we have $\mathcal{B}_k(s_n) = 1$ for $k \leq n$ and $\mathcal{B}_k(s_n) = 0$ otherwise.

Since dimensions, kernels, and ranges respect direct sums, we see that $\mathcal{B}_k$ counts the number of blocks of size at least $k + 1$ in the Jordan Canonical Form of any nilpotent transformation.

More generally, if $p(a)$ is nilpotent for some irreducible polynomial p, then $\mathcal{B}_k(a)$ can be computed by counting the blocks in the Classical Canonical Form of a having size at least $(\deg p)(k + 1)$ and multiplying the result by $\deg p$.

In particular, block invariants form a complete set of similarity invariants for primary transformations acting on finite–dimensional spaces. They do not, however, distinguish between $\bigoplus s_n$ and $\bigoplus s_{2n}$. For such tasks, one needs to consider dimensions of quotients of the spaces appearing in Definition 4.1, and we postpone such considerations to Section 6.

Proposition 4.3. *Suppose $a \in L(V)$ is primary and $k \in \mathbb{N}$.*

1. *Each finite block number is divisible by the degree of p.*
2. *$\mathcal{B}_k(a)$ is the dimension of the quotient space $\ker p^{k+1}(a)/\ker p^k(a)$.*
3. *In order that $A = \operatorname{pol} a$ act transitively on $\ker p^{k+1}(a)/\ker p^k(a)$, it is necessary and sufficient that $\mathcal{B}_k(a) \leq \deg p$.*

PROOF. Write K for the field $F[x]/pF[x]$. We make $\ker a$ into a vector space over K by defining $\bar{f}v = f(a)v$ for each $f \in F[x]$ and $v \in V$. By this definition, an F–closed subspace of $\ker a$ is invariant under a if and only if it is closed under muliplication by K. Thus, (1) follows from the fact that every a–invariant subspace M of $\ker a$ satisfies $\dim M = (\deg p)(\dim_K M)$. (Unqualified references to dimension are taken with respect to F.)

Since $p^k(a)$ maps $\ker p^{k+1}(a)$ onto $\ker p(a) \cap \operatorname{ran} p^k(a)$, the first isomorphism theorem tells us that $\ker p(a) \cap \operatorname{ran} p^k(a)$ has the same dimension (over F) as the quotient space $Q \equiv \ker p^{k+1}(a)/\ker p^k(a)$. This serves to establish (2).

Finally, we get (3) by making the quotient space Q into a vector space over K, noting that K acts transitively on Q iff $\dim_K Q = 1$. □

The *order* of a vector $v \in V$ is the smallest $n \in \mathbb{N}$ satisfying $p^n(a)v = 0$; when $a = 0$, the order of the zero vector is taken to be zero, but the order of every other vector is taken to be one. (This usage conflicts with [12] where the order of a vector refers to the minimal polynomial annihilating it.)

Proposition 4.4. *Suppose a is primary and $\dim[\ker p(a) \cap \operatorname{ran} p^k(a)] > \deg p$. Then given v of order $k+1$, there is a vector w of order $k+1$ with $Av \cap Aw = \{0\}$.*

Proof. Write $Q \equiv \ker p^{k+1}(a)/\ker p^k(a)$. In view of Proposition 4.3, the hypothesis means $\dim_K Q > 1$ so the proof is completed by choosing $w \in \ker p^{k+1}(a)$ such that $w + \ker p^k(a)$ is independent of $v + \ker p^k(a)$ over K. □

Proposition 4.5. *Let $a \in L(V)$ be primary. Then the following are equivalent.*

(1) $\dim[\ker p(a)] \leq \deg p$.
(2) $Av = \ker p^{\operatorname{ord} v}(a)$ *for some non–zero vector* $v \in V$.
(3) $Av = \ker p^{\operatorname{ord} v}(a)$ *for every vector* $v \in V$.
(4) *The invariant subspace lattice of a is totally ordered by inclusion.*

Proof. We leave it to the reader to check that the conventions have been arranged to make these conditions equivalent for the zero operator. We restrict attention to non–zero a in the rest of the proof.

Assuming (1), Proposition 4.3 tells us that A acts transitively on $\ker p(a)$, so (2) holds.

Assume (2) so that $Aw = \ker p^k(a)$ for some vector w of order $k > 0$. Given $y \in \ker p(a)$, there must be some polynomial f satisfying $f(a)v = y$. Comparing orders, we see that f is divisible by p^{k-1} whence $Ap^{k-1}(a)w = \ker a$, i.e., we may as well assume that the original vector w has order one. In particular, we see that $\dim \ker p(a) = \dim Aw = \deg p$ so (1) and (2) are equivalent.

We now establish (1) implies (3) by arguing inductively on the order of v. There is nothing to do for order zero. Given v of order $n+1$, we apply the inductive hypothesis to $p(a)v$, concluding that $\ker p^n(a) \subset Av$. But $\mathcal{B}_n(a) \leq \mathcal{B}_0(a)$ so A acts transitively on $\ker p^{n+1}(a)/\ker p^n(a)$. In other words Av contains representatives of each coset in this quotient space and we have $\ker p^{n+1}(a) \subset Av$. Since the opposite inclusion is automatic, we have completed the inductive argument.

(3) means that every proper invariant subspace for a takes the form $\ker p^k(a)$ and since these are totally ordered by inclusion, we get (3) implies (4).

Finally, (4) forces the invariant subspaces of $\ker p(a)$ to be totally ordered by inclusion, which means $\dim_K \ker p(a) \leq 1$ so (4) implies (1) and the proof is complete. □

Example 4.6. When V is finite dimensional, one can add two additional conditions to the preceding proposition:

(5) a admits a cyclic vector.

(6) The canonical form of a has a single block.

On infinite–dimensional spaces, however, no locally algebraic transformation can satisfy (5). On the other hand, there is a backward shift on such a space satisfying (1) through (4). More precisely, let F^ω denote the vector space of F–valued sequences having finite support, and define s_ω by

$$(S_\omega v)_n = v_{n+1}, \qquad v \in F^\omega, \quad n \in \mathbb{N}.$$

As we will see in Section 6, the appropriate version of (6) for infinite–dimensional V is "*a admits a canonical form* and that form has a single block".

The next result reduces to Proposition 4.4 when $n \leq k$ and to Proposition 4.5 when $k = 0$.

Corollary 4.7. *Suppose a is primary and write k for the smallest member of* $\mathbb{N} \cup \{\infty\}$ *satisfying* $\dim[\ker p(a) \cap \operatorname{ran} p^k(a)] \leq \deg p$. *For each $n \in \mathbb{N}$ we have*

$$\bigcap_{\operatorname{ord} y = n} Ay = \begin{cases} \{0\}, & \text{if } n \leq k \\ p^k(a) \ker p^n(a), & \text{if } n > k. \end{cases} \tag{4.1}$$

PROOF. Fix n and write M for the intersection appearing in Display 4.1.

If $n \leq k$, Proposition 4.4 yields vectors v, w of order n with $Av \cap Aw = \{0\}$ so $M = \{0\}$ as desired.

Suppose now that $n > k$. (Actually the proof only requires $k < \infty$.) Given z of order n, the set of polynomials q for which $q(a)z \in M$ is an ideal in $F[x]$ so $M = \langle p^m(a)z \rangle$ for some integer $m \leq n$. In particular, $\dim M = (n-m)d$, so m is in fact indepedent of z whence $M = p^m(a) \ker p^n(a)$.

It remains to show that $m = k$. Fix $v \in V$ of order n. Set $\hat{V} = \operatorname{ran} p^k(a)$, write $\hat{a}$ for the restriction of a to $\hat{V}$, and take $\hat{A} = \operatorname{pol} \hat{a}$. Then the vector $p^k(a)v$ will have order $n-k$ as a member of $\hat{V}$ so Proposition 4.5 yields $\hat{A}p^k(a)v = \ker p^{n-k}(\hat{a})$. Translating back to V, this implies

$$Ap^k(a)v = \ker p^{n-k}(a) \cap \operatorname{ran} p^k(a) = p^k(a) \ker p^n(a).$$

In particular, this implies that $Av \supset p^k(a) \ker p^n(a)$ so $m \leq k$. This completes the proof if $k = 0$.

On the other hand, if $k > 0$, we know that $\mathcal{B}_{k-1}(a) > \deg p$. Applying Proposition 4.5 to the restriction of a to $\operatorname{ran} p^{k-1}(a)$, we then conclude

$$Ap^{k-1}(a)v \not\supseteq p^{k-1}(a) \ker p^n(a). \tag{4.2}$$

Now if $f(a)v$ belongs to the right hand side of Display 4.2, its order cannot exceed $n-(k-1)$ and thus f must be divisible by p^{k-1}. Thus Display 4.2 is equivalent to

$$Av \not\supseteq p^{k-1}(a) \ker p^n(a)$$

whence $m > k-1$ and the proof is complete. □

Example 4.8. We investigate the meaning of Corollary 4.7 for a nilpotent transformation a acting on a finite–dimensional space. Suppose a is in (upper–triangular) Jordan canonical form relative to the standard basis $e_1, e_2, \ldots$, with block sizes arranged in non–increasing order. Write $m_1 \geq m_2$ for the sizes of the two largest blocks. ($m_2 = 0$ if there is only one block.) We have $p(x) = x$ and $k = m_2$. Thus Display 4.1 becomes

$$\bigcap_{\operatorname{ord} y = n} Ay = \begin{cases} \operatorname{span}\{e_1, \ldots, e_{n-m_2}\}, & \text{if } m_2 < n \leq m_1 \\ \{0\}, & \text{otherwise.} \end{cases} \tag{4.2}$$

In particular, for each n, we have

$$\bigcap\{Ay : \operatorname{ord} y = n\} = \begin{cases} \{0\} & \text{when } m_2 = m_1, \text{ and} \\ \ker a^n & \text{when } m_2 = 0. \end{cases}$$

These are consistent with Propositions 4.4 and 4.5, respectively.

5. Transformations of rank one

Throughout this section, A denotes the algebra generated by a transformation $a \in L(V)$. In view of Theorem 3.3(3) and Proposition 3.4, we concentrate on primary a. After finding the rank one members of $\operatorname{str} a$ in Proposition 5.3, we characterize the rank one members of $\operatorname{ref} a$ in Theorem 5.4. The pieces are then assembled in Theorem 5.8, which states that any excess of $\operatorname{str} a$ over $\operatorname{ref} a$ must be due to transformations of rank one.

The pioneering result concerning reflexive transformations is due to J. A. Deddens and P. A. Fillmore [5]; this can be paraphrased as follows.

Proposition 5.1. *Suppose $a \in L(V)$ is a nilpotent transformation acting on a finite–dimensional space and write $m_1 \geq m_2$ for the sizes of the two largest blocks in its Jordan Form. Then a is reflexive if and only if either $m_2 = m_1$ or $m_2 = m_1 - 1$.*

Successive generalizations in [10] and [6] allow $p(a)$ to be locally nilpotent for any irreducible polynomial p, remove the dimensionality restriction on V, and finally allow $p(a)$ to be locally nilpotent. An important distinction between [5] and its successors is that [5] catalogues all members of $\operatorname{ref} a$ regardless of whether a is reflexive. The goal of the present section is a unified analysis of this type for all members of $L(V)$. Our main result states that the excess of $\operatorname{ref} a$ over $\operatorname{pol} a$ can always be accounted for by rank one operators.

Proposition 5.2. *Suppose a is primary and consider the formal power series*

$$\sum_{i=0}^{\infty} q_i(a)p^i(a), \quad \deg q_i < \deg p, \quad i \in \mathbb{N}. \tag{5.1}$$

(1) *Every series of the form* (5.1) *defines a member of* $\operatorname{str} a$, *and every member of* $\operatorname{str} a$ *admits such a representation.*
(2) *The series* (5.1) *represents the zero operator if and only if each of its terms is zero.*

PROOF. The first assertion of (1) is clear since each finite–dimensional subspace of V is annihilated by a power of $p(a)$.

Conversely, given $b \in \operatorname{str} a$ and $v \in V$ of order n, there is a polynomial of the form $\sum_{i=0}^{n-1} q_i p^i$ whose value at a agrees with b on v. By requiring the degrees of the $\{q_i\}$ to be smaller than d, we guarantee uniqueness of the polynomial associated with the given vector v.

To see that the $\{q_i\}$ are independent of v, suppose $w \in V$ has order $m \geq n$ and write $\sum_{i=0}^{m-1} q_i' p^i$ for the associated polynomial. Since $b \in \operatorname{str} a$, there is a polynomial r simultaneously satisfying $bv = r(a)v$ and $bw = r(a)w$. Then we must have

$$r \equiv \sum_{i=0}^{n-1} q_i p^i \bmod p^n, \qquad r \equiv \sum_{i=0}^{m-1} q_i' p^i \bmod p^m,$$

whence $\sum_{i=0}^{n-1} q_i p^i \equiv \sum_{i=0}^{n-1} q_i' p^i \bmod p^n$. This forces $q_i = q_i'$ for $i < n$. Thus the $\{q_i\}$ associated with different vectors are consistent, and they can be assembled into a single series of the form (5.1).

For (2), let $q_n(a)(p(a))^n$ be the lowest degree non–vanishing term in (5.1). Then V contains vectors of order $n+1$ and such vectors will not lie in the kernel of (5.1). □

Proposition 5.3. *For primary a, the only possible rank one members of* $\operatorname{str} a$ *are scalar multiples of $p^{n-1}(a)$ where $p(a)$ is nilpotent of order n. In particular,* $\operatorname{str} a$ *can only have rank one members if p is linear and a is algebraic.*

PROOF. Let $b = \sum q_i(a)p^i(a)$ be a rank one member of $\operatorname{str} a$. Since zero is the only eignevalue of $p(a)$, we see that $p(a)b = 0$; in view of Proposition 5.2, b has a single non–zero term, $q_{n-1}(a)p^{n-1}(a)$, with $p^n(a) = 0$ and $\deg q_{n-1} < \deg p$. Moreover, if v is a non–zero vector in the range of b, then v, av must be dependent, which makes p of first degree and q_{n-1} constant. □

We now proceed to describe the rank one members of $\operatorname{ref} a$ in terms of its block numbers.

Theorem 5.4. *Suppose a is primary, $v \otimes \phi$ is a rank one member of $L(V)$, and n is the smallest possible order for vectors not belonging to* $\ker \phi$. *Then the following are equivalent.*

(1) $v \otimes \phi \in \operatorname{ref} a$.
(2) *$v \in Aw$ for each w not belonging to* $\ker \phi$.
(3) *$v \in Aw$ for each w of order n.*
(4) *$v \in p^k(a) \ker p^n(a)$ for some $k < n$ with* $\dim[\ker p(a) \cap \operatorname{ran} p^k(a)] = \deg p$.

PROOF. We begin by noting that there are non–zero vectors in $\ker p(a) \cap \operatorname{ran} p^{n-1}(a)$, so (4) $\iff$ (3) is the content of Corollary 4.7.

Assume next that (3) is satisfied and let $y \in V$. If the order of y is less than n, we have $(v \otimes \phi)y = 0 \in Ay$. On the other hand, if the order of y is at least n, then there is a polynomial f for which $f(a)y$ has order equal to n whence $(v \otimes \phi)y \in Af(a)y \subset Ay$. This establishes (3) implies (1).

Suppose (1) holds and $\phi(w) \neq 0$. Then $(v \otimes \phi)w$ is a non–zero scalar multiple of v and it must belong to Aw. Thus (1) implies (2).

We complete the proof by establishing (2) implies (3). Assume (2) and fix z of order n with $\phi(z) = 1$. We consider two cases. If $\dim[p^{n-1}(a) \ker p^n(a)] = d$, then Proposition 4.3 tells us that A acts transitively on $\ker p^n(a)/\ker p^{n-1}(a)$. Thus given w of order n, we can write $z = f(a)w + y$ for some $y \in \ker p^{n-1}(a)$. Since $\phi(z - y) = 1$, this yields $v \in A(z - y) \subset Aw$, as required by (3).

The remaining possibility is $\dim[p^{n-1}(a) \ker p^n(a)] > d$. Here, Proposition 4.4 yields a vector w of order n with $Aw \cap Az = \{0\}$. We can in fact arrange $\phi(w) \neq 0$. Indeed it $\phi(w) = 0$, replace w by $w + \lambda z$ where the non–zero scalar λ is chosen so that the latter vector still has order n. But this means (2) is not satisfied by any non–zero vector v so this case does not actually arise. □

Example 5.5. We informally investigate the meaning of Theorem 5.4 for a nilpotent transformation a acting on a finite–dimensional space. Continuing with the notation of Example 4.8, assume a is in (upper–triangular) Jordan canonical form relative to the standard basis $e_1, e_2, \ldots$, with block sizes arranged in non–increasing order. Write $m_1 \geq m_2$ for the sizes of the two largest blocks.

Suppose $v \otimes \phi$ is a rank one member of $\operatorname{ref} a$ and n is as in the statement of Theorem 5.4. The presence of vectors of order n in V forces $m_1 \geq n$. Because ϕ kills all vectors of smaller order, it must be supported on the span of $e_n \ldots e_{m_1}$. On the other hand, from Example 4.8, we know that Condition (4) is equivalent to having $v \in \operatorname{span}\{e_1 \ldots e_{n-m_2}\}$.

Thus (4) is equivalent to demanding that $v \otimes \phi$ be supported on rows $1 \ldots n - m_2$ and columns $n \ldots m_1$.

We note some special cases.

(1) If $m_2 = 0$ (only one block), then (4) reduces to the requirement that $v \otimes \phi$ be upper–triangular.
(2) At the other extreme, if $m_2 = m_1$, then $\operatorname{ref} a$ has no rank one members.
(3) If $m_2 = m_1 - 1$, the rank one members of $\operatorname{ref} a$ are supported on the single position $(1, m_1)$.
(4) If $m_2 < m_1 - 1$, then $\operatorname{ref} a$ contains an independent pair of rank one members.

In particular, the condition $m_2 \geq m_1 - 1$ is equivalent to having the rank one members of $\operatorname{ref} a$ and $\operatorname{str} a$ coincide. This reveals Proposition 5.1 as a special case of Theorem 5.8 below.

Lemma 5.6. *Suppose the order of x does not exceed the order of v and write k for the smallest integer satisfying $p^k(a)x \in \langle v \rangle$. Then $\langle v, x \rangle = \langle v \rangle \oplus \langle y \rangle$ for some vector y of order k.*

PROOF. Write $p^k(a)x = f(a)p^l(a)v$ with f relatively prime to p. Comparing orders, we see that $l \geq k$. Set $y = x - f(a)p^{l-k}(a)v$. Clearly, $\langle v, x \rangle = \langle v, y \rangle$. Since the order of y is k, we also have $\langle v \rangle \cap \langle y \rangle = \{0\}$ as desired. □

Proposition 5.7. *Given a primary, $b \in \operatorname{ref} a$, and $n \in \mathbb{N}$, there is a polynomial f and a finite linear combination c of rank one members of $\operatorname{ref} a$ such that b agrees with $f(a) + c$ on $\ker p^n(a)$.*

PROOF. We argue inductively on n. There is nothing to do for $n = 0$. Assuming we can implement the construction on $\ker p^n(a)$, we show how to adapt the decomposition to $\ker p^{n+1}(a)$. The procedure depends on $\ell \equiv \dim[\ker a \cap \operatorname{ran} p^n(a)]$. If $\ell = 0$, then $\ker p^{n+1}(a) = \ker p^n(a)$ and no adjustment is necessary.

Assume next that $\ell > d$ and apply Proposition 4.4 to find vectors v, w of order $n+1$ with $\langle v \rangle \cap \langle w \rangle = \{0\}$. By definition of ref, there is a polynomial g such that $b - g(a)$ vanishes on v. Now let x be an arbitrary member of $\ker p^{n+1}(a)$ and apply Lemma 5.6 to express $\langle v, x\rangle = \langle v\rangle \oplus \langle y\rangle$ for some y of order at most $n+1$. We now appeal to Proposition 3.2 three times to conclude that $b - g(a)$ vanishes on w, $\langle v\rangle$, and $\langle y\rangle$, respectively. In particular, $bx = g(a)x$, and we have shown that b agrees with $g(a)$ throughout $\ker p^{n+1}(a)$.

It remains to consider the case $\ell = d$. We begin by invoking the inductive hypothesis to find a polynomial g and a finite linear combination c of rank one members of $\operatorname{ref} a$ such that $r \equiv b - c - g(a)$ vanishes on $\ker p^n(a)$. Next apply the dimensionality assumption to find vectors $v_1, \dots, v_d$ of order $n+1$ such that the $\{v_i + \ker p^n(a) : i \leq d\}$ form a basis for the quotient space $\ker p^{n+1}(a)/\ker p^n(a)$. Then we choose a dual set in V^*, that is, functionals $\phi_1, \dots, \phi_d$ which vanish on $\ker p^n(a)$ and satisfy $\phi_i(v_j) = \delta_{i,j}$ for $i, j \leq d$.

Fix i for the moment, and suppose w has order $n+1$. Our dimensionality assumption means that A acts transitively on $\ker p^{n+1}(a)/\ker p^n(a)$ so we can find a polynomial f with $v_i - f(a)w \in \ker p^n(a)$. Since $r \in \operatorname{ref} A$, this yields

$$rv_i = rf(a)w \in Af(a)w \subset Aw.$$

Applying the equivalence (1) $\iff$ (3) of Theorem 5.4, we therefore conclude that each $rv_i \otimes \phi_i \in \operatorname{ref} A$. The proof is thus completed by observing that $r - \sum_{i=1}^d rv_i \otimes \phi_i$ vanishes on $\ker p^{n+1}$ whence b agrees with $g(a) + \left[c + \sum_{i=1}^d rv_i \otimes \phi_i\right]$ on $\ker p^{n+1}(a)$. □

Theorem 5.8. *Let A be the algebra generated by a single linear transformation. Then $\operatorname{ref} A$ is the strict closure of the span of those of its members which either have rank one or belong to A.*

PROOF. In view of Theorem 3.3(3), we may assume the generator a of A to be locally algebraic; applying Propostion 3.4, we may also assume a to be primary. The proof is therefore completed by appealing to Proposition 5.7. □

We close this section by recovering the characterizations of reflexive transformations discovered by Hadwin [10] and Delaï [6].

Recall the block numbers from Section 4:

$$\mathcal{B}_k(a) \equiv \dim \left[\, \ker p(a) \cap \operatorname{ran} p^k(a) \,\right] = \dim \frac{\ker p^{k+1}(a)}{\ker p^k(a)}.$$

Corollary 5.9. *Suppose an operator a on V has the property that $p(a)$ is locally nilpotent for the irreducible polynomial p. Then a is reflexive if and only if either*

(1) *$\mathcal{B}_k(a)$ does not agree with the degree of p for any integer k, or*
(2) *p is of first degree and $\mathcal{B}_k(a) = 1$ for a unique value of n.*

PROOF. In view of Theorem 5.8, we know that $\operatorname{str} A = \operatorname{ref} A$ precisely when all rank one members of $\operatorname{ref} A$ belong to $\operatorname{str} A$. If $\deg p > 1$, Proposition 5.3 tells us that $\operatorname{str} A$ has no rank one members, while Theorem 5.4 tells us that the same is true of $\operatorname{ref} A$ iff the condition of (1) is satisfied.

Thus we may as well assume that p is of first degree. If $\mathcal{B}_k(a)$ is never one, then $\operatorname{str} A = \operatorname{ref} A$ as in the preceding paragraph. On the other hand, if $\mathcal{B}_k(a) = 1$ for more than one value of k, then Theorem 5.4 guarantees an independent pair of rank one members in $\operatorname{ref} A$, so Proposition 5.2 precludes equality of $\operatorname{str} A$ and $\operatorname{ref} A$.

Suppose finally that $\mathcal{B}_k(a) = 1$ for a unique integer k. Then $\mathcal{B}_{k+1}(a) = 0$, so $p^{k+1}(a) = 0$ and $p^k(a)$ has rank one. Given a rank one member $v \otimes \phi$ of $\operatorname{ref} A$, Theorem 5.4 yields $\ker p^k(a) \subset \ker \phi$, whence $v \otimes \phi$ is in fact a scalar multiple of $p^k(a)$. This means all rank one members of $\operatorname{ref} A$ belong to A, and $\operatorname{str} A = \operatorname{ref} A$ by Theorem 5.8. □

To recover Proposition 5.1, note that the condition $m_2 \geq m_1 - 1$ appearing there is equivalent to requiring $\mathcal{B}_k(a) = 1$ for *at most* one value of k.

Several alternate ways of expressing Corollary 5.9 in the nilpotent case can be found in Lemma 2.5 of Hadwin and Nordgren's paper [11]; in fact, it was our reading of [11] which first led to the considerations of the present paper.

6. REFLEXIVITY VERSUS CANONICAL FORMS

In this section, we apply ideas from I. Kaplansky's monograph [12] to explain the title of this paper. In order to simplify the notation, we concentrate on locally nilpotent transformations; the discussion is easily adapted to the more general setting of primary transformations.

Definition 6.1. *A locally algebraic operator in $L(V)$ is a nest operator* if its lattice of invariant subspaces is totally ordered by inclusion. An operator is said to *admit a canonical form* if it is similar to some (possibly infinite) direct sum of nest operators.

Example 6.2. Recall the backward shifts s_n acting on F^{n+1} and s_ω acting on F^ω studied in Examples 4.2 and 4.6, respectively. These have one–dimensional kernels, so Proposition 4.5 tells us they are nest transformations.

In the preceding section, we saw that block numbers determine whether a locally nilpotent operator is reflexive. These simple similarity invariants do not distinguish between the operators $\bigoplus_{n\in\mathbb{N}} s_n$ and $\bigoplus_{n\in\mathbb{N}} s_{2n}$. To do that, we consider dimensions of quotient spaces; these appear in Display 6.1 below.

Given an operator $b \in L(V)$ and a cardinal number n, we write $b^{(n)}$ for the direct sum of n copies of b. This operator can also be realized as the tensor product of b with the identity operator acting on an n–dimensional space and is usually referred to as the *n–fold ampliation* of b.

Proposition 6.3. *Suppose the locally nilpotent operator $a \in L(V)$ admits a canonical form.*

(1) *If a is a nest operator, it must be similar to one of the operators from Example 6.2.*
(2) *a is similar to a unique operator of the form $\bigoplus_{0 \le k \le \omega} s_k^{(n_k)}$.*
(3) *a maps $\bigcap_{k \in \mathbb{N}} \operatorname{ran} a^k$ onto itself and n_ω is the dimension of the intersection of this space with $\ker a$.*
(4) *For each finite k, the cardinal number n_k equals*

$$\dim \left[\frac{\ker a \cap \operatorname{ran} a^k}{\ker a \cap \operatorname{ran} a^{k+1}} \right]. \tag{6.1}$$

Proof. For (1), suppose a is a nest operator. Any non–zero vector $e_1 \in \ker a$ is a basis for $\ker a$. We need to extend this to a (finite or countable) basis $\{e_n\}$ for V satisfying $ae_{i+1} = e_i$ for all i. Assume such a basis $e_1, \dots, e_n$ for $\ker a^n$ has been constructed. If $\ker a^{n+1} = \ker a^n$, we're done. Otherwise, Proposition 4.5 tells us that $\ker a^n$ has codimension one in $\ker a^{n+1}$. In fact, the rank–nullity theorem tells us that a maps the latter space *onto* the former, so we can choose e_{n+1} satisfying $ae_{n+1} = e_n$ to get the desired basis $e_1, \dots, e_{n+1}$ for $\ker a^{n+1}$.

Existence in (2) is a matter of gathering similar direct summands. (3) and (4) are easily verified for the operators s_n, s_ω of Example 6.2; they extend to arbitrary direct sums of such operators because ranges, kernels and dimensions respect direct sums of subspaces. Finally, once (3) and (4) are established, they yield uniqueness in (2). □

Example 6.4. There is a locally nilpotent operator which is reflexive, but does not admit a canonical form.

Construction. Take W to be the space of all lower–triangular infinite matrices having finite support and write $\{e_{i,j}\}_{i \ge j}$ for its standard basis. Let b act on W by "shifting one column to the left", i.e.,

$$be_{i,1} = 0, \quad be_{i,j} = e_{i,j-1} \qquad \text{for } i, j \in \mathbb{N}, \quad j > 1.$$

(b is similar to the transformation $\bigoplus_{n \in \mathbb{N}} s_n$.)

Now take M to be the set of matrices in $\ker b$ whose non–zero entries sum to zero, i.e.,

$$M = \Big\{ \sum_i c_i e_{i,1} : \; c_i = 0 \text{ for all but finitely many } i; \; \sum c_i = 0 \Big\}.$$

Then M is invariant under b, so b induces a locally nilpotent transformation a on the quotient space $V = W/M$.

Given n, note that

$$\begin{aligned} e_{1,1} + M &= e_{n+1,1} + M = a^n(e_{n+1,n+1} + M) \quad \text{and} \\ e_{n+2,2} - e_{n+3,2} + M &= a^n(e_{n+2,n+2} - e_{n+3,n+2} + M) \end{aligned}$$

are independent members of $\ker a \cap \operatorname{ran} a^n$. Thus none of these spaces has dimension one and reflexivity of a follows from Corollary 5.9.

We next compute $\bigcap_{n\in\mathbb{N}} \operatorname{ran} a^n$. We already know that $e_{1,1}+M$ belongs to this space. On the other hand, $\operatorname{ran} b^n = \operatorname{span}\{e_{i,j} : i \geq j+n\}$ so $\bigcap_{n\in\mathbb{N}}[M + \operatorname{ran} b^n] = \operatorname{span}\{e_{i,1} : i \in \mathbb{N}\} = M + \operatorname{span} e_{1,1}$. This means that $\bigcap_{n\in\mathbb{N}} \operatorname{ran} a^n$ is the one–dimensional space spanned by $e_{1,1}+M$. Since a does not map this space onto itself, we see that Proposition 6.3(3) is violated whence a does not admit a canonical form.

The main point of the present section is that the behavior of Example 6.4 is typical. The reader familiar with Kaplansky's monograph [12] will recognize Proposition 6.6 below as a consequence of his Theorems 6, 2, and 4.

Lemma 6.5. *Suppose $a \in L(V)$, and $n, k \in \mathbb{N}$. If $\ker a \cap \operatorname{ran} a^k = \ker a \cap \operatorname{ran} a^{n+k}$, then $\ker a^n \cap \operatorname{ran} a^k \subset \operatorname{ran} a^{k+1}$.*

PROOF. There is nothing to do when $n = 0$. When $n > 0$, the hypothesis implies that the transformation a^{n-1} maps $\ker a^n \cap \operatorname{ran} a^k$ and $\ker a^n \cap \operatorname{ran} a^{k+1}$ onto the same space. It follows that

$$\ker a^n \cap \operatorname{ran} a^k \subset \ker a^n \cap \operatorname{ran} a^{k+1} + \ker a^{n-1} \cap \operatorname{ran} a^k$$

from which point an inductive argument yields the desired result. □

Proposition 6.6. *If $a \in L(V)$ is locally nilpotent and $\ker a \cap \operatorname{ran} a^k$ is finite–dimensional for some $k \in \mathbb{N}$, then a admits a canonical form.*

PROOF. Suppose first $a \in L(V)$ is nilpotent. Using bases for succesive quotients of $\ker a$, $\ker a \cap \operatorname{ran} a$, $\ker a \cap \operatorname{ran} a^2$, ..., we can write $\ker a = \bigoplus_{i=0}^n K_i$ where $K_i \subset \operatorname{ran} a^i$ and $K_i \cap \operatorname{ran} a^{i+1} = \{0\}$ for $i \leq n$. For each i, we then find a subspace $M_i \subset V$ which a^i maps injectively onto K_i and set $V_i = \langle M_i \rangle$. By construction, the restriction of a to V_i is similar to a direct sum of $\dim K_i$ copies of the operator s_i from Example 6.2. The proof of this case is then completed by verifying that $V = \bigoplus V_i$.

In the general case, fix k so that

$$K_\omega \equiv \bigcap_{n\in\mathbb{N}} [\ \ker a \cap \operatorname{ran} a^n\ ] = \ker a \cap \operatorname{ran} a^k.$$

Applying Lemma 6.5, we see that $\ker a^n \cap \operatorname{ran} a^k \subset \operatorname{ran} a^{k+1}$ for each n, so a maps the range of a^k onto itself. Proceeding inductively, we can then construct subspaces $W_0 = K_\omega$, $W_1, W_2, \ldots$ such that for each i, the operator a maps W_{i+1} injectively onto W_i. Set $V_\omega = \bigcup_{n\in\mathbb{N}} W_i$. By construction, the restriction of a to V_ω is similar to a direct sum of finitely many copies of s_ω. One then uses a Zorn's lemma argument to find an invariant subspace M complementary to V_ω. Write b for the restriction of a to M, and observe that $\bigcap_{n\in\mathbb{N}}[\ker b \cap \operatorname{ran} b^n] = \{0\}$. Since one of the intersectands is finite–dimensional by hypothesis, it follows that $\ker b \cap \operatorname{ran} b^n = \{0\}$ for some n, and the proof is completed by applying the first case to b. □

Corollary 6.7. *Every locally nilpotent operator which is not reflexive admits a canonical form.*

PROOF. If a is locally nilpotent and non–reflexive, then Corollary 5.9 tells us that $\ker a \cap \operatorname{ran} a^k$ is one–dimensional for some $k \in \mathbb{N}$, and Proposition 6.6 applies. □

The reader may object to the arbitrary nature of Definition 6.1. Thus while Corollary 6.7 is fairly definitive in saying that non–reflexive local nilpotents are well–behaved, Example 6.4 is scant evidence for the converse. The following discussion of Ulm invariants reveals Example 6.4 as "the tip of the iceberg" concerning intractability of reflexive local nilpotents.

Fix a locally nilpotent operator $a \in L(V)$. Following [12], we construct a family of subspaces of V indexed by ordinal numbers (starting at 0), by setting

$$\begin{aligned} &V_0 = V, \\ &V_{\alpha+1} = aV_\alpha, \text{ and} \\ &V_\alpha = \textstyle\bigcap_{\beta<\alpha} V_\beta \text{ when } \alpha \text{ is a limit ordinal.} \end{aligned}$$

For finite n we have $V_n = \operatorname{ran} a^n$. There must be a smallest ordinal λ for which $V_\lambda = V_{\lambda+1}$ and we avoid set theoretic difficulties by halting the construction at that point. Note that V_λ is in fact the largest subspace of V mapped onto itself by a.

Definition 6.8. *The Ulm invariants* for a locally nilpotent transformation a are given by

$$\mathcal{U}_\alpha(a) = \begin{cases} \dim \dfrac{\ker a \cap V_\alpha}{\ker a \cap V_{\alpha+1}}, & \text{for } \alpha < \lambda \\ \dim [\ \ker a \cap V_\lambda\ ], & \text{for } \alpha = \lambda. \end{cases}$$

The ordinal λ is referred to as the *length* of a.

For finite α, these similarity invariants agree with those of Display 6.1. Our earlier block invariants can be recovered from the formula

$$\dim [\ \ker a \cap V_\beta\ ] = \sum_{\alpha\ge\beta} \mathcal{U}_\alpha(a), \qquad \beta \le \lambda.$$

Kaplansky proves that the Ulm invariants are complete if and only if the underlying space V has countable dimension. More to the point of the present discussion, even on F^ω, there are local nilpotents of each length $< \Omega$. Moreover, Ulm invariants of such operators can be arbitrarily specified subject to the mild restriction that there be infinitely many non–vanishing invariants between any two limit ordinals not exceeding λ (see Remark (d) on page 31 of [12]). From this perspective, we see that Example 6.4 is tame indeed with its length of $\omega + 1$ and its Ulm invariants being one for $\alpha \leq \omega$.

We conclude the paper by observing that Ulm invariants lead to a simple characterization of those operators on F^ω which admit canonical forms; a variant of Example 6.4 shows that the dimensionality restriction cannot be dropped.

Theorem 6.9. *In order for a local nilpotent operator to admit a canonical form it is necessary that its length not exceed ω. If V has countable dimension, the condition is also sufficient.*

PROOF. Necessity follows from Proposition 6.3(3) since $V_\omega = \bigcap_{n\in\mathbb{N}} \operatorname{ran} a^n$. For sufficiency, we have only to note that a has the same Ulm invariants as $\bigoplus_{0\leq k\leq\lambda} s_k^{(\mathcal{U}_k)}$. □

Example 6.10. There is a locally nilpotent operator of length ω which does not admit a canonical form.

Construction. Take V to be the space of all lower–triangular infinite matrices which are supported on finitely many columns. Let a act on V by "shifting one column to the left", i.e.,

$$(av)_{i,j} = v_{i,j+1}, \quad v \in V \quad i, j \in \mathbb{N}.$$

For each $k \in \mathbb{N}$, we have

$$\ker a \cap \operatorname{ran} a^k = \{\, v \in V : v \text{ is supported on the positions } (i,1),\ i > k\}.$$

It follows that $V_\omega \equiv \cap_{k\in\mathbb{N}} \operatorname{ran} a^k = \{0\}$ and all the dimensions of Display 6.1 equal one. In view of Proposition 6.3, this means that if a had a canonical form, it would have to be $\bigoplus_{k\in\mathbb{N}} s_k$. This however is incompatible with the fact that $\dim V > \aleph_0$.

REFERENCES

[1] E. A. AZOFF, *On finite rank operators and preannihilators*, Memoirs Amer. Math. Soc. **357** (1986).

[2] E. A. AZOFF AND H. A. SHEHADA, *Literal embeddings of linear spaces of operators*, Indiana Univ. Math. J. **22** (1993), 571–589.

[3] H. Bercovici, C. Foias, and C. Pearcy, *Dual algebras with applications to invariant subspaces*, Regional Conference Series in Mathematics **56**, American Mathematical Society (1985).

[4] L. Brickman and P. A. Fillmore, *The invariant subspace lattice of a linear transformation*, Canad. J. Math. **19** (1967), 810–822.

[5] J. A. Deddens and P. A. Fillmore, *Reflexive linear transformations*, Linear Algebra Appl. **10** (1975), 89–93.

[6] M. B. Delaï, *Sur la réflexivité des opérateurs linéaires*, Linear and Multilinear Algebra **38** (1994), 39–43 .

[7] L. Ding, *Separating vectors and reflexivity*, Linear Algebra Appl. **174** (1992), 37–52.

[8] L. Ding, *On a pattern of reflexive operator spaces*, Proc. Amer. Math. Soc. (to appear).

[9] F. Gilfeather, A. Hopenwasser, and D. R. Larson, *Reflexive algebras with finite width lattices: Tensor products, cohomology, compact perturbations*, J. Funct. Analysis **55** (1984), 176–199.

[10] D. Hadwin, *Algebraically reflexive linear transformations*, Linear and Multilinear Algebra **14** (1983), 225–233.

[11] D. Hadwin and E. A. Nordgren, *Reflexivity and direct sums*, Acta Sci. Math. **55** (1991), 181–197.

[12] I. Kaplansky, *Infinite Abelian Groups*. University of Michigan Press, Ann Arbor, 1969.

[13] A. Katavolos, M. S. Lambrou, and M. Papadakis, *On some algebras diagonalized by M-bases of ℓ^2*, Integral Equations Operator Theory **17** (1993), 68–94.

[14] W. R. Wogen, *Some counterexamples in nonselfadjoint algebras*, Ann. of Math. **126** (1987), 415–427.

Edward A. Azoff
Department of Mathematics
University of Georgia
Athens, GA 30602-7403
E-mail: azoff@alpha.math.uga.edu

Lifeng Ding
Dept. of Mathematics and Computer Science
Georgia State University
Atlanta, GA 30303-3083
E-mail: matlfd@gsusgi2.gsu.edu

Received: August 23rd, 1995.

Operator Theory:
Advances and Applications, Vol. 104

On Skew Toeplitz Operators, II

H. Bercovici, C. Foias, and A. Tannenbaum*

To our dear friend and inspiring colleague Carl Pearcy on the occasion of his sixtieth birthday

0. Introduction

Robust feedback control ideas crystalized in the 1980's under the form of H^∞ control. An important issue in this theory is the effective calculation of the norms of certain operators. In many cases of interest these operators can be written as (scalar or more general) functions of a given contraction T, and this makes it possible to bring into the picture ideas from the dilation theory of contractions. These ideas were used for the first time in [7] for the calculation of $\|f(T)\|$, where f is a rational function, and T is a contraction with defect indices equal to one. A more general approach was introduced in Part I of this paper [2]. In [2] the operator T was allowed to have finite defect indices, and the calculation of $\|f(T)\|$ (with f no longer a scalar function) was replaced by the study of invertibility for skew Toeplitz operators. The work in [2] was given an explicitly algorithmic form for scalar Toeplitz operators in [6]. A unified presentation of these results is given in [3].

In the present work we take up again the calculation of $\|f(T)\|$ with f rational, but we now allow T to be an arbitrary contraction of class $C_{\cdot 0}$. In addition, we remove a certain condition which appears in our earlier work (see condition (1) in Section 4. of [2]). This condition is generically satisfied in the context of [2], but not in the framework of this paper.

For the case of defect indices equal to one (corresponding to a scalar characteristic function), various skew Toeplitz algorithms were presented by Gu [8]. Gu's algorithms are more inolved but they cover some nongeneric situations.

1. The Problem

Let T be a bounded operator on a Hilbert space $\mathcal{H}$, and let $f(\lambda) = p(\lambda)/q(\lambda)$ be a rational function with poles off the spectrum $\sigma(T)$ of T, i.e., $q(\lambda) \neq 0$ for $\lambda \in \sigma(T)$. Further, denote $A = f(T) = p(T)q(T)^{-1}$. We will be interested in the effective calculation of the norm $\|A\|$ in the case when T is a contraction represented as a functional model, and q has no zeros in the closed unit disk. However, some simple

*The authors were partially supported by grants from the National Science Foundation, Air Force Office of Scientific Research, and Army Research Office.

observations can be made in the general case. Thus, for instance, $\|A\|$ is greater than the spectral radius $|A|_{\text{sp}}$, hence

$$|A|_{\text{sp}} = \sup\left\{\left|\frac{p(\lambda)}{q(\lambda)}\right| : \lambda \in \sigma(T)\right\} \leq \|A\|.$$

Next, if ρ denotes $\|A\|$, then the operator $\rho^2 - AA^*$ is positive definite but not invertible, and hence it has zero as an approximate eigenvalue. Since

$$\rho^2 - AA^* = q(T)^{-1}(\rho^2 q(T)q(T)^* - p(T)p(T)^*)q(T)^{*-1},$$

we deduce that the operator

$$Q = \rho^2 q(T)q(T)^* - p(T)p(T)^*$$

is positive definite and not invertible. If $p(\lambda) = \sum_{j=0}^n p_j\lambda^j$ and $q(\lambda) = \sum_{j=0}^n q_j\lambda^j$, then Q can be written as

$$Q = \sum_{i,j=0}^{n} c_{ij}T^iT^{*j},$$

where the coefficients $c_{ij} = \rho^2 q_i\bar{q}_j - p_i\bar{p}_j$ satisfy the condition $c_{ij} = \bar{c}_{ji}$, $0 \leq i, j \leq n$. Now, given an arbitrary polynomial in two variables

$$\omega(\lambda, \mu) = \sum_{i,j=0}^{n} c_{ij}\lambda^i\mu^j,$$

one can introduce an operator

$$Q_\omega = \omega(T, T^*) = \sum_{i,j=0}^{n} c_{ij}T^iT^{*j}.$$

We have seen above how deciding whether $\rho^2 - AA^*$ has zero as an approximate eigenvalue is equivalent to the corresponding question for an operator of the form Q_ω such that $c_{ij} = \bar{c}_{ji}$, $0 \leq i, j \leq n$. Since the calculation of $\rho = \|A\|$ is only a problem when $|A|_{\text{sp}} < \rho$, we may restrict ourselves to the case in which $\omega(\lambda, \bar{\lambda}) \neq 0$ for every $\lambda \in \sigma(T)$. We arrive thus at the study of the following problem.

Problem 1.1. *Given a polynomial $\omega(\lambda, \mu) = \sum_{i,j=0}^n c_{ij}\lambda^i\mu^j$ such that*

(i) *$c_{ij} = \bar{c}_{ji}$, $0 \leq i, j \leq n$; and*
(ii) *$\omega(\lambda, \bar{\lambda}) \neq 0$ for every $\lambda \in \sigma(T)$,*

decide whether zero is an approximate eigenvalue for Q_ω.

Clearly, an effective answer to Problem 1.1 will allow one to calculate $\|A\| = \|f(T)\|$ numerically by testing the various operators Q_ω corresponding to values $\rho > |A|_{\mathrm{sp}}$. The norm of A is the largest ρ for which zero is an approximate eigenvalue of Q_ω.

In the cases of interest more information is available about T and f. More precisely, T is a contraction with inner characteristic function, and f belongs to the algebra H^∞ of bounded analytic functions in the unit disk $\mathbf{D} = \{\lambda : |\lambda| < 1\}$. This means that q has no zeros in the closure $\overline{\mathbf{D}}$ of $\mathbf{D}$, and then von Neumann's inequality implies that

$$|A|_{\mathrm{sp}} = \sup\{|f(\lambda)| : \lambda \in \sigma(T)\} \leq \|A\| \leq \sup\{|f(\zeta)| : |\zeta| = 1\}.$$

These inequalities become equalities if $\sigma(T)$ contains the entire unit circle $\partial\mathbf{D} = \{\zeta : |\zeta| = 1\}$. Hence we will assume throughout that $\sigma(T)$ does not contain the unit circle. We arrive thus at the problem which we will actually study in this paper.

Problem 1.2. *We are given a contraction T and a polynomial $\omega(\lambda,\mu) = \sum_{i,j=0}^n c_{ij}\lambda^i\mu^j$ such that*

(i) *the characteristic function of T is inner;*
(ii) *$\sigma(T)$ does not contain the unit circle;*
(iii) *$c_{ij} = \bar{c}_{ji}$, $0 \leq i,j \leq n$; and*
(iv) *$\omega(\lambda,\bar{\lambda}) \neq 0$ for every $\lambda \in \sigma(T)$.*

Determine whether zero is an approximate eigenvalue of $Q_\omega = \omega(T,T^*) = \sum_{i,j=0}^n c_{ij}T^iT^{*j}$.

We recall that the operators Q_ω considered in Problem 1.2 are exactly the scalar skew Toeplitz operators considered in [2], [6], and [3].

2. The Main Result

In order to study Problem 1.2, we will represent the operator T as a functional model. We refer to [5] (see also [1] and [4]) for the relevant notation and results used below. Let $\mathcal{E}$ be a separable Hilbert space, and let $\Theta : \mathbf{D} \to \mathcal{L}(\mathcal{E})$ be an inner function (which "coincides" with the characteristic function of T). Thus Θ is bounded, analytic, and the boundary values $\Theta(\zeta)$, $|\zeta| = 1$, are isometric for almost every ζ. The function Θ can be viewed as a multiplication operator on the $\mathcal{E}$-valued Hardy space $H^2(\mathcal{E})$, and the Hilbert space $\mathcal{H}$ will be identified with $\mathcal{H}(\Theta) = H^2(\mathcal{E}) \ominus \Theta H^2(\mathcal{E})$. Further, let us denote by U_+ the unilateral shift on $H^2(\mathcal{E})$, i.e.,

$$(U_+h)(\lambda) = \lambda h(\lambda), \quad h \in H^2(\mathcal{E}), \quad \lambda \in \mathbf{D}.$$

The operator T will be given by $T = P_{\mathcal{H}(\Theta)}U_+|\mathcal{H}(\Theta)$ or, equivalently, $T^* = U_+^*|\mathcal{H}(\Theta)$. We are allowed to restrict our attention to such operators because every contraction T such that $T^{*n} \to 0$ in the strong topology as $n \to \infty$ is unitarily equivalent to one of them. Recall that $H^2(\mathcal{E})$ can also be viewed as a subspace of $L^2(\mathcal{E})$, the space of all $\mathcal{E}$-valued square integrable functions defined on $\partial\mathbf{D}$. The function Θ also defines a multiplication operator on $L^2(\mathcal{E})$. A vector $h \in H^2(\mathcal{E})$ has a Fourier expansion of the form $h(\zeta) = \sum_{j=0}^{\infty} \zeta^j h_j$, $|\zeta| = 1$, with coefficients $h_j \in \mathcal{E}$. Furthermore, if the vector h belongs to $\mathcal{H}(\Theta)$ then $\Theta^* h$ is orthogonal to $H^2(\Theta)$, and hence it has a Fourier expansion of the form

$$(\Theta^* h)(\zeta) = \sum_{j=1}^{\infty} \zeta^{-j} h_{-j}, \quad \zeta \in \partial\mathbf{D}.$$

Thus, with each element $h \in \mathcal{H}(\Theta)$ we can associate a doubly infinite sequence $\{h_j\}_{j=-\infty}^{\infty}$ of elements of $\mathcal{E}$. Since the sequence $\{h_j\}_{j=0}^{\infty}$ entirely determines h, it will also determine h_{-j} for $j \geq 1$. The relevant formula is

$$h_{-j} = \sum_{i=j}^{\infty} \Theta_i^* h_{i-j}, \quad j \geq 1,$$

where the Θ_i are the Taylor coefficients of Θ, i.e., $\Theta(\lambda) = \sum_{i=0}^{\infty} \lambda^i \Theta_i$, $|\lambda| < 1$.

Before returning to Problem 1.2, we want to establish a useful notational convention. If F is a function of a Hilbert space variable k, we will write $F(k) = o(k)$ if there exists a sequence of nonzero vectors k_j such that $\lim_{j\to\infty} F(k_j)/\|k_j\| = 0$. Thus, for instance, $Qk = o(k)$ simply means that zero is an approximate eigenvalue for Q.

From this point on Θ and ω will be fixed, T will be given as above, and Q will denote the skew Toeplitz operator Q_ω. All the conditions of Problem 1.2 will be assumed to hold. We will also consider the polynomials

$$C(\lambda) = \lambda^n \omega\left(\lambda, \frac{1}{\lambda}\right) = \sum_{i=0}^{2n} c_i \lambda^i,$$

and

$$C_\ell(\lambda) = \sum_{i=1}^{n} \sum_{j=l+1}^{n} c_{ij} \lambda^{n+i-j+\ell}, \quad \ell = 0, 1, 2, \ldots, n-1.$$

Observe that the degree of C_ℓ is at most $2n - 1$. The reason for introducing these polynomials is the following result.

Lemma 2.1. *For every vector $h \in \mathcal{H}(\Theta)$ we have $Qh = P_{\mathcal{H}(\Theta)} u$, where*

$$u(\lambda) = \lambda^{-n}\Big[C(\lambda)h(\lambda) - \sum_{\ell=0}^{n-1} C_\ell(\lambda) h_\ell\Big], \quad \lambda \in \mathbf{D}.$$

PROOF. Observe that $T^iT^{*j}h = P_{\mathcal{H}(\Theta)}U_+^iU_+^{*j}h$, and

$$(U_+^iU_+^{*j}h)(\lambda) = \lambda^{i-j}\Big(h(\lambda) - \sum_{\ell=0}^{j-1}\lambda^\ell h_\ell\Big).$$

The lemma follows now immediately because $\sum_{i,j=1}^n c_{ij}\lambda^{i-j}h(\lambda) = \lambda^{-n}C(\lambda)h(\lambda)$ and

$$\sum_{i,j=1}^n c_{ij}\lambda^{i-j}\sum_{\ell=0}^{j-1}\lambda^\ell h_\ell = \lambda^{-n}\sum_{\ell=0}^{n-1}C_\ell(\lambda)h_\ell. \qquad \square$$

The function u in the preceding lemma clearly belongs to $H^2(\mathcal{E})$ since $u = \omega(U_+, U_+^*)h$. Thus the first n Fourier coefficients of $Ch - \sum_{\ell=0}^{n-1} C_\ell h_\ell$ must vanish. Equivalently, the first n Fourier coefficients of $\sum_{\ell=0}^{n-1}(\lambda^\ell C(\lambda) - C_\ell(\lambda))h_\ell$ vanish. Since this is true regardless of the function Θ, we must have that $\lambda^\ell C(\lambda) - C_\ell(\lambda)$ is divisible by λ^n. This can be established by direct computation as well. For further reference we include this fact in the next result.

Lemma 2.2. (i) *$\lambda^{2n}\overline{C(1/\bar\lambda)} = C(\lambda)$ for $\lambda \neq 0$.*
(ii) *$c_{2n-i} = \bar c_i$.*
(iii) *λ^n divides $\lambda^\ell C(\lambda) - C_\ell(\lambda)$ for $\ell = 0, 1, \ldots, n-1$.*
(iv) *$C(\zeta) \neq 0$ for $\zeta \in \sigma(T) \cap \partial\mathbf{D}$.*

PROOF. If $\omega(\lambda, \mu) = c\lambda^i\mu^j + \bar c\lambda^j\mu^i$ then $C(\lambda) = c\lambda^{n+i-j} + \bar c\lambda^{n+j-i}$, and this polynomial satisfies (ii). The general case of (ii) follows from this, and (i) is equivalent to (ii). For (iii), we calculate

$$\lambda^\ell C(\lambda) - C_\ell(\lambda) = \sum_{i=0}^{n}\sum_{j=0}^{\ell} c_{ij}\lambda^{n+i-j+\ell},$$

and clearly $i - j + \ell \geq 0$ in the summation range. Finally, (iv) follows because $\omega(\zeta, \bar\zeta) \neq 0$, and $\bar\zeta = 1/\zeta$ for $\zeta \in \partial\mathbf{D}$. $\square$

Using the convention established earlier, we can now reformulate Problem 1.2.

Corollary 2.3. *We have $Qh = o(h)$ if and only if $Ch - \sum_{\ell=0}^{n-1} C_\ell h_\ell = U_+^n\Theta g + o(h)$, $h \in \mathcal{H}(\Theta)$, $g \in H^2(\mathcal{E})$.*

In the second relation above we used the notational convention established earlier. In other words there are vectors $h_N \in \mathcal{H}(\Theta)$, $h_N \neq 0$, and $g_N \in H^2(\mathcal{E})$ such that

$$\lim_{N\to\infty}\Big(Ch_N - \sum_{\ell=0}^{n-1} C_\ell(h_N)_\ell - U_+^n\Theta g_N\Big)/\|h_N\| = 0.$$

PROOF. By Lemma 2.1, for every $h \in \mathcal{H}(\Theta)$ there exists $g \in H^2(\mathcal{E})$ such that $U_+^{*n}(Ch - \sum_{\ell=0}^{n-1} C_\ell h_\ell) = Qh + \Theta g$. As observed before, $Ch - \sum_{\ell=0}^{n-1} C_\ell h_\ell$ belongs to the range of U_+^n, so the last relation is equivalent to

$$Ch - \sum_{\ell=0}^{n-1} C_\ell h_\ell = U_+^n Qh + U_+^n \Theta g.$$

Thus, if $Qh = o(h)$, then the second relation in the statement must hold. Conversely, if $Ch - \sum_{\ell=0}^{n-1} C_\ell h_\ell = U_+^n \Theta g + o(h)$ with $h \in \mathcal{H}(\Theta)$ and $g \in H^2(\mathcal{E})$, then

$$U_+^{*n}\left(Ch - \sum_{\ell=0}^{n-1} C_\ell h_\ell\right) = \Theta g + o(h)$$

and hence $Qh = P_{\mathcal{H}(\Theta)} U_+^{*n}(Ch - \sum_{\ell=0}^{n-1} C_\ell h_\ell) = o(h)$. □

Let us denote by $\alpha_1, \alpha_2, \ldots, \alpha_s$ the roots of C which also belong to $\sigma(T)$, repeated according to their multiplicities. Observe that these roots belong to $\mathbf{D}$. We can then factor C as $C = RS$, where $S(\lambda) = (\lambda - \alpha_1)(\lambda - \alpha_2)\cdots(\lambda - \alpha_s)$, while R has no zeros in $\sigma(T)$, and therefore $R(T)$ is an invertible operator. Observe that C may have no zeros in $\sigma(T)$, in which case $S = 1$ and $R = C$. In order to better understand the approach in this paper, let us consider in more detail the case when $S = 1$. The relation $Ch - \sum_{\ell=0}^{n-1} C_\ell h_\ell = U_+^n \Theta g + o(h)$ projected onto $\mathcal{H}(\Theta)$ yields $C(T)h - \sum_{\ell=0}^{n-1} C_\ell(T) P_{\mathcal{H}(\Theta)} h_\ell = o(h)$. Since $C(T)$ is invertible in this case, we have $h = \sum_{\ell=0}^{n-1} C(T)^{-1} C_\ell(T) P_{\mathcal{H}(\Theta)} h_\ell + o(h)$. This suggests that the condition $Qh = o(h)$ can be translated into a set of conditions on the Fourier coefficients $\{h_\ell\}_{\ell=0}^{n-1}$, thus reducing Problem 1.2 to a linear algebra problem (at least when $\mathcal{E}$ is a finite dimensional space; see [6]).

If $S \neq 1$ the situation is more involved. We begin by introducing linear operators $X_\ell : \mathcal{E} \to H^2(\mathcal{E})$ and $Y_\ell : \mathcal{E} \to \mathcal{H}(\Theta)$ by

$$Y_\ell \xi = R(T)^{-1} C_\ell(T) P_{\mathcal{H}(\Theta)} \xi, \quad \xi \in \mathcal{E}, 0 \le \ell \le n,$$

and

$$X_\ell \xi = \Theta^*(RY_\ell \xi - C_\ell \xi), \quad \xi \in \mathcal{E}, 0 \le \ell \le n.$$

Observe that $X_\ell \xi$ is indeed in $H^2(\mathcal{E})$ because

$$P_{\mathcal{H}(\Theta)}(RY_\ell \xi - C_\ell \xi) = R(T) Y_\ell \xi - C_\ell(T) \xi = 0.$$

Using these operators, Problem 1.2 can be reformulated as follows.

Proposition 2.4. *We have $Qh = o(h)$ if and only if the following system is satisfied*

$$Sh - \sum_{\ell=0}^{n-1} Y_\ell h_\ell = \Theta k + o(h)$$

$$U_+^n g - \sum_{\ell=0}^{n-1} X_\ell h_\ell - Rk = o(h)$$

with $h \in \mathcal{H}(\Theta)$ *and* $g, k \in H^2(\mathcal{E})$.

PROOF. Assume first that $Qh = o(h)$. We will analyze the equivalent condition given by Corollary 2.3. Since $C_\ell \xi = RY_\ell \xi - \Theta X_\ell \xi$, that condition is equivalent to

$$RSh - \sum_{\ell=0}^{n-1} RY_\ell h_\ell = \Theta\Big(U_+^n g - \sum_{\ell=0}^{n-1} X_\ell h_\ell\Big) + o(h).$$

Upon projecting on $\mathcal{H}(\Theta)$ we obtain $R(T)\Big(S(T)h - \sum_{\ell=0}^{n-1} Y_\ell h_\ell\Big) = o(h)$ and, since $R(T)$ is invertible, $S(T)h - \sum_{\ell=0}^{n-1} Y_\ell h_\ell = o(h)$. We must have therefore

$$Sh - \sum_{\ell=0}^{n-1} Y_\ell h_\ell = \Theta k + o(h)$$

with $k \in H^2(\mathcal{E})$, which is the first condition in the statement. Next observe that

$$\begin{aligned}
\Theta\Big(U_+^n g - \sum_{\ell=0}^{n-1} X_\ell h_\ell - Rk\Big) &= \Theta\Big(U_+^n g - \sum_{\ell=0}^{n-1} X_\ell h_\ell\Big) - R\Theta k \\
&= \Theta\Big(U_+^n g - \sum_{\ell=0}^{n-1} X_\ell h_\ell\Big) - \Big(RSh - \sum_{\ell=0}^{n-1} RY_\ell h_\ell\Big) + o(h) \\
&= o(h),
\end{aligned}$$

and hence the second condition also follows because Θ is inner. Conversely, assume that the two conditions in the statement are satisfied. Then

$$\begin{aligned}
Ch - \sum_{\ell=0}^{n-1} C_\ell h_\ell - U_+^n \Theta g &= Ch - \sum_{\ell=0}^{n-1} (RY_\ell h_\ell - \Theta X_\ell h_\ell) - U_+^n \Theta g \\
&= R\Big(Sh - \sum_{\ell=0}^{n-1} Y_\ell h_\ell\Big) + \Theta\Big(\sum_{\ell=0}^{n-1} X_\ell h_\ell - U_+^n g\Big) \\
&= \Theta\Big(Rk + \sum_{\ell=0}^{n-1} X_\ell h_\ell - U_+^n g\Big) + o(h) \\
&= o(h),
\end{aligned}$$

so that $Qh = o(h)$ by Corollary 2.3. □

With the notation of Proposition 2.4, let us observe that

$$S\Theta^* h - \Theta^* \sum_{\ell=0}^{n-1} Y_\ell h_\ell = k + o(h).$$

Since $Y_\ell h_\ell \in \mathcal{H}(\Theta)$ we have that $\Theta^* Y_\ell h_\ell$ is orthogonal to $H^2(\mathcal{E})$. Therefore we have

$$k = P_{H^2(\mathcal{E})}(S\Theta^* h) + o(h).$$

If we write $S(\lambda) = \sum_{j=0}^{s} \beta_j \lambda^j$, we have

$$(S\Theta^* h)(\zeta) = \sum_{j=0}^{s} \beta_j \zeta^j \sum_{\ell=1}^{\infty} \zeta^{-\ell} h_{-\ell},$$

and therefore

$$P_{H^2(\mathcal{E})}(S\Theta^* h) = \sum_{\ell=1}^{s} X_{-\ell} h_{-\ell},$$

where $X_{-\ell}(\lambda) = \sum_{j=\ell}^{s} \beta_j \lambda^{j-\ell}$ is a polynomial of degree exactly $s - \ell$ for $\ell = 1, 2, \ldots, s$; indeed, $\beta_s = 1$.

At this point it will be expedient to make two additional genericity assumptions.

Assumption 2.5. (i) $C(0) \neq 0$.
(ii) *The zeros of S are simple.*

Lemma 2.6. *If Assumption 2.5 is satisfied, then:*

(i) *$s \leq n$; and*
(ii) *the matrix $[X_{-\ell}(\alpha_j)]_{1 \leq \ell, j \leq s}$ is invertible.*

PROOF. Lemma 2.2.(i) shows that $C(1/\bar{\alpha}_j) = 0$ for $j = 1, 2, \ldots, s$. Thus (i) follows because C has at most $2n$ roots. To prove (ii), assume that there exist numbers δ_ℓ, $1 \leq \ell \leq s$, such that $\sum_{\ell=1}^{s} X_{-\ell}(\alpha_j)\delta_\ell = 0$ for $1 \leq j \leq s$. These equations can be written as $P(\alpha_j) = 0$, $1 \leq j \leq s$, where $P = \sum_{\ell=1}^{s} X_{-\ell}\delta_\ell$ is a polynomial of degree at most $s - 1$. We conclude that P must be identically zero. Now, the degree of $X_{-\ell}$ is exactly $s - \ell$, so the polynomials $X_{-1}, X_{-2}, \ldots, X_{-\ell}$ are linearly independent. We conclude that $\delta_\ell = 0$ for all ℓ, and (ii) follows. □

We can now state the main result of this paper.

Theorem 2.7. *If Assumption 2.5 is satisfied then we have $Qh = o(h)$ if and only if $\Omega\xi = o(\xi)$, $\xi = [\xi_\ell]_{\ell=-s}^{n-1} \in \mathcal{E}^{n+s}$, where $\Omega \in \mathcal{L}(\mathcal{E}^{n+s})$ is given by*

$$(\Omega\xi)_{-j} = \Theta(\alpha_j) \sum_{l=1}^{s} (X_{-\ell}\xi_{-\ell})(\alpha_j) + \sum_{\ell=0}^{n-1} (Y_\ell \xi_\ell)(\alpha_j)$$

for $j = 1, 2, \ldots, s$, *and*

$$(\Omega\xi)_j = \sum_{\ell=1}^{n-1} (X_\ell \xi_\ell)_j + (R \sum_{\ell=0}^{s} X_{-\ell}\xi_{-\ell})_j$$

for $j = 0, 1, \ldots, n-1$, *where the subscript* j *in the right hand side indicates the* j*th Fourier coefficient.*

PROOF. Assume first that $Qh = o(h)$. We will use the equivalent form of this condition given by Proposition 2.4. As we saw before the statement of Assumption 2.5, the vector k in the formulas of Proposition 2.4 can be replaced by $\sum_{\ell=1}^{s} X_{-\ell}h_{-\ell}$, and hence we get the equivalent conditions

$$\sum_{\ell=0}^{n-1} Y_\ell h_\ell + \Theta \sum_{\ell=1}^{s} X_{-\ell}h_{-\ell} = Sh + o(h),$$

and

$$\sum_{\ell=0}^{n-1} X_\ell h_\ell + R \sum_{\ell=1}^{s} X_{-\ell}h_{-\ell} = U_+^n g + o(h).$$

Evaluating the first expression above at $\lambda = \alpha_j$, and considering the first n Fourier coefficients in the second expression, we get that $\Omega\xi = o(h)$, with $\xi_j = h_j$, $-s \leq j \leq n-1$. Thus, to prove the first implication it will suffice to show that $h = O(\xi)$, i.e. $\|h\|/\|\xi\|$ is bounded. Observe that for almost every $\zeta \in \partial\mathbf{D}$ we have

$$h(\zeta) = \Big(\sum_{\ell=0}^{n-1} (Y_\ell h_\ell)(\zeta) + \Theta(\zeta) \sum_{\ell=1}^{s} X_{-\ell}(\zeta)h_{-\ell} + o(h)\Big)/S(\zeta),$$

and since $S(\zeta)$ is bounded away from zero on $\partial\mathbf{D}$, we get $\|h\| \leq \gamma\|\xi\| + o(h)$ for some constant γ. This clearly establishes that $h = O(\xi)$.

Conversely, assume that $\Omega\xi = o(\xi)$. This condition is equivalent to the existence of vectors $h', g \in H^2(\mathcal{E})$ such that

$$Sh' - \sum_{\ell=0}^{n-1} Y_\ell \xi_\ell = \Theta k + o(\xi), \tag{2.8}$$

and

$$U_+^n g - \sum_{\ell=0}^{n-1} X_\ell \xi_\ell - Rk = o(\xi), \tag{2.9}$$

where $k = \sum_{\ell=1}^{s} X_{-\ell}\xi_{-\ell}$. If we define $h = P_{\mathcal{H}(\Theta)}h'$ then h' can be written as $h' = h + \Theta\psi$ for some $\psi \in H^2(\mathcal{E})$. We have then by (2.8)

$$Sh + S\Theta\psi - \sum_{\ell=0}^{n-1} Y_\ell \xi_\ell = \Theta k + o(\xi),$$

and therefore

$$S\Theta^* h + S\psi - \Theta^* \sum_{\ell=0}^{n-1} Y_\ell \xi_\ell = k + o(\xi).$$

Projecting on $H^2(\mathcal{E})$ we obtain $P_{H^2(\mathcal{E})}(S\Theta^* h) + S\psi = k + o(\xi)$. Recalling that

$$P_{H^2(\mathcal{E})}(S\Theta^* h) = \sum_{\ell=1}^{s} X_{-\ell} h_{-\ell},$$

we obtain

$$\sum_{\ell=1}^{s} X_{-\ell}(h_{-\ell} - \xi_{-\ell}) + S\psi = o(\xi). \tag{2.10}$$

If we evaluate this at the points α_j we see that $\sum_{\ell=1}^{s} X_{-\ell}(\alpha_j)(h_{-\ell} - \xi_{-\ell}) = o(\xi)$, and Lemma 2.6 implies now that $h_{-\ell} - \xi_{-\ell} = o(\xi)$ for $\ell = 1, 2, \ldots, s$. Combining this with (2.10) we see that $S\psi = o(\xi)$, and hence (2.8) can be rewritten as

$$Sh - \sum_{\ell=0}^{n-1} Y_\ell \xi_\ell = \Theta k + o(\xi), \tag{2.11}$$

where now $h \in \mathcal{H}(\Theta)$. By Proposition 2.4, to conclude the proof it suffices to show that $\xi_\ell = h_\ell + o(\xi)$, $\ell = 0, 1, \ldots, n-1$, and $\xi = O(h)$. Note however that the second relation follows from the first, and from the fact proved earlier that $h_{-\ell} - \xi_{-\ell} = o(\xi)$ for $\ell = 1, 2, \ldots, s$.

If we multiply (2.11) by R and (2.9) by Θ we get

$$Ch - \sum_{\ell=0}^{n-1} RY_\ell \xi_\ell = R\Theta k + o(\xi),$$

and

$$\Theta U_+^n g - \sum_{\ell=0}^{n-1} \Theta X_\ell \xi_\ell - R\Theta k = o(\xi),$$

These two relations can be combined to yield

$$\Theta U_+^n g - \sum_{\ell=0}^{n-1} \Theta X_\ell \xi_\ell - Ch + \sum_{\ell=0}^{n-1} RY_\ell \xi_\ell = o(\xi).$$

Since $\Theta X_\ell \xi_\ell$ - $RY_\ell \xi_\ell = -C_\ell \xi_\ell$, we obtain

$$Ch - \sum_{\ell=0}^{n-1} C_\ell \xi_\ell = \Theta U_+^n g + o(\xi).$$

From Lemma 2.2.(iii) we deduce now that

$$C(\lambda)\Big(h(\lambda)-\sum_{\ell=0}^{n-1}\lambda^\ell\xi_\ell\Big)=\lambda^n g_1(\lambda)+o(\xi),\quad \lambda\in\mathbf{D},$$

with $g_1\in H^2(\mathcal{E})$. Setting $\lambda=0$, we obtain $C(0)(h_0-\xi_0)=o(\xi)$ and hence $h_0-\xi_0=o(\xi)$ because $C(0)\neq 0$. We conclude the proof by showing inductively that $h_\ell-\xi_\ell=o(\xi)$ for $\ell\le n-1$. Assume that this has been established for $\ell<k\le n-1$. Then we conclude that

$$C(\lambda)\Big(h(\lambda)-\sum_{\ell=0}^{k-1}\lambda^\ell h_\ell-\sum_{\ell=k}^{n-1}\lambda^\ell\xi_\ell\Big)=\lambda^n g_k(\lambda)+o(\xi),$$

with $g_k\in H^2(\mathcal{E})$. We now divide both sides of this equality by λ^k and set $\lambda=0$ to get $C(0)(h_k-\xi_k)=o(\xi)$, and hence $h_k-\xi_k=o(\xi)$ as desired. □

3. The Singular System

The equations $\Omega\xi=o(\xi)$ in Theorem 2.7 form the *singular system* associated with the operator Q. For numerical calculations it is important to find explicit formulas for the operator coefficients appearing in the singular system. These coefficients are the operators $\Theta(\alpha_j)$, $j=1,2,\dots,s$, $X_{-\ell}(\alpha_j)$, j, $\ell=1,2,\dots,s$, and X_ℓ,Y_ℓ $\ell=0,1,\dots n-1$. Thus, in order to calculate the singular system we must find the singular factor S of C, i.e., the roots $\alpha_1,\alpha_2,\dots,\alpha_s$. This gives us immediately the coefficients $\Theta(\alpha_j)$ as well as the polynomials $X_{-\ell}$ via the equation

$$X_{-\ell}(\lambda)=\sum_{j=\ell}^{s}\beta_j\lambda^{j-\ell},$$

where $S(\lambda)=\sum_{j=0}^{s}\beta_j\lambda^j$. We would like to show that the operators X_ℓ and Y_ℓ can also be calculated explicitly. To do this we recall the defining equation of these operators:

$$C_\ell\xi=RY_\ell\xi-\Theta X_\ell\xi,\quad \xi\in\mathcal{E},\tag{3.1}$$

with $Y_\ell\xi\in\mathcal{H}(\Theta)$ and $X_\ell\xi\in H^2(\mathcal{E})$. Thus

$$X_\ell\xi=R\Theta^*Y_\ell\xi-C_\ell\Theta^*\xi,\tag{3.2}$$

and hence

$$X_\ell\xi=P_{H^2(\mathcal{E})}R\Theta^*Y_\ell\xi-P_{H^2(\mathcal{E})}C_\ell\Theta^*\xi.$$

Since C_ℓ is a polynomial of degree at most $2n-1$, and $\Theta^*\xi$ has zero Fourier coefficients of order ≥ 1, the function $P_{H^2(\mathcal{E})}C_\ell\Theta^*\xi$ is a polynomial of degree

$\leq 2n-1$. Similarly, $\Theta^* Y_\ell \xi$ is orthogonal onto $H^2(\mathcal{E})$, and R has degree $\leq 2n-s$. Thus $P_{H^2(\mathcal{E})}(R\Theta^* Y_\ell \xi)$ is a polynomial of degree $\leq 2n-s-1$. We conclude that $X_\ell \xi$ itself is a polynomial of degree at most $2n-1$, say

$$X_\ell(z)\xi = \sum_{j=0}^{2n-1} z^j X_{\ell,j}\xi,$$

where $X_{\ell,j} \in \mathcal{L}(\mathcal{E})$ are linear operators. Observe that relation (3.1) also yields s of the coefficients of X; indeed, $X_{\ell,j}\xi$ must coincide with the corresponding coefficient in $-C_\ell \Theta^* \xi$ for $j = 2n-s, 2n-s+1, \ldots, 2n-1$. Thus we have the formulas

$$X_{\ell,j} = -\sum_{\alpha-\beta=j} C_{\ell,\alpha}\Theta^*_\beta$$

for $2n-s \leq j \leq 2n-1$. Here, of course, $C_\ell(z) = \sum_{\alpha=0}^{2n-1} C_{\ell,\alpha} z^\alpha$ and $\Theta(z) = \sum_{\beta=0}^{\infty} z^\beta \Theta_\beta$. Now, the polynomial

$$X_\ell(z)\xi - \sum_{j=2n-s}^{2n-1} z^j X_{\ell,j}\xi$$

is of degree $\leq 2n-s-1$, and could be determined by Lagrange interpolation once we determine its values at $2n-s$ distinct points. Equivalently, we must find the values of $X_\ell(z)\xi$ at $2n-s$ points. The points in question will be just the zeros of R. Indeed assume that $R(\alpha) = 0$. We consider first the case $|\alpha| \leq 1$, in which case, since α is not in the spectrum of T, the function Θ is analytic in a neighborhood of α and $\Theta(\alpha)$ is invertible ($\Theta(\alpha)$ is unitary if $|\alpha| = 1$). If we evaluate now (3.1) at $z = \alpha$ we obtain

$$X_\ell(\alpha) = -C_\ell(\alpha)\Theta(\alpha)^{-1}.$$

If $|\alpha| > 1$ we will use an equivalent version of (3.2) for values outside the unit disk. Observing that $1/\bar{z} = z$ for $|z| = 1$, we can rewrite (3.2) as

$$(X_\ell \xi)(z) = R(z)\Theta(1/\bar{z})^*(Y_\ell \xi)(1/\bar{z}) - C_\ell(z)\Theta(1/\bar{z})^*\xi$$

for $|z| > 1$. Setting $z = \alpha$ in this relation we obtain the interpolation condition

$$X_\ell(\alpha) = -C_\ell(\alpha)\Theta(1/\bar{\alpha})^*$$

for such zeros of R. Thus, if the zeros of R are distinct, the operators X_ℓ can be calculated numerically, and then Y_ℓ is simply determined by $Y_\ell = (C_\ell - \Theta X_\ell)/R$.

One must observe that the generic assumption that R has simple zeros is not essential in these considerations. If R has a multiple zero α, one can still derive the required number of conditions on X_ℓ by differentiating (3.1) with respect to z, and then setting $z = \alpha$. Then X is obtained by Hermite interpolation.

References

[1] H. Bercovici, *Operator Theory and Arithmetic in H^∞*, Math. Surveys and Monographs No. 56, Amer. Math. Soc., Providence, Rhode Island, 1988.

[2] H. Bercovici, C. Foias, and A. Tannenbaum, *On skew Toeplitz operators. I*, Operator Theory: Advances and Applications **29** (1988), 21–43.

[3] C. Foias, *Commutant lifting techniques for computing H^∞ controlers*. In: 'H_∞-control Theory,' Lecture Notes in Mathematics, No. 1496. Springer-Verlag, New York, 1991.

[4] C. Foias and A. Frazho, *The Commutant Lifting Approach to Interpolation Problems*. Operator Theory: Advances and Applications, Vol. 44. Birkhäuser, Boston, 1990.

[5] C. Foias and B. Sz.-Nagy, *Harmonic Analysis of Operators in Hilbert Space*. North Holland, Amsterdam, 1970.

[6] C. Foias and A. Tannenbaum, *Some remarks on optimal interpolation*, System and Control Letters **11** (1988), 259–265.

[7] C. Foias, A. Tannenbaum, and G. Zames, *On the H^∞-optimal sensitivity problem for systems with delays*, SIAM J. Control and Optimization **25** (1987), 686–705.

[8] C. Gu, *Eliminating the genericity conditions in the skew Toeplitz operator algorithm*, SIAM J. Math. Anal. **23** (1992), 1623–1638.

H. Bercovici & C. Foias
Department of Mathematics
Indiana University
Bloomington, IN 47405
E-mail: bercovic@indiana.edu,
foias@indiana.edu

A. Tannenbaum
Department of Electrical Engineering
University of Minnesota
Minneapolis, MN 55455
E-mail: tannenba@ee.umn.edu

Received: August 23rd, 1995.

Operator Theory:
Advances and Applications, Vol. 104

Regularity Questions for Free Convolution

HARI BERCOVICI AND DAN VOICULESCU*

For Carl Pearcy on his sixtieth birthday

1. INTRODUCTION

Free additive convolution is a binary operation on the set $\mathcal{M}$ of all probability measures on the real line $\mathbf{R}$. This operation was first defined in [7] for measures with finite moments of all orders (in particular for compactly supported measures). Maassen [5] extended this operation to measures with finite variance, and the extension to arbitrary measures was done in [1]. The free convolution of μ, $\nu \in \mathcal{M}$ is denoted $\mu \boxplus \nu$. Unlike classical convolution, free convolution is a highly non-linear operation, and therefore it is not obvious that various regularity properties of μ (like absolute continuity, differentiability, etc.) should be passed on to $\mu \boxplus \nu$. In some respects however free convolution has a stronger regularizing effect than classical convolution. It is our purpose in this paper to examine a few instances in which regularity properties of $\mu \boxplus \nu$ can be inferred. Some of the earlier results in this direction were only proved for measures with compact support. We will extend these results to general probability measures, giving as much of the technical detail as necessary. Among the new results, we show that there may be a loss of smoothness under free convolution. We also give a complete description of the atoms of a free convolution of probability measures.

2. THE CALCULATION OF FREE CONVOLUTION

An analogue of the Fourier transform for free convolution was discovered in [7] (and extended in [5] and [1]). This analogue involves the Cauchy transform of a measure $\mu \in \mathcal{M}$ which is the analytic function G_μ defined in the upper half-plane $\mathbf{C}^+ = \{z = x + iy : y > 0\}$ by the formula

$$G_\mu(z) = \int_{-\infty}^{\infty} \frac{d\mu(t)}{z - t}.$$

The function $F_\mu(z) = 1/G_\mu(z)$ is also analytic in $\mathbf{C}^+$ and

$$\Im F_\mu(z) \geq \Im z, \quad z \in \mathbf{C}^+.$$

*The authors were partially supported by grants from the National Science Foundation.

(cf. [5] and [1]; see also Lemma 4.2 in [9].) In addition, $(F_\mu(z)-z)/z$ tends to zero if $z \to \infty$ nontangentially to $\mathbf{R}$, i.e., such that $\Re z/\Im z$ stays bounded. It follows that F_μ is conformal in a truncated cone of the form

$$\Gamma_{\alpha,\beta} = \{z = x+iy : y \geq \beta,\ |x| \leq \alpha y\},$$

provided that $\alpha > 0$ and $\beta = \beta(\alpha)$ is sufficiently large. The inverse F_μ^{-1} is also defined in a similar truncated cone, and

$$\Im F_\mu^{-1}(z) \leq \Im z.$$

When more is known about the growth of μ, the domain of F_μ^{-1} can be suitably extended. Thus, if μ has compact support, F_μ^{-1} is defined in a neighborhood of ∞ [7], while for μ with finite variance F_μ^{-1} is defined in a half-plane $\{z : \Im z \geq \beta\}$ [5].

We can now define on the domain of F_μ^{-1} a new function

$$\phi_\mu(z) = F_\mu^{-1}(z) - z;$$

observe that $\Im\phi_\mu(z) \leq 0$. The function ϕ_μ makes it possible to handle free convolutions because

$$\phi_{\mu\boxplus\nu} = \phi_\mu + \phi_\nu, \quad \mu,\nu \in \mathcal{M},$$

where the equality holds in any truncated cone where all the three functions $\phi_{\mu\boxplus\nu}$, ϕ_μ, and ϕ_ν are defined (cf. [7], [5], and [1]).

3. Preliminaries about subordination in the upper half-plane

The subordination of analytic functions in the unit disk was first studied by J. Littlewood. If f and g are defined in the unit disk, f is subordinate to g if $f = g \circ \omega$, where ω is analytic in the unit disk and $|\omega(z)| \leq |z|$. If f is subordinate to g then one can make estimates of f (e.g., Hardy space norm estimates) in terms of g. A version of subordination appropriate for the upper half-plane is useful in the study of free convolution. A good reference for the facts reviewed below is [4].

For $y > 0$ the Poisson kernel $P_y \in L^1(\mathbf{R})$ is defined by

$$P_y(t) = \frac{1}{\pi}\frac{y}{t^2+y^2}, \quad t \in \mathbf{R}.$$

Given a (generally complex) measure σ on $\mathbf{R}$ such that $\int_{-\infty}^{\infty} 1/(t^2+1)\,d|\sigma|(t) < \infty$, the Poisson integral of σ is the harmonic function $P[\sigma]$ defined by

$$P[\sigma](x+iy) = (P_y * \sigma)(x) = \int_{-\infty}^{\infty} P_y(x-t)\,d\sigma(t)$$

in the upper half-plane. If σ is a positive measure then $P[\sigma]$ is a positive function. Conversely, an arbitrary positive harmonic function u in $\mathbf{C}^+$ can be written as

$$u(x+iy) = \beta y + P[\sigma]$$

for some $\beta > 0$ and some positive measure σ such that $\int_{-\infty}^{\infty} 1/(t^2+1)\,d\sigma(t) < \infty$. The measure σ can be recovered from u as the weak*-limit as $y \to 0$ of the measures $u(t+iy)\,dt$, while $\beta = \lim_{y\to\infty} u(iy)/y$. Thus u is a Poisson integral if and only if $\lim_{y\to\infty} u(iy)/y = 0$. The measure σ is finite if and only if $\sup_{y>0} yP[\sigma](iy) < \infty$. In this case

$$\sigma(\mathbf{R}) = \pi \lim_{y\to+\infty} yP[\sigma](iy) = \pi \sup_{y>0} yP[\sigma](iy) = \pi \sup_{\Im z>0} (\Im z)P[\sigma](z),$$

and the L^1 norms $\int_{-\infty}^{\infty} P[\sigma](t+iy)\,dt$ are all equal to $\sigma(\mathbf{R})$ for $y > 0$.

We are now ready to prove a version of the Littlewood subordination theorem. A slightly different version appears in [9], and in any case the result is probably known.

Proposition 3.1. *Let σ be a finite positive measure on $\mathbf{R}$, set $u = P[\sigma]$, and let $\omega : \mathbf{C}^+ \to \mathbf{C}^+$ be an analytic function such that $\Im\omega(z) \geq \Im z$ for $z \in \mathbf{C}^+$. Then the harmonic function $u \circ \omega$ is of the form $P[\sigma']$, where σ' is a positive meaasure on $\mathbf{R}$ with $\sigma'(\mathbf{R}) \leq \sigma(\mathbf{R})$.*

PROOF. Let us set $v(z) = u(\omega(z))$. It will suffice to show that $\pi y v(iy) \leq \sigma(\mathbf{R})$ for $y > 0$. Indeed, if this is shown, the condition $\lim_{y\to\infty} v(iy)/y = 0$ is automatically satisfied. Thus we just estimate

$$\pi y v(iy) = \frac{y}{\Im\omega(iy)}\pi(\Im\omega(iy))u(\omega(iy)) \leq \frac{y}{\Im\omega(iy)}\sigma(\mathbf{R}) \leq \sigma(\mathbf{R}),$$

where we used in the last inequality the fact that $y \leq \Im\omega(iy)$. □

This subordination result can also be used to estimate L^p norms. Do do this, assume that $d\sigma(t) = f(t)\,dt$, with $f \in L^p(\mathbf{R})$, $1 \leq p \leq \infty$. In this case $u = P[\sigma]$ is also denoted $P[f]$. The functions u_y defined by $u_y(x) = u(x+iy)$ belong to $L^p(\mathbf{R})$, $\|u_y\|_p \leq \|f\|_p$, and $\|f\|_p = \lim_{y\to 0} \|u_y\|_p$. Moreover, $f(x) = \lim_{y\to 0} u(x+iy)$ for almost every $x \in \mathbf{R}$. Conversely, if $p \in (1,\infty)$ and u is a harmonic function in $\mathbf{C}^+$ such that $\sup_{y>0} \|u_y\|_p < \infty$, then u is the Poisson integral of a function $f \in L^p(\mathbf{R})$. This is also true for $p = \infty$ provided that $\sup_{y>0} y\|u_y\|_\infty < \infty$. The corresponding statement for $p = 1$ is not true because the Poisson integral u of a finite complex measure always satisfies $\sup_{y>0} \|u_y\|_1 < \infty$.

The following result was proved in [9, Lemma 4.1] under a somewhat more restrictive hypothesis.

Proposition 3.2. *Fix a function $f \in L^p(\mathbf{R})$, $1 < p \leq \infty$, and an analytic function $\omega : \mathbf{C}^+ \to \mathbf{C}^+$ such that $\Im\omega(z) \geq \Im z$, $z \in \mathbf{C}^+$. Denote by F the Poisson integral of f. Then the function $G = F \circ \omega$ is the Poisson integral of a function $g \in L^p(\mathbf{R})$ with $\|g\|_p \leq \|f\|_p$.*

PROOF. Denote by u the Poisson integral of the function $|f|^p$, and observe that Jensen's inequality implies that $|F(z)|^p \le u(z)$, $z \in \mathbf{C}^+$. We conclude that $|G(z)|^p \le u(\omega(z))$, $z \in \mathbf{C}^+$. By Proposition 3.1, $u \circ \omega$ is the Poisson integral of a measure with total mass $\le \int_{-\infty}^{\infty} |f(t)|^p \, dt = \|f\|_p^p$, so that $\int_{-\infty}^{\infty} u(\omega(x+iy))\, dx \le \|f\|_p^p$, $y > 0$. Therefore $\int_{-\infty}^{\infty} |G(x+iy)|^p \, dx \le \|f\|_p^p$, $y > 0$, and this implies the result in view of the remarks preceding the statement of the proposition. □

A consequence of Proposition 3.2 is that we also have (with the notation of the proposition) $\|G_y\|_p \le \|F_y\|_p$ for all $y > 0$. Indeed, Proposition 3.2 can be applied in the half-plane $\mathbf{C}^+ + iy$, and $F|\mathbf{C}^+ + iy$ is the Poisson integral of F_y.

Another type of inequality can be obtained by considering the energies

$$I_\alpha[\mu] = \int_{-\infty}^{\infty}\int_{-\infty}^{\infty} |x-y|^{\alpha-2}\, d\mu(x)\, d\mu(y)$$

defined for a positive measure μ and for $0 < \alpha < 2$. These are the energies associated with the M. Riesz kernels. Another quantity of interest is the logarithmic potential

$$I_2[\mu] = -\int_{-\infty}^{\infty}\int_{-\infty}^{\infty} \log|x-y|\, d\mu(x)\, d\mu(y).$$

If $d\mu = f\,dt$ we will also write $I_\alpha[f] = I_\alpha[\mu]$. In order for $I_2[\mu]$ to be defined (possibly $+\infty$) we will assume that $\iint_{|x-y|>1} \log|x-y|\, d\mu(x)\, d\mu(y) < \infty$. This condition is satisfied if, for instance, μ has compact support. The quantity $\Sigma(\mu) = -I_2[\mu]$ was shown in [9] to be a good substitute for entropy in the context of free probability theory. The following result appears as Lemma 6.1 in [9] for compactly supported measures.

Lemma 3.3. *Let $\mu \in \mathcal{M}$, let $\omega : \mathbf{C}^+ \to \mathbf{C}^+$ be an analytic function such that $\Im\omega(z) \ge \Im z$, $z \in \mathbf{C}^+$, and set $F = P[\mu]$. Then we have*

$$I_\alpha[(F \circ \omega)_y] \le I_\alpha[F_y]$$

for every $y > 0$ and $\alpha \in (0,2)$. If $\lim_{y\to\infty} \omega(iy)/y = 1$, then the inequality holds for $\alpha = 2$ as well.

4. REGULARITY RESULTS DERIVED FROM SUBORDINATION

The following result was proved in [9] for measures with compact support (under a genericity condition) and by Biane [3] for the general case (see Proposition 4.4 in [9] and Theorem 3.1 in [3]).

Proposition 4.1. *Given $\mu, \nu \in \mathcal{M}$, there exists an analytic function $\omega : \mathbf{C}^+ \to \mathbf{C}^+$ such that*

(i) $G_{\mu \boxplus \nu} = G_\mu \circ \omega$;

(ii) $\Im\omega(z) \geq \Im z$ *for* $z \in \mathbf{C}^+$; *and*
(iii) $\omega(z) = z(1 + o(1))$ *when* $z \to \infty$ *nontangentially to* $\mathbf{R}$.

We will not prove here the existence of ω for which we refer to Biane [3]. We will just argue that (ii) and (iii) are true. Recall first that ω must have a Nevanlina representation, say

$$\omega(z) = \alpha + \beta z + \int_{-\infty}^{\infty} \frac{1+tz}{t-z}\, d\sigma(t),$$

where $\alpha \in \mathbf{R}$, $\beta \geq 0$, and σ is a finite positive measure on $\mathbf{R}$. From this representation it follows easily that $\omega(z) = \beta z(1+o(1))$ as z tends to infinity nontangentially to $\mathbf{R}$. By Proposition 5.2 of [1] we similarly have $F_\mu(z) = z(1 + o(1))$. Combining this with the identity $F_{\mu\boxplus\nu} = F_\mu \circ \omega$ we conclude that $\beta = 1$. Statements (ii) and (iii) follow immediately.

Proposition 4.1 might be restated by saying that $G_{\mu\boxplus\nu}$ is subordinated to G_μ (and, by symmetry, to G_ν) in a sense appropriate for the upper half-plane. Proposition 3.2 can now be used to obtain some regularity results. The following two results were first proved for measures with compact support in [9]. The general case is proved in the same way from Proposition 4.1 and Proposition 3.2.

Proposition 4.2. *For* μ, $\nu \in \mathcal{M}$, $y > 0$, *and* $1 < p \leq \infty$, *we have*

$$\begin{aligned} \|(G_{\mu\boxplus\nu})_y\|_p &\leq \|(G_\mu)_y\|_p, \\ \|(\Re G_{\mu\boxplus\nu})_y\|_p &\leq \|(\Re G_\mu)_y\|_p, \text{ and} \\ \|(\Im G_{\mu\boxplus\nu})_y\|_p &\leq \|(\Im G_\mu)_y\|_p . \end{aligned}$$

If μ *is absolutely continuous relative to Lebesgue measure, and its density* f *is in* $L^p(\mathbf{R})$, *then* $\mu \boxplus \nu$ *is also absolutely continuous, and its density* g *satisfies* $\|g\|_p \leq \|f\|_p$.

We only comment briefly on the proof. For the last statement of the proposition above, one only needs to note that $-\Im G_\mu = \pi P[\mu]$ and apply Proposition 3.2. The inequality involving $\Im G_\mu$ also follows from the remark immediately following Proposition 3.2. The other two inequalities can be deduced from the fact that, if $(\Re G_\mu)_y \in L^p(\mathbf{R})$ then $(\Im G_\mu)_y \in L^p(\mathbf{R})$ by a well-known result of M. Riesz; Indeed, $(\Im G_\mu)_y$ is (up to sign) the Hilbert transform of $(\Re G_\mu)_y$. It follows then that G_μ and $\Re G_\mu$, restricted to a half-plane of the form $\mathbf{C}^+ + i\varepsilon$, $\varepsilon > 0$, are Poisson integrals of functions in $L^p(\mathbf{R})$, and Proposition 3.2 can be applied again.

An analogue of Proposition 4.2 for the Riesz energies is as follows (see [9] for the case of compact supports).

Proposition 4.3. *With the notation of Proposition 4.2, we have*

$$I_\alpha[(G_{\mu\boxplus\nu})_y] \leq I_\alpha[(G_\mu)_y]$$

for $\alpha \in (0,2)$ *and* $y > 0$.

The preceding results were stated asymmetrically, but μ could be replaced by ν in the right hand sides of all inequalities. For $\alpha = 2$ a more powerful inequality was proved in [6] for compactly supported measures. It is as follows (see Theorem 2.1′ in [6]).

Theorem 4.4. *If μ, $\nu \in \mathcal{M}$ are compactly supported, then*

$$\exp(-2I_2[\mu]) + \exp(-2I_2[\nu]) \le \exp(-2I_2[\mu \boxplus \nu]).$$

5. Regularity results derived from the continuity of convolution

It was shown in [1] that free convolution is Lipschitz continuous with respect to two natural metrics on $\mathcal{M}$. These metrics are defined using the distribution function $\mathcal{F}_\mu$ associated with a measure $\mu \in \mathcal{M}$:

$$\mathcal{F}_\mu(t) = \mu((-\infty, t)), \quad t \in \mathbf{R}.$$

Given $\mu, \nu \in \mathcal{M}$ one defines the uniform distance

$$d_\infty(\mu, \nu) = \sup_{t \in \mathbf{R}} |\mathcal{F}_\mu(t) - \mathcal{F}_\nu(t)|,$$

and the Lévy distance

$$d(\mu, \nu) = \inf\{\varepsilon > 0 : \mathcal{F}_\mu(t - \varepsilon) - \varepsilon \le \mathcal{F}_\nu(t) \le \mathcal{F}_\mu(t + \varepsilon) + \varepsilon \text{ for all } t\}.$$

Then we have the following result (Proposition 4.13 in [1]).

Proposition 5.1. *If μ, μ', ν, $\nu' \in \mathcal{M}$, then*

$$d(\mu \boxplus \nu, \mu' \boxplus \nu') \le d(\mu, \mu') + d(\nu, \nu')$$

and

$$d_\infty(\mu \boxplus \nu, \mu' \boxplus \nu') \le d_\infty(\mu, \mu') + d_\infty(\nu, \nu').$$

The proof of this result is based on a remarkable realization of close measures as distributions of noncommutative random variables which coincide on a 'set' of large measure. We digress a little in order to give the statement of one such result. Let A be a finite von Neumann algebra with normal trace state τ. If X is a selfadjoint operator affiliated with A, say $X = \int_{-\infty}^{\infty} t\, de_X(t)$, where e_X is A-valued, one defines the distribution μ_X of X by setting $\mu_X(\sigma) = \tau(e_X(\sigma))$ for every Borel subset σ of $\mathbf{R}$. The following result is Theorem 3.9 in [1].

Theorem 5.2. *Let A be a finite von Neumann algebra with normal trace state τ, let X, Y be selfadjoint operators affiliated with A, and let δ be a positive number. If there exists a projection $p \in A$ such that $pXp = pYp$ and $\tau(p) \ge 1 - \delta$, then $d_\infty(\mu_X, \mu_Y) \le \delta$. Conversely, if $\mu, \nu \in \mathcal{M}$ and $d_\infty(\mu, \nu) \le \delta$, then there exist a finite von Neumann algebra A with normal trace state τ, selfadjoint operators X, Y affiliated with A, and a projection $p \in A$ such that $\mu_X = \mu$, $\mu_Y = \nu$, $pXp = pYp$, and $\tau(p) \ge 1 - \delta$.*

Proposition 5.1 can be used to infer some regularity results for the distribution function of $\mu \boxplus \nu$. We recall that a function f defined on $\mathbf{R}$ is α-Hölder with constant c if $|f(x)-f(y)| \leq c|x-y|^{\alpha}$ for all $x, y \in \mathbf{R}$. This notion is only useful for $\alpha \in (0,1]$.

Proposition 5.3. *Let μ, $\nu \in \mathcal{M}$ and $\alpha \in (0,1]$ be such that $\mathcal{F}_{\mu}$ is α-Hölder with constant c. Then $\mathcal{F}_{\mu \boxplus \nu}$ is also α-Hölder with constant $\leq c$.*

PROOF. It is easy to see that for $h \in \mathbf{R}$ $\mu \boxplus \delta_h$ is just the translate of μ by h, i.e., $\mathcal{F}_{\mu \boxplus \delta_h}(x) = \mathcal{F}_{\mu}(x-h)$ for all x. Thus, the fact that $\mathcal{F}_{\mu}$ is α-Hölder with constant c can be rewritten as

$$d_{\infty}(\mu \boxplus \delta_h, \mu) \leq ch^{\alpha}, \quad h > 0.$$

Proposition 4.1 implies now that

$$\begin{aligned} d_{\infty}((\mu \boxplus \nu) \boxplus \delta_h, \mu \boxplus \nu) &\leq d_{\infty}((\mu \boxplus \delta_h) \boxplus \nu, \mu \boxplus \nu) \\ &\leq d_{\infty}(\mu \boxplus \delta_h, \mu), \end{aligned}$$

and this implies the result. □

Remark that Proposition 4.2 for $p = \infty$ can also be deduced from Proposition 5.3. Indeed, μ is absolutely continuous with bounded derivative if and only if $\mathcal{F}_{\mu}$ is Lipschitz (=1-Hölder), and the smallest Lipschitz constant coincides with the L^{∞}-norm of the derivative.

6. Irregularity phenomena

We have seen in the previous sections some instances when the regularity of μ (Hölder continuity, absolute continuity) is passed on to $\mu \boxplus \nu$. One may ask whether differentiability properties are passed on as well. We will see however that this is not the case. It was shown in [2], in relation to the free central limit theorem, that for every compactly supported measure μ with nonzero variance, the n-fold convolution

$$\mu^{\boxplus n} = \underbrace{\mu \boxplus \mu \boxplus \cdots \boxplus \mu}_{n \text{ times}}$$

is absolutely continuous for large n, and its density looks like a semicircle. For such values of n, $\mu^{\boxplus n}$ has a density which is differentiable at all but two points where a one-sided derivative is infinite. Thus the density is not of class C^1, even though the original measure μ might have had a density of class C^{∞}.

Corollary 6.1. *There exist compactly supported measures $\mu, \nu \in \mathcal{M}$ such that μ has density of class C^{∞} and ν has density of class C^1, but the density of $\mu \boxplus \nu$ is not of class C^1.*

For the proof choose $\nu = \mu^{\boxplus n}$, where n is the largest integer so that ν has density of class C^1.

One might remark that there exist measures μ with unbounded support such that $\mu^{\boxplus n}$ has density of class C^∞ for all n. The Cauchy distribution is an example because $\mu^{\boxplus n}$ are also Cauchy distributions.

7. FREE CONVOLUTION AND ATOMS

Let μ, $\nu \in \mathcal{M}$, and assume that α, $\beta \in \mathbf{R}$ are atoms of μ, ν, respectively, such that $\mu(\{\alpha\}) + \nu(\{\beta\}) > 1$. It is easy to see in this case that $\alpha + \beta$ is an atom of $\mu \boxplus \nu$ with mass at least $\mu(\{\alpha\}) + \nu(\{\beta\}) - 1$. The easiest way to argue is to note that there exists a finite von Neumann algebra A with normal trace state τ, and selfadjoint operators X, Y affiliated with A, such that μ, ν, $\mu + \nu$ are the distributions of X, Y, $X + Y$, respectively. The fact that α (resp. β) is an atom of μ (resp. ν) is equivalent to the existence of a projection p (resp. q) with $\tau(p) = \mu(\{\alpha\})$ (resp. $\tau(q) = \nu(\{\beta\})$) such that $Xp = \alpha p$ (resp. $Yq = \beta q$). Then $(X + Y)(p \wedge q) = (\alpha + \beta)(p \wedge q)$, and $1 - \tau(p \wedge q) \leq (1 - \tau(p)) + (1 - \tau(q))$.

We will see in this section that this is the only way in which atoms of $\mu \boxplus \nu$ arise. The tool will be again subordination. We begin with two simple results which help identify an atom of μ using the Cauchy transform G_μ.

Lemma 7.1. *Fix $\mu \in \mathcal{M}$ and $\alpha \in \mathbf{R}$. Then $(z - \alpha)G_\mu(z) \to \mu(\{\alpha\})$ as $z \to \alpha$ nontangentially to $\mathbf{R}$.*

PROOF. It suffices to prove the lemma for $\alpha = 0$. Also, considering the measure $\mu - \mu(\{0\})\delta_0$ we can reduce to the case when 0 is not an atom of μ, in which case we must show that $zG_\mu(z) \to 0$ as $z \to 0$ nontangentially. Indeed,

$$|zG_\mu(z)| \leq \int_{-\infty}^{\infty} \frac{|z|}{|z - t|}\, d\mu(t),$$

and $|z|/|z - t|$ converges to zero μ almost everywhere as $z \to 0$. If $z \to 0$ nontangentially, the integrand stays bounded as well because

$$\frac{|z|}{|z - t|} = \frac{|\frac{x}{y} + i|}{|\frac{x-t}{y} + i|} \leq \left|\frac{x}{y} + i\right|,$$

where $z = x + iy$, and x/y is bounded. □

A sequence $\{z_n\}_{n=0}^\infty \subset \mathbf{C}^+$ converges tangentially to $\mathbf{R}$ if

$$\lim_{n \to \infty} \frac{y_n}{x_n - t} = 0$$

for all $t \in \mathbf{R}$, where $x_n = \Re z_n$ and $y_n = \Im z_n$. Observe that, if $\{z_n\}_{n=0}^\infty$ does not converge tangentially to $\mathbf{R}$, then a subsequence of $\{z_n\}_{n=0}^\infty$ is contained in an angle of the form $\{x+iy : |x-t| \le \gamma y\}$ for some $t \in \mathbf{R}$ and some $\gamma > 0$. The following result shows that sequences which converge tangentially to $\mathbf{R}$ do not yield enough information about the atoms of a measure in $\mathcal{M}$.

Lemma 7.2. *If $\mu \in \mathcal{M}$, and $\{z_n\}_{n=0}^{\infty} \subset \mathbf{C}^+$ converges tangentially to $\mathbf{R}$, then $\lim_{n\to\infty}(\Im z_n)G_\mu(z_n) = 0$.*

PROOF. Write $z_n = x_n + iy_n$, so that

$$(\Im z_n)G_\mu(z_n) = \int_{-\infty}^{\infty} \frac{y_n}{(x_n - t) + iy_n}\, d\mu(t).$$

Since clearly

$$\left|\frac{y_n}{(x_n - t) + iy_n}\right| \le 1,$$

the result will follow from the dominated convergence theorem once we verify that $\lim_{n\to\infty} y_n/(x_n - t) + iy_n = 0$ for $t \in \mathbf{R}$. This however is clear because

$$\left|\frac{y_n}{(x_n - t) + iy_n}\right| \le \frac{y_n}{|x_n - t|}. \qquad \square$$

For the remainder of this section we fix measures $\mu, \nu \in \mathcal{M}$, and we consider the analytic functions $\omega_\mu, \omega_\nu : \mathbf{C}^+ \to \mathbf{C}^+$ provided by Proposition 4.1, satisfying the identities $G_{\mu\boxplus\nu} = G_\mu \circ \omega_\mu$, $G_{\mu\boxplus\nu} = G_\nu \circ \omega_\nu$.

Proposition 7.3. *We have $\omega_\mu(z) + \omega_\nu(z) = z + F_{\mu\boxplus\nu}(z)$ for every $z \in \mathbf{C}^+$.*

PROOF. Observe that we also have $F_{\mu\boxplus\nu} = F_\mu \circ \omega_\mu$, and this implies that we can write $\omega_\mu(z) = F_\mu^{-1}(F_{\mu\boxplus\nu}(z))$ in a convenient truncated cone of the form $\Gamma_{\alpha,\beta}$ (see Section 2). Since $F_\mu^{-1}(z) + F_\nu^{-1}(z) = z + \phi_\mu(z) + z + \phi_\nu(z) = z + z + \phi_{\mu\boxplus\nu}(z) = z + F_{\mu\boxplus\nu}^{-1}(z)$, the relation in the statement follows immediately for z in a sufficiently small truncated cone. The validity of the equation for all z follows by analytic continuation. $\square$

We are now ready for the main result of this section.

Theorem 7.4. *Let μ, $\nu \in \mathcal{M}$, and let γ be a real number. The following are equivalent:*

(i) *γ is an atom of $\mu \boxplus \nu$;*
(ii) *there exist atoms α, β for μ, ν, respectively, such that $\gamma = \alpha + \beta$ and $\mu(\{\alpha\}) + \nu(\{\beta\}) > 1$.*

If (ii) is satisfied then $(\mu \boxplus \nu)(\{\gamma\}) = \mu(\{\alpha\}) + \nu(\{\beta\}) - 1$.

PROOF. We have already seen that (ii) implies (i). Assume therefore that (i) holds. Replacing μ by $\mu \boxplus \delta_{-\gamma}$ we may assume that $\gamma = 0$. By Proposition 7.3 we have

$$\frac{\omega_\mu(iy)}{iy} + \frac{\omega_\nu(iy)}{iy} = 1 + \frac{1}{iyG_{\mu\boxplus\nu}(iy)}$$

for $y > 0$. Moreover, by Lemma 7.1, the right hand side of this equation converges to the real number $1 + 1/(\mu \boxplus \nu)(\{0\})$ as $y \to 0$. We conclude that $\Im\omega_\mu(iy) = O(y)$ as $y \to 0$, and hence $\omega_\mu(iy)$ approaches the real line as $y \to 0$. Since $\Im\omega_\mu(iy) \geq y$, the identity

$$(\Im\omega_\mu(iy))G_\mu(\omega_\mu(iy)) = (\Im\omega_\mu(iy))G_{\mu\boxplus\nu}(iy) = \frac{\Im\omega_\mu(iy)}{iy} iyG_{\mu\boxplus\nu}(iy)$$

shows then that $(\Im\omega_\mu(iy))G_\mu(\omega_\mu(iy))$ stays bounded away from zero as $y \to 0$. If $y_n > 0$ is a sequence converging to zero, Lemma 7.2 implies that $\omega_\mu(iy_n)$ does not converge tangentially to $\mathbf{R}$. Passing to a subsequence, we may assume that there exist $\alpha \in \mathbf{R}$ and $\gamma > 0$ such that $\omega_\mu(iy_n)$ belongs to the cone $\{x + iy : |x - \alpha| \leq \gamma y\}$. Thus $\omega_\mu(iy_n)$ converges to α nontangentially to $\mathbf{R}$. Dropping to a further subsequence, we may also assume that $\omega_\nu(iy_n)$ converges nontangentially to some $\beta \in \mathbf{R}$ as $n \to \infty$. Since

$$\omega_\mu(iy_n) + \omega_\nu(iy_n) = iy_n \left(1 + \frac{1}{iy_nG_{\mu\boxplus\nu}(iy_n)}\right),$$

we conclude easily that $\alpha + \beta = 0$. Finally we note that

$$(\omega_\mu(iy_n) - \alpha)G_{\mu\boxplus\nu}(iy_n) + (\omega_\nu(iy_n) - \beta)G_{\mu\boxplus\nu}(iy_n) = iy_nG_{\mu\boxplus\nu}(iy_n) + 1$$

or, equivalently,

$$(\omega_\mu(iy_n) - \alpha)G_\mu(\omega_\mu(iy_n)) + (\omega_\nu(iy_n) - \beta)G_\nu(\omega_\nu(iy_n)) = iy_nG_{\mu\boxplus\nu}(iy_n) + 1.$$

Let now $n \to \infty$ and recall that $\omega_\mu(iy_n) \to \alpha$ nontangentially. Lemma 7.1 gives now the desired identity

$$\mu(\{\alpha\}) + \nu(\{\beta\}) = (\mu \boxplus \nu)(\{0\}) + 1.$$ □

We note a few immediate consequences of this result. The proofs are left to the interested reader.

Corollary 7.5. (i) *If μ and ν are not Dirac measures then $\mu \boxplus \nu$ has at most a finite number of atoms.*
(ii) *$\mu \boxplus \mu$ has at most one atom.*
(iii) *If μ is $\boxplus$-infinitely divisible then μ has at most one atom.*

Part (iii) was observed before in [1].

REFERENCES

[1] H. BERCOVICI AND D. VOICULESCU, *Free convolution of measures with unbounded support*, Indiana Univ. Math. J. **42** (1993), 733–773.

[2] H. BERCOVICI AND D. VOICULESCU, *Superconvergence to the central limit and failure of the Cramér theorem for free random variables*, Probability Theory Related Fields **102** (1995), 215–222.

[3] PH. BIANE, *On processes with free increments*, (Preprint).

[4] P. DUREN, *H^p Spaces*, Academic Press, New York, 1972.

[5] H. MAASSEN, *Addition of freely independent random variables*, J. Funct. Anal. **106** (1992), 409–438.

[6] S. SZAREK AND D. VOICULESCU, *Volumes of restricted Minkowski sums and the free analogue of the entropy power inequality*, (Preprint).

[7] D. VOICULESCU, *Symmetries of some reduced free product C^*-algebras*, in: "Operator Algebras and Their Connections with Topology and Ergodic Theory." Springer-Verlag, New York, 1985. (Lecture Notes in Mathematics, No. 1132, pp. 556–588.)

[8] D. VOICULESCU, *Addition of certain non-commuting random variables*, J. Funct. Anal. **66** (1986), 323–346.

[9] D. VOICULESCU, *The analogues of entropy and Fisher's information measure in free probability theory. I*, Commun. Math. Phys. **155** (1993), 71–92.

[10] D. VOICULESCU, K. DYKEMA, AND A. NICA, *Free random variables*, CRM Monograph Series No. 1. Amer. Math. Soc., Providence, Rhode Island, 1992.

H. BERCOVICI
Department of Mathematics
Indiana University
Bloomington, IN 47405
E-MAIL: bercovic@indiana.edu

D. VOICULESCU
Department of Mathematics
University of California
Berkeley, CA 94720
E-MAIL: dvv@math.berkeley.edu

Received: August 23rd, 1995.

Operator Theory:
Advances and Applications, Vol. 104

Operators of Putinar Type

SCOTT W. BROWN AND EUNGIL KO

This paper is dedicated to Carl Pearcy on the event of his 60th birthday

ABSTRACT. Mihai Putinar has shown that if T is a hyponormal operator acting on a Hilbert space $\mathbf{H}$, then there exists a (generalized) scalar operator S acting on a Hilbert space $\mathbf{K}$ and a map $V : \mathbf{H} \longrightarrow \mathbf{K}$ such that $VT = SV$. Furthermore, he has shown that the map V arising in this construction is bounded below. The construction can be applied to any operator. By slightly modifying Putinar's proof, it is shown that the construction yields an operator V which is nonzero (and sometimes one–to–one) for operators closely related to hyponormal operators.

1. INTRODUCTION

Let $\mathbf{H}$ and $\mathbf{K}$ be separable, complex Hilbert spaces and $\mathcal{L}(\mathbf{H}, \mathbf{K})$ denote the space of all linear, bounded operators from $\mathbf{H}$ to $\mathbf{K}$. If $\mathbf{H} = \mathbf{K}$, we write $\mathcal{L}(\mathbf{H})$ in place of $\mathcal{L}(\mathbf{H}, \mathbf{K})$.

Let D be an open disc in $\mathbf{C}$ the complex plane. Let $\bar{\mathrm{D}}$ denote the closure of D in $\mathbf{C}$. And let $C^2(\bar{\mathrm{D}})$ denote the space of continuous complex valued functions on $\bar{\mathrm{D}}$ with first and second order partials that have continuous extensions from D to $\bar{\mathrm{D}}$. The notation $\|f\|_\infty$ will be used to denote the sup norm of a function f on $\bar{\mathrm{D}}$. Then define for $f \in C^2(\bar{\mathrm{D}})$,

$$\|f\| \equiv \|f\|_\infty + \|f_x\|_\infty + \|f_y\|_\infty + \|f_{xy}\|_\infty + \|f_{xx}\|_\infty + \|f_{yy}\|_\infty$$

where for example f_x denotes the partial with respect to the coordinate variable x. This makes $C^2(\bar{\mathrm{D}})$ into a Banach space. Note that the pointwise product of two functions in $C^2(\bar{\mathrm{D}})$ is again in $C^2(\bar{\mathrm{D}})$ (in fact $C^2(\bar{\mathrm{D}})$ with this norm is a topological algebra).

An operator S is called scalar of order two (or 2-scalar) if for some open disc D in $\mathbf{C}$ there exists a map

$$\Phi : C^2(\bar{\mathrm{D}}) \longrightarrow \mathcal{L}(\mathbf{K})$$

such that 1) Φ is an algebra homomorphism, and 2) Φ is continuous when the above norm is used on $C^2(\bar{\mathrm{D}})$ and the operator norm is placed on $\mathcal{L}(\mathbf{K})$, and 3) $\Phi(z) = S$, where z stands for the identity function on $\mathbf{C}$, and 4) the constant one function is mapped to the identity operator on $\mathbf{K}$. The map Φ is called a spectral resolution for S. This paper will deal with the following notions.

Definition 1.1. *An operator is* subscalar *if it is the restriction of a scalar operator to a closed invariant subspace.*

We now define the weaker form of a subscalar operator:

Definition 1.2. *An operator $T \in \mathcal{L}(\mathbf{H})$ is B–*quasisubscalar *if there exists a nonzero V in $\mathcal{L}(\mathbf{H}, \mathbf{K})$ such that $VT = SV$ where S ($=\Phi(z)$ in the above definition) is a scalar operator. If the map V is one–to–one then T is said to be* quasisubscalar.

Definition 1.3. *An operator T in $\mathcal{L}(\mathbf{H})$ is said to be an operator of* Putinar type *if for some fixed non-zero $y \in \mathbf{H}$ and some $A, X \in \mathcal{L}(\mathbf{H})$ with $X^*y \neq 0$,*

$$|\langle (A-z)^* Xx, y\rangle| \leq \|(T-z)x\|$$

for all $x \in \mathbf{H}$ and all $z \in \mathbf{C}$.

This paper has been divided into four sections. Section two deals with some preliminary facts. In section three, we investigate the properties of operators of Putinar type. In section four, we show that an operator of this type is B-quasisubscalar, and if in addition the operator has no nontrivial invariant subspaces then it is quasisubscalar.

2. FURTHER PRELIMINARIES

An operator T in $\mathcal{L}(\mathbf{H})$ is called hyponormal if $TT^* \leq T^*T$, or equivalently, if $\|T^*h\| \leq \|Th\|$ for every h in $\mathbf{H}$. A subspace $\mathcal{M}$ of $\mathbf{H}$ is called invariant under T if $Tx \in \mathcal{M}$ for any $x \in \mathcal{M}$.

Let $d\mu(z)$, or simply $d\mu$, denote the planar Lebesgue measure. Let $\mathbf{H}$ be a complex separable Hilbert space, and let D be an open disc in $\mathbf{C}$. For $p = 1$ or 2, we shall denote by $L^p(\mathrm{D}, \mathbf{H})$ the Banach space of measurable functions $f : \mathrm{D} \to \mathbf{H}$, such that

$$\|f\|_{p,\mathrm{D}} \equiv \left(\int_{\mathrm{D}} \|f(z)\|^p \, d\mu(z) \right)^{1/p} < \infty .$$

If E is a Borel subset of D, then replacing D with E in the expression above also defines $\|f\|_{1,\mathrm{E}}$ and $\|f\|_{2,\mathrm{E}}$. Let $A^2(\mathrm{D}, \mathbf{H})$ denote the closure in $L^2(\mathrm{D}, \mathbf{H})$ of the analytic functions (i.e., $\bar{\partial} f = 0$) on D that have finite $\| \ \|_{2,\mathrm{D}}$ norm. Now $A^2(\mathrm{D}, \mathbf{H})$ is called the Bergman space for D, and every element in it corresponds to a function analytic on D.

Let P denote the orthogonal projection of $L^2(\mathrm{D}, \mathbf{H})$ onto $A^2(\mathrm{D}, \mathbf{H})$. Let $L^\infty(\mathrm{D}, \mathbf{H})$ denote the Banach space of essentially bounded $\mathbf{H}$-valued functions on D.

We will use the following version of Green's formula for the plane, also known as the Cauchy-Pompeiu formula. Define $C^2(\bar{\mathrm{D}}, \mathbf{H})$ in exactly the same way as $C^2(\bar{\mathrm{D}})$ except that the functions in the space are now $\mathbf{H}$ valued.

2.1. Green's Formula. Let D be an open disc in the plane, let $z \in \mathrm{D}$ and $f \in C^2(\bar{\mathrm{D}}, \mathbf{H})$. Then

$$f(z) = \frac{1}{2\pi i}\int_{\partial \mathrm{D}} \frac{f(s)}{s-z}\, ds - \frac{1}{\pi}\int_{\mathrm{D}} \frac{\bar{\partial} f(s)}{s-z}\, d\mu(s).$$

Remark 2.2. *The function $g(z) = \int_{\partial \mathrm{D}}(f(s)/(s-z))\, ds$ appearing in Cauchy-Pompeiu formula is analytic in D and extends continuously to $\bar{\mathrm{D}}$ as can be seen by examining the $\int_{\mathrm{D}}$ term. So, $g \in A^2(\mathrm{D}, \mathbf{H})$ for $f \in C^2(\bar{\mathrm{D}}, \mathbf{H})$.*

Let us define now a special Sobolev type space, called $W^2(\mathrm{D}, \mathbf{H})$ where, as before, D is a bounded disc in $\mathbf{C}$. For $f \in C^2(\bar{\mathrm{D}}, \mathbf{H})$, let

$$\|f\|_{W^2}^2 \equiv \sum_{i=0}^{2} \|\bar{\partial}^i f\|_{2,\mathrm{D}}^2.$$

Then let $W^2(\mathrm{D}, \mathbf{H})$ be the completion of $C^2(\bar{\mathrm{D}}, \mathbf{H})$ under this norm. Note that $W^2(\mathrm{D}, \mathbf{H})$ is a Hilbert space contained continuously in $L^2(\mathrm{D}, \mathbf{H})$.

Now for $f \in C^2(\bar{\mathrm{D}}, \mathbf{H})$, let M_f denote the operator on $W^2(\mathrm{D}, \mathbf{H})$ given by multiplication by f. This defines a natural map $\Phi : C^2(\bar{\mathrm{D}}) \longrightarrow \mathcal{L}(W^2(\mathrm{D}, \mathbf{H}))$. Hence M_z is 2-scalar by the definition given in section one. In fact, it can be shown [Pu] that M_z is subnormal.

3. Class of operators of Putinar type

We present some properties about operators of Putinar type in this section.

Definition 3.1. *An operator $C \in \mathcal{L}(\mathbf{H})$ is called M-hyponormal if there exists $M > 0$ such that $\|(C-z)^*x\| \leq M\|(C-z)x\|$ for all $z \in \mathbf{C}$ and for all x in $\mathbf{H}$.*

Proposition 3.2. *Every M-hyponormal operator C is an operator of Putinar type.*

Proof. Let $w \in \mathbf{H}$ be any unit vector and set $y = (1/M)w$. Then

$$\begin{aligned} |\langle (C-z)^*x, y\rangle| &= \frac{1}{M}|\langle (C-z)^*x, w\rangle| \\ &\leq \frac{1}{M}\|(C-z)^*x\| \\ &\leq \|(C-z)x\|. \end{aligned}$$ □

Proposition 3.3. *Let A be an operator of Putinar type.*

(a) *If $\alpha \in \mathbf{C}$, then αA is an operator of Putinar type.*
(b) *If $\beta \in \mathbf{C}$, $A - \beta$ is an operator of Putinar type.*

Proof. The proof is left as an exercise for the reader. □

Example 3.4. *If $T \in \mathcal{L}(\mathbf{H})$ has nonempty compression spectrum (i.e. if there exists $\beta \in \mathbf{C}$ such that $T - \beta$ has nondense range), then T is of Putinar type.*

PROOF. Suppose $T - \beta$ has nondense range for some $\beta \in \mathbf{C}$. Without loss of generality assume $\beta = 0$. Let Q be the projection of $\mathbf{H}$ onto the null space of T^*. Choose unit vector v in the null space of T^*. Given $x \in \mathbf{H}$ and $\lambda \in \mathbf{C}$, note that $(T - \lambda)x = -\lambda Qx + w$ for some $w \in \mathbf{H}$ orthogonal to the null space of T^*. Now,

$$|\langle (T-\lambda)^* Qx, v\rangle| \leq |\bar{\lambda}|\, |\langle Qx, v\rangle| \leq |\lambda|\, \|Qx\| \leq \|(T-\lambda)x\|.$$

The proof is complete. □

Proposition 3.5. *Let $T \in \mathcal{L}(\mathbf{H})$ be an operator of Putinar type and let $A \in \mathcal{L}(\mathbf{H})$ be any operator such that $YA = TY$ where Y is one-to-one and has a dense range. Then A is an operator of Putinar type.*

PROOF. Since T is an operator of Putinar type, there exists a fixed nonzero $r \in \mathbf{H}$ and some $R, X \in \mathcal{L}(\mathbf{H})$ with $X^* r \neq 0$ such that

$$|\langle (R-z)^* Xh, r\rangle| \leq \|(T-z)h\|$$

for any $h \in \mathbf{H}$ and any $z \in \mathbf{C}$. Set $y = (1/\|Y\|)r$. Then $(XY)^* y \neq 0$ and for any $h \in \mathbf{H}$,

$$\begin{aligned} |\langle (R-z)^* XYh, y\rangle| &= \frac{1}{\|Y\|} |\langle (R-z)^* XYh, r\rangle| \\ &\leq \frac{1}{\|Y\|} \|(T-z)Yh\| \\ &= \frac{1}{\|Y\|} \|Y(A-z)h\| \\ &\leq \|(A-z)h\|. \end{aligned}$$

□

Theorem 3.6. *If T and $T^* \in \mathcal{L}(\mathbf{H})$ are both B-quasisubscalar, then T has a nontrivial invariant subspace.*

PROOF. Suppose $T^* \in \mathcal{L}(\mathbf{H})$ is B-quasisubscalar implemented by V_1 and S_1 (i.e. $V_1 T^* = S_1 V_1$ with S_1 scalar of order 2). Suppose T is B-quasisubscalar implemented by V_2 and S_2. If $\sigma(S_1)$ consists of only one point, assume without loss of generality that this point is the origin, then S_1 is nilpotent (see [C-F] p. 106 lemma 3.5), and the result is easily seen to follow. Therefore, it may be assumed that there exist $\lambda_1 \in \sigma(S_1^*)$ and $\lambda_2 \in \sigma(S_2)$ with $\lambda_1 \neq \lambda_2$. Choose disjoint discs

D_1 and D_2 with $\lambda_1 \in D_1$ and $\lambda_2 \in D_2$. Using the fact that S_1 is 2-scalar, it can be shown that there exists $y_1 \in \mathbf{H}$ such that $(T-\lambda)f_1(\lambda) = y_1$ for some function $f_1(\lambda)$ analytic on $\mathbf{C} - D_1$. Also, there exists $y_2 \in \mathbf{H}$ such that $(T^* - \lambda)f_2(\lambda) = y_2$ for some function $f_2(\lambda)$ analytic on $\mathbf{C} - D_2^*$ (where D_2^* is the complex conjugate of D_2). Note that $(T^* - \bar{\lambda})f_2(\bar{\lambda}) = y_2$ for $\lambda \in \mathbf{C} - D_2$. Now on $\mathbf{C} - \sigma(T)$, the two analytic functions $\langle y_1, f_2(\bar{\lambda})\rangle$ and $\langle f_1(\lambda), y_2\rangle$ agree. It follows that each of these functions can be extended to a bounded entire function, and therefore is constant (and equal to zero). Therefore, y_1 is not cyclic. □

3.7 Nonexample: The *backward* unilateral shift is not of Putinar type. The proof of this is left to the reader. In fact, it can be shown (using "spectral" reasoning similar to that used in 3.6) that the backward shift is not B-quasisubscalar.

4. Operators of Putinar type as quasisubscalar operators

Recall from 1.2 that an operator $T \in \mathcal{L}(\mathbf{H})$ is said to be B-quasisubscalar if there exists a nonzero $V \in \mathcal{L}(\mathbf{H}, \mathbf{K})$ such that

$$VT = SV$$

where S is scalar. In order to obtain operators V and S that might satisfy this definition, the following construction from [Pu] is presented.

4.1 Construction of V and S: Let $T \in \mathcal{L}(\mathbf{H})$. The operator $T - z$ naturally acts on the space $W^2(D, \mathbf{H})$ via pointwise multiplication. Let us consider the space

$$\mathbf{K} \equiv W^2(D, \mathbf{H}) \ominus \overline{(T-z)W^2(D, \mathbf{H})}$$

and let Q denote the orthogonal projection of $W^2(D, \mathbf{H})$ onto $\mathbf{K}$. Let M be the operator of multiplication by z on $W^2(D, \mathbf{H})$. As noted at the end of section 2, M is a 2-scalar (subnormal) operator and has a spectral distribution Φ. Let $S = QM|_{\mathbf{K}}$ be the compression of the operator M to the space $\mathbf{K}$. Since $\overline{(T-z)W^2(D, \mathbf{H})}$ is invariant under every operator M_f, $f \in C^2(D)$, we infer that S is a 2-scalar operator with spectral distribution Φ_S given by $\Phi_S(f) = Q\Phi(f)|_{\mathbf{K}}$.

Consider the natural map $V : \mathbf{H} \longrightarrow \mathbf{K}$ defined by $Vh = Q(1 \otimes h)$, for $h \in \mathbf{H}$, where $1 \otimes h$ denotes the constant function identically equal to h. Note that $VT = SV$. Indeed, for every $h \in \mathbf{H}$ we have

$$\begin{aligned} VTh &= Q(1 \otimes Th) \\ &= Q(z \otimes h) \\ &= QM(1 \otimes h) \\ &= QMQ(1 \otimes h) \\ &= SVh. \end{aligned}$$

□

The following together with 4.5 is the main result.

Theorem 4.2. *Let $T \in \mathcal{L}(\mathbf{H})$ be an operator of Putinar type. Then T is a B-quasisubscalar operator.*

PROOF. First, choose and fix a disc D in the complex plane centered at the origin with area less than one. Let R be the radius of this disc. By 3.3, it can be assumed without loss of generality that $\|T\| < a < R$ where a is also fixed in what follows. Let $\mathrm{D}(a, R) \equiv \{z \in \mathbf{C} : a < |z| < R\}$. The proof consists of showing that (under these constraints placed on $\|T\|$ and R) the map V constructed in 4.1 is nonzero. This will be done in the two lemmas presented below. We repeat (often times verbatim) the reasoning of [Pu]. □

Lemma 4.3. *Let $A \in \mathcal{L}(\mathbf{H})$. There exists a constant C_{D} such that given any $f \in W^2(\mathrm{D}, \mathbf{H})$ there exists $g \in A^2(\mathrm{D}, \mathbf{H})$ such that*

$$\|\langle f - g, y\rangle\|_{2,\mathrm{D}} \le C_{\mathrm{D}}(\|\langle (A-z)^*\bar{\partial} f, y\rangle\|_{2,\mathrm{D}} + \|\langle (A-z)^*\bar{\partial}^2 f, y\rangle\|_{2,\mathrm{D}})$$

for all $y \in \mathbf{H}$.

PROOF. It can be assumed that $f \in C^2(\bar{\mathrm{D}}, \mathbf{H})$ since $C^2(\bar{\mathrm{D}}, \mathbf{H})$ is dense in $W^2(\mathrm{D}, \mathbf{H})$. Clearly

$$\bar{\partial}(f(z) + (A-z)^*\bar{\partial} f(z)) = (A-z)^*\bar{\partial}^2 f(z).$$

By the Cauchy-Pompeiu formula, 2.1, we have

$$\begin{aligned} f(z) + (A-z)^*\bar{\partial} f(z) = \frac{1}{2\pi i}\int_{\partial \mathrm{D}} (f(s) + (A-s)^*\bar{\partial} f(s))(s-z)^{-1}\, ds \\ -\frac{1}{\pi}\int_{\mathrm{D}} ((A-s)^*\bar{\partial}^2 f(s))(s-z)^{-1}\, d\mu(s). \qquad (*) \end{aligned}$$

Set

$$g(z) = \frac{1}{2\pi i}\int_{\partial \mathrm{D}} (f(s) + (A-s)^*\bar{\partial} f(s))(s-z)^{-1}\, ds.$$

Now $\langle g(z), y\rangle$ is analytic on D. And, from $(*)$,

$$\begin{aligned} &\langle f(z) - g(z), y\rangle \\ &= -\langle (A-z)^*\bar{\partial} f(z), y\rangle \\ &\quad -\left\langle \frac{1}{\pi}\int_{\mathrm{D}} ((A-s)^*\bar{\partial}^2 f(s))(s-z)^{-1}\, d\mu(s), y\right\rangle \\ &= -\langle (A-z)^*\bar{\partial} f(z), y\rangle \qquad (1) \\ &\quad -\frac{1}{\pi}\int_{\mathrm{D}} \langle (A-s)^*\bar{\partial}^2 f(s), y\rangle (s-z)^{-1}\, d\mu(s). \end{aligned}$$

The proof will be completed by using the triangle inequality on (1) and estimating the norm of the last term of the sum in (1). Let E be the open disc centered at the origin with radius $2R$. Let $*$ denote the convolution of two functions on $\mathbf{C}$, and let $\|\ \|_2$ denote $\|\ \|_{2,\mathrm{U}}$ in the case that the disc U is the entire complex plane $\mathbf{C}$ (and similarly for $\|\ \|_1$). Let $a(s) = \frac{1}{s}\chi_{\mathrm{E}}(s)$ where χ_{E} is the characteristic function of E, and let

$$b(s) = \langle (A-s)^*\bar{\partial}^2 f(s), y\rangle \chi_{\mathrm{D}}(s).$$

Then the norm of the last term in (1) is given by

$$\begin{aligned}
&\frac{1}{\pi}\Big\|\int_{\mathbf{C}} \langle (A-s)^*\bar{\partial}^2 f(s)\chi_{\mathrm{D}}(s), y\rangle \; ((s-z)^{-1}\chi_{\mathrm{E}}(s-z))\, d\mu(s)\Big\|_{2,\mathrm{D}} \\
\leq\ & \frac{1}{\pi}\|a * b\|_2 \leq \frac{1}{\pi}\|a\|_1\|b\|_2 \leq \frac{1}{\pi}4\pi R\|\langle (A-z)^*\bar{\partial}^2 f, y\rangle\|_{2,\mathrm{D}}
\end{aligned}$$

From this the result follows. □

Notice that since the area of R is less than one, $\|f\|_{1,\mathrm{D}(a,R)} < \|f\|_{2,\mathrm{D}(a,R)}$ for all $f \in C^2(\bar{\mathrm{D}}, \mathbf{H})$. For $r > 0$ we let Γ_r denote the positively oriented circular path of radius r centered at the origin. Now for any $f \in C^2(\bar{\mathrm{D}}, \mathbf{H})$,

$$\begin{aligned}
\left\|\int_a^R \int_{\Gamma_r} f(z)\, dz\, dr\right\| &= \left\|\int_a^R \int_0^{2\pi} f(re^{i\theta}) ire^{i\theta}\, d\theta\, dr\right\| \\
&\leq \int_a^R \int_0^{2\pi} \|f(re^{i\theta}) re^{i\theta}\|\, d\theta\, dr \\
&= \|f\|_{1,\mathrm{D}(a,R)} < \|f\|_{2,\mathrm{D}(a,R)}\,.
\end{aligned} \tag{2}$$

The next lemma provides the final step in the proof of 4.2.

Lemma 4.4. *For the operator T of Putinar type, the map V is nonzero.*

Proof. Assume that $Vh = \tilde{0}$ for every $h \in \mathbf{H}$. A contradiction will be found. Now given $h \in \overline{(T-z)W^2(\mathrm{D}, \mathbf{H})}$, there exists a sequence $\{f_n\}$ in $C^2(\bar{\mathrm{D}}, \mathbf{H})$ (recall that this set is dense in $W^2(\mathrm{D}, \mathbf{H})$) such that

$$\lim_{n\to\infty} \|(T-z)f_n + h\|_{W^2} = 0 \tag{3}$$

which implies

$$\lim_{n\to\infty} (\|(T-z)\bar{\partial} f_n\|_{2,\mathrm{D}} + \|(T-z)\bar{\partial}^2 f_n\|_{2,\mathrm{D}}) = 0\,.$$

From the definition of an operator of Putinar type, for some fixed nonzero $y \in \mathbf{H}$, and some $A, X \in \mathcal{L}(\mathbf{H})$ where $X^*y \neq 0$,

$$\lim_{n\to\infty} (\|\langle (A-z)^* X\bar{\partial} f_n, y\rangle\|_{2,\mathrm{D}} + \|\langle (A-z)^* X\bar{\partial}^2 f_n, y\rangle\|_{2,\mathrm{D}}) = 0.$$

Given any $\varepsilon > 0$, select n sufficiently high so that

$$\|\langle (A-z)^* X\bar{\partial} f_n, y\rangle\|_{2,\mathrm{D}} + \|\langle (A-z)^* X\bar{\partial}^2 f_n, y\rangle\|_{2,\mathrm{D}} < \varepsilon, \tag{4}$$

and (using (3) and the fact that $(T-z)^{-1}$ is bounded on $\mathrm{D}(a,R)$)

$$\|f_n(z) + (T-z)^{-1}h\|_{2,\mathrm{D}(a,R)} < \varepsilon\,. \tag{5}$$

Now by Lemma 4.3 and (4), there exists $g_n \in A^2(\mathrm{D}, \mathbf{H})$ such that

$$\|\langle X f_n - g_n\,,\, y\rangle\|_{2,\mathrm{D}} < C_{\mathrm{D}}\varepsilon\,. \tag{6}$$

Therefore (as explained just below), letting $\delta = 1/(R-a)$, it follows that

$$\begin{aligned}
|\langle h, X^* y\rangle| &= \left|\delta \int_a^R \frac{1}{2\pi i}\int_{\Gamma_r} \langle (T-z)^{-1}h, X^*y\rangle\, dz\, dr\right| \\
&\le \left|\delta \int_a^R \frac{1}{2\pi i}\int_{\Gamma_r} \langle (T-z)^{-1}h + f_n(z), X^*y\rangle\, dz\, dr\right| \\
&\qquad + \left|\delta \int_a^R \frac{1}{2\pi i}\int_{\Gamma_r} \langle f_n(z), X^*y\rangle\, dz\, dr\right| \\
&= \left|\frac{\delta}{2\pi}\left\langle \int_a^R \int_{\Gamma_r} (T-z)^{-1}h + f_n(z)\, dz\, dr\,,\, X^*y\right\rangle\right| \\
&\qquad + \left|\frac{\delta}{2\pi}\int_a^R \int_{\Gamma_r} \langle X f_n(z) - g_n(z)\,,\, y\rangle\, dz\, dr\right| \\
&\le \frac{\delta}{2\pi}\|(T-z)^{-1}h + f_n(z)\|_{2,\mathrm{D}(a,R)}\, \|X^*y\| + \frac{\delta}{2\pi}\|\langle X f_n - g_n\,,\, y\rangle\|_{2,\mathrm{D}} \\
&\le \frac{\delta}{2\pi}\varepsilon\|X^*y\| + \frac{\delta}{2\pi}C_{\mathrm{D}}\varepsilon\,.
\end{aligned}$$

The first equality holds by the functional calculus. The first inequality holds by the triangle inequality. The second equality holds since g_n is analytic. The second inequality follows by applying (2) to each of the two terms in the sum (note that the second term only involves scalar valued functions). The final inequality holds by (5) and (6). Since $\varepsilon > 0$ and h can be chosen arbitrarily, this shows that $X^*y = 0$, which is a contradiction. Therefore, we conclude V is nonzero. □

Corollary 4.5. *If T is an operator of Putinar type and T has no nontrivial invariant subspace, then T is quasisubscalar.*

Proof. Since T is an operator of Putinar type, T has a nonzero V such that $VT = SV$ as in the proof of Lemma 4.4. Since $\ker V \in \operatorname{Lat} T$ and T has no non-trivial invariant subspaces, $\ker V = \{0\}$. Thus T is quasisubscalar. □

This material represents the starting point of the second named author's dissertation which involves a much more general study of the Putinar map V in this situation and in the case of Sobolev spaces using higher order partials (in order to force V to be highly nontrivial for many operators). In [Ko], by using an infinite number of partials, a true (but complicated) model is found.

References

[Br] S. W. Brown. *Hyponormal operators with thick spectrum have invariant subspaces.* Ann. of Math., **125**, 93–103, 1987.

[C-F] I. Colojoară and C. Foiaş. *Theory of Generalized Spectral Operators.* Gordon and Breach, New York, 1968.

[Co] J. B. Conway. *Subnormal Operators.* Pitman, London, 1981.

[Ko] Eungil Ko. *Subscalar and quasisubscalar operators.* Ph. D. Thesis, Indiana University, 1993.

[MP] M. Martin and M. Putinar. *Lectures on Hyponormal Operators.* Birkhäuser, Basel-Boston, MA., 1989.

[Pu] M. Putinar. *Hyponormal operators are subscalar.* J. Operator Theory **12** (1984), 385-395.

S. W. Brown
Department of Mathematics
Indiana University
Bloomington, IN 47405

Eungil Ko
Department of Mathematics
Seoul National University
Seoul 151–742, Korea

Received: August 23rd, 1995.

Operator Theory:
Advances and Applications, Vol. 104

Flat Extensions of Positive Moment Matrices: Relations in Analytic or Conjugate Terms*

RAÚL E. CURTO AND LAWRENCE A. FIALKOW

Dedicated to our teacher and friend, Carl M. Pearcy, on the occasion of his sixtieth birthday

1. INTRODUCTION

Given a doubly indexed finite sequence of complex numbers $\gamma \equiv \gamma^{(2n)}$: γ_{00}, γ_{01}, $\gamma_{10}, \dots, \gamma_{0,2n}, \dots, \gamma_{2n,0}$, with $\gamma_{00} > 0$ and $\gamma_{ij} = \overline{\gamma_{ji}}$, the *truncated complex moment problem* entails finding a positive Borel measure μ supported in the complex plane $\mathbf{C}$ such that

$$\gamma_{ij} = \int \bar{z}^i z^j \, d\mu \qquad (0 \leq i + j \leq 2n); \tag{1.1}$$

γ is called a *truncated moment sequence* (*of order* $2n$) and μ is called a *representing measure* for γ. The truncated complex moment problem is closely related to several other moment problems: the *full moment problem* prescribes moments of *all* orders, i.e., $\gamma = (\gamma_{ij})_{i,j\geq 0}$, $\gamma_{00} > 0$, $\gamma_{ij} = \overline{\gamma_{ji}}$; the K*-moment problem* (truncated or full) prescribes a closed set $K \subseteq \mathbf{C}$ which is to contain the support of the representing measure ([Atz], [BM], [Cas], [CP], [P3], [Sch2], [StSz], [Sza]); and the *multidimensional moment problem* extends each of these problems to measures supported in $\mathbf{C}^k$ ([Ber], [BCJ], [Cas], [Fug], [Hav1], [Hav2], [McG], [P1], [P2], [P4]); moreover, the k-dimensional complex moment problem is equivalent to the $2k$-dimensional real moment problem [CF4, Section 6]. All of these problems generalize classical power moment problems on the real line, whose study was initiated by Stieltjes, Riesz, Hamburger, and Hausdorff (cf. [AK], [Akh], [Hau], [KrN], [Lan], [Sar], [ShT]). Recently, J. Stochel [Sto] proved that a solution to the multidimensional truncated K-moment problem actually implies a solution to the corresponding full moment problem. For $k = 1$, we may informally paraphrase Stochel's result as follows: If $K \subseteq \mathbf{C}$ is closed, if $\gamma = (\gamma_{ij})_{i,j\geq 0}$ is a full moment sequence, and if for each $n \geq 1$ there exists a representing measure μ_n for $\{\gamma_{ij}\}_{0\leq i+j\leq 2n}$ such that $\operatorname{supp} \mu_n \subseteq K$, then there exists a subsequence of $\{\mu_n\}$

*Both authors were partially supported by NSF research grants

that converges (in an appropriate weak topology) to a representing measure μ for γ with $\operatorname{supp}\mu \subseteq K$.

In [CF4] we initiated an approach to the truncated complex moment problem based on positivity and extension properties of the *moment matrix* $M(n) \equiv M(n)(\gamma)$ associated to a truncated moment sequence γ (see below for notation). If μ is any representing measure for γ, then $\operatorname{card}\operatorname{supp}\mu \geq \operatorname{rank} M(n)$ (see (1.5) below); the main results of [CF4] characterize the existence of representing measures μ for which $\operatorname{card}\operatorname{supp}\mu = \operatorname{rank} M(n)$.

Theorem 1.1. [CF4, Corollary 5.14] *If $M(n) \geq 0$ and $M(n)$ is flat, i.e.,* $\operatorname{rank} M(n) = \operatorname{rank} M(n-1)$, *then γ has a unique representing measure, which is* $\operatorname{rank} M(n)$*-atomic.*

Theorem 1.2. [CF4, Theorem 5.13] *γ has a* $\operatorname{rank} M(n)$*-atomic representing measure if and only if $M(n) \geq 0$ and $M(n)$ admits a flat extension $M(n+1)$, i.e., $M(n)$ can be extended to a moment matrix $M(n+1)$ satisfying* $\operatorname{rank} M(n+1) = \operatorname{rank} M(n)$.

In [CF4] we conjectured that if γ has *any* representing measure, then it has a $\operatorname{rank} M(n)$-atomic representing measure; this conjecture remains open. In the present note we study concrete sufficient conditions for the existence of flat moment matrix extensions of positive moment matrices; in view of Theorem 1.2, each such condition leads to the solution of a corresponding truncated moment problem.

To explain our results we require some additional notation. For $m \geq 1$, let $M_m(\mathbf{C})$ denote the $m \times m$ complex matrices. For $n \geq 1$, let $m \equiv m(n) := (n+1)(n+2)/2$; we introduce the following lexicographic order on the rows and columns of matrices in $M_{m(n)}(\mathbf{C})$: $1, Z, \bar{Z}, Z^2, Z\bar{Z}, \bar{Z}^2, \ldots, Z^n, \ldots, \bar{Z}^n$; rows or columns indexed by $1, Z, Z^2, \ldots, Z^n$ are said to be *analytic*. Let $\gamma : \gamma_{00}, \ldots, \gamma_{0,2n}, \ldots, \gamma_{2n,0}$ be a truncated moment sequence; given $0 \leq i,j \leq n$, we define the $(i+1) \times (j+1)$ matrix B_{ij} whose entries are the moments of order $i+j$:

$$B_{ij} \equiv \begin{pmatrix} \gamma_{ij} & \gamma_{i+1,j-1} & \cdots & \gamma_{i+j,0} \\ \gamma_{i-1,j+1} & \gamma_{ij} & \gamma_{i+1,j-1} & \cdots \\ \vdots & \gamma_{i-1,j+1} & & \vdots \\ \gamma_{0,j+i} & \cdots & & \gamma_{ji} \end{pmatrix}; \tag{1.2}$$

B_{ij} has the Toeplitz-like property of being constant on each diagonal. We now define the *moment matrix* $M(n) \equiv M(n)(\gamma)$ via the block decomposition $M(n) = (B_{ij})_{0 \leq i,j \leq n}$. For example, if $n = 1$, the *quadratic moment problem* for $\gamma : \gamma_{00}, \gamma_{01}, \gamma_{10}, \gamma_{02}, \gamma_{11}, \gamma_{20}$ corresponds to

$$M(1) = \begin{pmatrix} B_{00} & B_{01} \\ B_{10} & B_{11} \end{pmatrix} = \begin{pmatrix} \gamma_{00} & \gamma_{01} & \gamma_{10} \\ \gamma_{10} & \gamma_{11} & \gamma_{20} \\ \gamma_{01} & \gamma_{02} & \gamma_{11} \end{pmatrix}.$$

Note that for $0 \le i+j \le n$, $0 \le k+\ell \le n$, the entry in row $\bar{Z}^k Z^\ell$, column $\bar{Z}^i Z^j$ of $M(n)$ is equal to $\gamma_{i+\ell,j+k}$.

Let $\mathcal{P}_n \subseteq \mathbf{C}[z,\bar{z}]$ denote the complex polynomials in z, $\bar{z}$ of total degree $\le n$. For $p \in \mathcal{P}_n$, $p(z,\bar{z}) \equiv \sum_{0 \le i+j \le n} a_{ij}\bar{z}^i z^j$, let $\bar{p}(z,\bar{z}) \equiv \sum \overline{a_{ij}} z^i \bar{z}^j$ and let $\hat{p} \equiv (a_{00}, a_{01}, a_{10}, \dots, a_{0n}, \dots, a_{n0})^T \in \mathbf{C}^{m(n)}$. The basic connection between $M(n)(\gamma)$ and any representing measure μ is provided by the identity

$$\int f\bar{g}\, d\mu = (M(n)\hat{f}, \hat{g}) \qquad (f, g \in \mathcal{P}_n); \tag{1.3}$$

in particular $(M(n)\hat{f}, \hat{f}) = \int |f|^2\, d\mu \ge 0$, so $M(n) \ge 0$. For the quadratic moment problem ($n = 1$), positivity of $M(1)$ implies the existence of $\operatorname{rank} M(1)$-representing measures [CF4, Theorem 6.1], but in general positivity of $M(n)$ does not by itself imply the existence of representing measures.

We next recall from [CF4] some additional necessary conditions for the existence of representing measures. Let $\mathcal{C}_{M(n)}$ denote the column space of $M(n)$, i.e., $\mathcal{C}_{M(n)} = \langle 1, Z, \bar{Z}, \dots, Z^n, \dots, \bar{Z}^n \rangle \subseteq \mathbf{C}^{m(n)}$. For $p \in \mathcal{P}_n$, $p \equiv \sum a_{ij}\bar{z}^i z^j$, we define $p(Z,\bar{Z}) \in \mathcal{C}_{M(n)}$ by $p(Z,\bar{Z}) := \sum a_{ij}\bar{Z}^i Z^j$; note that if $p(Z,\bar{Z}) = 0$, then $\bar{p}(Z,\bar{Z}) = 0$ [CF4, Lemma 3.10]. If μ is a representing measure for γ, then

$$\text{For } p \in \mathcal{P}_n,\ p(Z,\bar{Z}) = 0 \Leftrightarrow \operatorname{supp}\mu \subseteq \mathcal{Z}(p) := \{z \in \mathbf{C} : p(z,\bar{z}) = 0\} \tag{1.4}$$

[CF4, Prop. 3.1]. It follows from (1.4) that

$$\text{If } \mu \text{ is a representing measure for } \gamma, \text{ then } \operatorname{card}\operatorname{supp}\mu \ge \operatorname{rank} M(n) \tag{1.5}$$

[CF4, Cor. 3.5].

The following Structure Theorem for positive moment matrices provides a basic tool for constructing flat extensions.

Theorem 1.3. [CF4, Theorem 3.14] *Let $M(n)(\gamma) \ge 0$. If f, g, $fg \in \mathcal{P}_{n-1}$ and $f(Z,\bar{Z}) = 0$, then $(fg)(Z,\bar{Z}) = 0$.*

In view of Theorem 1.3 the following condition is *necessary* for the existence of a positive extension $M(n+1)$ of $M(n)(\gamma)$:

(RG) $\qquad f,\ g,\ fg \in \mathcal{P}_n, \quad f(Z,\bar{Z}) = 0 \Rightarrow (fg)(Z,\bar{Z}) = 0.$

A moment matrix satisfying (RG) is said to be *recursively generated.*

For the case of the truncated moment problems in one real variable, where the "moment matrix" associated to moments $\gamma : \gamma_0, \dots, \gamma_{2n}$ is the *Hankel matrix* $H(n) \equiv (\gamma_{i+j})_{0 \le i,j \le n}$, we have the following result.

Theorem 1.4. [CF3, Section 3] *The following are equivalent:*

(1) *There exists a positive Borel measure* μ, $\operatorname{supp}\mu \subseteq \mathbf{R}$, *such that* $\gamma_i = \int t^i\, d\mu(t)$ $(0 \le i \le 2n)$;
(2) γ *has a* rank $H(n)$*-atomic representing measure supported in* $\mathbf{R}$;
(3) $H(n) \ge 0$ *and* $H(n)$ *is recursively generated (in the one-variable sense);*
(4) $H(n) \ge 0$ *and* $H(n)$ *admits a flat (i.e., rank-preserving) extension* $H(n+1)$.

In [CF4] we presented several cases in which Theorem 1.4 admits the following analogue for the truncated complex moment problem:

$$\text{If } M(n) \text{ is positive and satisfies (RG), } M(n) \text{ admits a flat extension } M(n+1). \tag{1.6}$$

Of course, if (1.6) holds for a particular $M(n)(\gamma)$, then by Theorem 1.2, γ has a rank $M(n)$-atomic representing measure. Theorem 1.1 corresponds to the case of (1.6) in which $M(n) \ge 0$ and for all $i+j=n$, $\bar{Z}^i Z^j \in \langle \bar{Z}^\ell Z^m \rangle_{0 \le \ell+m \le n-1}$.

In the present note we establish (1.6) in the following two new cases:

$$\bar{Z} = \alpha 1 + \beta Z \text{ for some } \alpha,\ \beta \in \mathbf{C} \text{ (Theorem 2.1);} \tag{1.7}$$

$$Z^k = p(Z, \bar{Z}) \text{ for some } p \in \mathcal{P}_{k-1}, \text{ where } k \le \left[\tfrac{n}{2}\right] + 1 \text{ (Theorem 3.1).} \tag{1.8}$$

On the other hand, we show that (1.6) does not always hold. In Section 4 we use an example of Schmüdgen [Sch1] to construct a positive invertible (hence recursively generated) moment matrix $M(3)(\gamma)$ for which γ admits *no* representing measure; hence $M(3)(\gamma)$ admits no flat extension $M(4)$.

We conclude this section with some preliminaries concerning flat moment matrix extensions of positive moment matrices. For $n \ge 1$ and $A \in M_{m(n)}(\mathbf{C})$, $A = A^*$, we define an hermitian sesquilinear form $\langle \cdot, \cdot \rangle_A$ on $\mathcal{P}_n$ by $\langle p, q \rangle_A := (A\hat{p}, \hat{q})$; thus the entry in row $\bar{Z}^k Z^\ell$, column $\bar{Z}^i Z^j$ of A is given by

$$\begin{aligned} (\bar{Z}^i Z^j, \widehat{\bar{z}^k z^\ell}) &= (A\widehat{\bar{z}^i z^j}, \widehat{\bar{z}^k z^\ell}) \\ &= \langle \bar{z}^i z^j, \bar{z}^k z^\ell \rangle_A \qquad (0 \le i+j \le n,\ 0 \le k+\ell \le n). \end{aligned}$$

Moreover, if there exist $p,\ q \in \mathcal{P}_n$ such that $\bar{Z}^i Z^j = p(Z, \bar{Z})$ and $\bar{Z}^k Z^\ell = q(Z, \bar{Z})$, then $\langle \bar{z}^i z^j, \bar{z}^k z^\ell \rangle_A = \langle p, q \rangle_A$. The following intrinsic characterization of moment matrices provides a useful tool for constructing extensions.

Theorem 1.5. [CF4, Theorem 2.1] *Let* $n \ge 1$ *and let* $A \in M_{m(n)}(\mathbf{C})$. *There exists a truncated moment sequence* γ *such that* $A = M(n)(\gamma)$ *if and only if*

(1) $\langle 1, 1 \rangle_A > 0$;
(2) $A = A^*$;
(3) $\langle p, q \rangle_A = \langle \bar{q}, \bar{p} \rangle_A$ $(p, q \in \mathcal{P}_n)$ *(symmetric property);*
(4) $\langle zp, q \rangle_A = \langle p, \bar{z}q \rangle_A$ $(p, q \in \mathcal{P}_{n-1})$;
(5) $\langle zp, zq \rangle_A = \langle \bar{z}p, \bar{z}q \rangle_A$ $(p, q \in \mathcal{P}_{n-1})$ *(normality property).*

For $0 \le i, j \le n$, let $B[i,j] \in M_{(i+1)\times(j+1)}(\mathbf{C})$. Let Z^i, $Z^{i-1}\bar{Z}$, ..., $\bar{Z}^i$ denote the rows of $B[i,j]$ and let Z^j, $Z^{j-1}\bar{Z}$, ..., $\bar{Z}^j$ denote the columns of $B[i,j]$. For $0 \le r \le i$, $0 \le s \le j$, we denote the entry in row $\bar{Z}^r Z^{i-r}$, column $\bar{Z}^s Z^{j-s}$ of $B[i,j]$ by $\langle \bar{z}^s z^{j-s}, \bar{z}^r z^{i-r}\rangle_{B[i,j]}$. For $i = j$, this notation is consistent with our previous definition of $\langle \cdot,\cdot\rangle_{M(n)}$. We say that $B[i,j]$ is *symmetric* if

$$\langle \bar{z}^s z^{j-s}, \bar{z}^r z^{i-r}\rangle_{B[i,j]} = \overline{\langle z^s \bar{z}^{j-s}, z^r \bar{z}^{i-r}\rangle}_{B[i,j]} \qquad (0 \le r \le i,\ 0 \le s \le j), \tag{1.9}$$

and we say that $B[i,j]$ satisfies *normality* if it is constant on diagonals, i.e.,

$$\langle \bar{z}^s z^{j-s}, \bar{z}^r z^{i-r}\rangle_{B[i,j]} = \langle \bar{z}^{s+1} z^{j-s-1}, \bar{z}^{r+1} z^{i-r-1}\rangle_{B[i,j]} \tag{1.10}$$

$$(0 \le r < i,\ 0 \le s < j).$$

Note that if $B[i,j]$ is symmetric and constant on upper diagonals (i.e., where $r \le s$ in (1.10)), then $B[i,j]$ satisfies normality. More generally, given $B[i,j]$, there exist scalars $\{\gamma_{\ell,m}\}_{\ell+m=i+j}$, $\gamma_{\ell m} = \overline{\gamma_{m\ell}}$, such that for all $s+t=i$, $u+v=j$, we have $\langle \bar{z}^u z^v, \bar{z}^s z^t\rangle_{B[i,j]} = \gamma_{u+t,v+s}$ if and only if $B[i,j]$ is symmetric and satisfies normality.

Given $\gamma \equiv \gamma^{(2n)}$, in addition to $M(n)$ we may also define blocks $B_{0,n+1}$, ..., $B_{n-1,n+1}$ via (1.2). Given $B \equiv B[n,n+1]$, let

$$\tilde{B} := \begin{pmatrix} B_{0,n+1} \\ \vdots \\ B_{n-1,n+1} \\ B \end{pmatrix}.$$

Given $C := B[n{+}1\ ,\ n{+}1]$, let $M = \begin{pmatrix} M(n) & \tilde{B} \\ \tilde{B}^* & C \end{pmatrix}$; M is an *extension* of $M(n)$; M is a *flat extension* if $\operatorname{rank} M = \operatorname{rank} M(n)$. Note that if C is self-adjoint and constant on upper diagonals, then C is constant on all diagonals and is thus also symmetric.

In the sequel we seek to construct M so that it is a positive flat extension of the form $M(n+1)$. The structure theory of positive operator matrices (cf. [Fia], [Smu]) implies that if $M \ge 0$, then $\operatorname{Ran}\tilde{B} \subseteq \mathcal{C}_{M(n)}$; equivalently, there exists a matrix W such that $\tilde{B} = M(n)W$. Conversely, given $M(n) \ge 0$ and $\tilde{B} = M(n)W$, then $M \ge 0 \Leftrightarrow C \ge W^*M(n)W$; moreover, M is a flat extension of $M(n) \ge 0 \Leftrightarrow C = W^*M(n)W$. (In this case, C is independent of W.) Thus a flat extension of a positive moment matrix is positive. The block structures of $\tilde{B}$ and $M(n)$, Theorem 1.5, and the preceding remarks, imply that to construct a flat moment matrix extension $M(n+1)$ of $M(n) \ge 0$ it is necessary and sufficient to construct a block $B[n,n+1]$ such that

$B[n\ ,\ n{+}1]$ is symmetric and satisfies normality; (1.11)

$\operatorname{Ran}\tilde{B} \subseteq \mathcal{C}_{M(n)}$ (so that $\tilde{B} = M(n)W$ for some W); (1.12)

$W^*M(n)W$ is constant on upper diagonals. (1.13)

We next provide a sufficient condition for $B[n, n+1]$ to be symmetric. Assume $\operatorname{Ran}\tilde{B} \subseteq \mathcal{C}_{M(n)}$; thus, for $i+j = n+1$, there exist $p_{ij} \in \mathcal{P}_n$ such that $\bar{Z}^i Z^j = p_{ij}(Z, \bar{Z}) \in \mathcal{C}_{M(n)}$.

Lemma 1.6. *If $p_{ij} = \overline{p_{ji}}$ for all $i+j = n+1$, then $B \equiv B[n, n+1]$ is symmetric.*

PROOF. For $i+j = n+1$, $k+\ell = n$,

$$\begin{aligned}
\langle \bar{z}^i z^j, \bar{z}^k z^\ell \rangle_B &= (p_{ij}(Z, \bar{Z}), \widehat{\bar{z}^k z^\ell}) \\
&= \langle p_{ij}, \bar{z}^k z^\ell \rangle_{M(n)} \\
&= \langle z^k \bar{z}^\ell, \overline{p_{ij}} \rangle_{M(n)} \qquad \text{(by Theorem 1.5–(3))} \\
&= \overline{\langle \overline{p_{ij}}, z^k \bar{z}^\ell \rangle_{M(n)}} = \overline{(p_{ji}(Z, \bar{Z}), \widehat{z^k \bar{z}^\ell})} \\
&= \overline{\langle \bar{z}^j z^i, z^k \bar{z}^\ell \rangle_B}.
\end{aligned}$$

□

Let $[M(n)]_{n-1} := (B_{ij})_{0 \le i \le n-1, 0 \le j \le n}$, let $[\tilde{B}]_{n-1} := (B_{i,n+1})_{0 \le i \le n-1}$, and let $\{\bar{V}^i V^j\}_{0 \le i+j \le n+1}$ denote the columns of the block $S := ([M(n)]_{n-1} \; [\tilde{B}]_{n-1})$.

Lemma 1.7. *Suppose $M(n)$ is recursively generated. If $p, q \in \mathcal{P}_n$, $pq \in \mathcal{P}_{n+1}$ and $p(Z, \bar{Z}) = 0$ in $\mathcal{C}_{M(n)}$, then $(pq)(V, \bar{V}) = 0$ in $\mathcal{C}_S$.*

PROOF. Let $k := \deg p \ (\le n)$ and denote $p(z, \bar{z}) = \sum_{0 \le r+s \le k} a_{rs} \bar{z}^r z^s$. It suffices to prove that if $i+j \le n+1-k$, then $(\bar{z}^i z^j p)(V, \bar{V}) = 0$. Since $M(n)$ satisfies (RG), if $i+j \le n-k$, then $(\bar{z}^i z^j p)(Z, \bar{Z}) = 0$ in $\mathcal{C}_{M(n)}$, so clearly $(\bar{z}^i z^j p)(V, \bar{V}) = 0$ in $\mathcal{C}_S$. Suppose $i+j = n+1-k$ and assume $j \ge 1$. Then (RG) $\Rightarrow (\bar{z}^i z^{j-1} p)(Z, \bar{Z}) = 0$ in $\mathcal{C}_{M(n)}$, and an obvious adaptation of the proof of Theorem 1.3 shows that $(\bar{z}^i z^j p)(V, \bar{V}) = (z(\bar{z}^i z^{j-1} p))(V, \bar{V}) = 0$. The proof when $j = 0$, $i \ge 1$ is similar. □

Lemma 1.8. (cf. [CF4, Lemma 5.2–ii]) *Suppose $M(n+1)$ is a flat extension of $M(n) \ge 0$, and let $p \in \mathcal{P}_{n+1}$ be such that $p(Z, \bar{Z}) = 0$ in $\mathcal{C}_{[M(n+1)]_n}$. Then $p(Z, \bar{Z}) = 0$ in $\mathcal{C}_{M(n+1)}$.*

Lemma 1.9. *Suppose $M(n)$ is positive and recursively generated. If $M(n+1)$ is a flat extension of $M(n)$, then $M(n+1)$ is recursively generated.*

PROOF. Suppose p, q, $pq \in \mathcal{P}_{n+1}$ and $p(Z, \bar{Z}) = 0$. We seek to show that $(pq)(Z, \bar{Z}) = 0$, and we may assume $p \in \mathcal{P}_n$, for otherwise q is a constant function and the result is immediate. Using [CF4, Lemma 3.10], it suffices to consider the case $q(z, \bar{z}) \equiv z^i$. When $\deg p + i \le n$, a combination of $M(n)$ being recursively generated and Lemma 1.8 yields the desired result. Assume, therefore, $\deg p + i = n+1$. Since $p \in \mathcal{P}_n$ and $p(Z, \bar{Z}) = 0$ in $\mathcal{C}_{M(n+1)}$, the proof of Theorem 1.3 (applied to $M(n+1)$) shows that $[(z^i p)(Z, \bar{Z})]_n = 0$ (where $[\cdot]_n$ denotes truncation of a vector through rows corresponding to monomials of total degree n). The result now follows from Lemma 1.8. □

The authors wish to thank Professors Jim Agler, Edwin Franks, John McCarthy, Mihai Putinar, Bruce Reznick and Mark Stankus for helpful discussions concerning the subject of this paper. Some of the calculations in Section 4 were obtained with the help of the software tool *Mathematica* [Wol].

2. The Case $\bar{Z}=\alpha 1+\beta Z$

In this section we focus on a positive, recursively generated moment matrix $M(n) \equiv M(n)(\gamma)$ in which the third column, $\bar{Z}$, is a linear combination of the first and second columns, *1* and Z. We will show that $M(n)$ always admits a flat extension $M(n+1)$; thus there exists a rank $M(n)$-atomic representing measure μ for γ. As outlined in the Introduction, our method is to define a suitable block $\tilde{B}$ of the form $M(n)W$ so that properties (1.11)–(1.13) are satisfied. To motivate our construction, we consider first the quadratic moment problem ($n=1$).

We are given six numbers, $\gamma_{00}, \gamma_{01}, \gamma_{10}, \gamma_{02}, \gamma_{11}, \gamma_{20}$, ($\gamma_{00}>0$, $\gamma_{ij}=\overline{\gamma_{ji}}$) such that $M(1) \geq 0$ and $\bar{Z}=\alpha 1+\beta Z$ for some α, $\beta \in \mathbf{C}$, and we would like to find a 6×6 moment matrix $M(2)$ which is a flat extension of $M(1)$. Since the case rank $M(1)=1$ is straightforward, we focus on rank $M(1)=2$, that is, *1* and Z are linearly independent. By Lemma 1.8, the relation $\bar{Z}=\alpha 1+\beta Z$ and the flatness condition would imply $\bar{Z}Z=\alpha Z+\beta Z^2$ and $\bar{Z}^2=\alpha\bar{Z}+\beta\bar{Z}Z$ in $\mathcal{C}_{M(2)}$. To describe $M(2)$ it thus remains to define column Z^2; we focus on the case $\alpha \neq 0$. By (1.2), we know that $\langle z^2,1\rangle_{M(2)}$ must equal γ_{02}, and by Theorem 1.5, a necessary condition for the existence of an extension $M(2)$ is

$$\begin{aligned} \langle z^2,z\rangle_{M(2)} &= \langle z,\bar{z}z\rangle_{M(2)}=\langle z,(\alpha 1+\beta z)z\rangle_{M(2)} \\ &= \bar{\alpha}\langle z,z\rangle_{M(1)}+\bar{\beta}\langle z,z^2\rangle_{M(2)} \\ &= \bar{\alpha}\langle z,z\rangle_{M(1)}+\bar{\beta}\overline{\langle z^2,z\rangle}_{M(2)}. \end{aligned} \tag{2.1}$$

Let us briefly pause to establish a relation between α and β : Since $\bar{Z}=\alpha 1+\beta Z$, [CF4, Lemma 3.10] forces at once the relation $Z=\bar{\alpha}1+\bar{\beta}\bar{Z}=\bar{\alpha}1+\bar{\beta}(\alpha 1+\beta Z)=(\bar{\alpha}+\bar{\beta}\alpha)1+|\beta|^2 Z$. By the linear independence of *1* and Z, we must then have

$$\begin{cases} \bar{\alpha}+\bar{\beta}\alpha=0 \\ |\beta|=1 \end{cases}. \tag{2.2}$$

Thus, (2.1) becomes

$$\alpha\langle z^2,z\rangle_{M(2)}=|\alpha|^2\langle z,z\rangle_{M(1)}+\alpha\bar{\beta}\overline{\langle z^2,z\rangle}_{M(2)}=|\alpha|^2\langle z,z\rangle_{M(1)}-\bar{\alpha}\overline{\langle z^2,z\rangle}_{M(2)},$$

that is

$$2\operatorname{Re}(\alpha\langle z^2,z\rangle_{M(2)})=|\alpha|^2\langle z,z\rangle_{M(1)}.$$

Therefore, $\alpha\langle z^2,z\rangle_{M(2)}=\frac{1}{2}|\alpha|^2\langle z,z\rangle_{M(1)}+it$, for some $t\in\mathbf{R}$. Observe also that

$$\langle z^2,\bar{z}\rangle_{M(2)}=\langle z^2,\alpha 1+\beta z\rangle_{M(2)}=\bar{\alpha}\langle z^2,1\rangle_{M(2)}+\bar{\beta}\langle z^2,z\rangle_{M(2)},$$

so the choice of $\langle z^2, z\rangle_{M(2)}$ (and the flatness requirement) fully determines the remaining entries of $M(2)$. We shall now extend this idea to the general case $n \geq 1$.

Theorem 2.1. *Assume that $M(n) \geq 0$ satisfies (RG) and that $\bar{Z} = \alpha 1 + \beta Z$. Then $M(n)$ admits a flat extension $M(n+1)$.*

We note that Theorem 2.1 is independent of Theorems 1.1 and 3.1. Indeed, [CF4, Section 6] contains the case when $M(1)$ is positive, recursively generated, $\bar{Z} = \alpha 1 + \beta Z$ and $\{1, Z\}$ is linearly independent; thus $M(1)$ is not flat (independence from Theorem 1.1) and $Z \notin \langle 1\rangle$ (independence from Theorem 3.1). A more ambitious example is contained in [Fia, Section 5], wherein $M(2)$ is positive, recursively generated, $\bar{Z} = \alpha 1 + \beta Z$, and $\{1, Z, Z^2\}$ is linearly independent.

First, let us show that the analytic columns of $M(n)$ can always be assumed to be linearly independent.

Proposition 2.2. ([CF4, Corollary 5.15]) *Assume $M(n) \geq 0$ and that the analytic columns of $M(n)$ are linearly dependent. Let $r := \min\{k \geq 1 : Z^k \in \langle 1, Z, \ldots, Z^{k-1}\rangle\}$. Then γ has a representing measure if and only if $\{1, Z, \ldots, Z^{r-1}\}$ spans $\mathcal{C}_{M(n)}$. In this case, the representing measure is unique, and is r ($= \operatorname{rank} M(n)$)-atomic.*

In Theorem 2.1, (RG) and $\bar{Z} = \alpha 1 + \beta Z$ imply $\bar{Z}^i Z^j = ((\alpha + \beta z)^i z^j)(Z, \bar{Z})$ in $\mathcal{C}_{M(n)}$ $(0 \leq i + j \leq n,\ i \geq 1)$, whence $\{Z^i\}_{i=0}^n$ spans $\mathcal{C}_{M(n)}$. Since $M(n) \geq 0$, it follows from (RG), Proposition 2.2 and Theorem 1.2 that if $\{Z^i\}_{i=0}^n$ is dependent, then $M(n)$ has a flat extension $M(n+1)$. In the sequel we thus assume $\{Z^i\}_{i=0}^n$ is independent.

The proof of Theorem 2.1 will be a consequence of a series of lemmas. Our first goal is to define column Z^{n+1} of the block $\tilde{B}$. If $\alpha = 0$, then $\bar{Z} = \beta Z$ with $|\beta| = 1$, say $\beta = e^{i\psi}$. The requirement $\langle z^{n+1}, z^n\rangle_{M(n+1)} = \langle z^n, \bar{z} z^n\rangle_{M(n+1)} = \bar{\beta}\overline{\langle z^{n+1}, z^n\rangle}_{M(n+1)}$ forces us to define $\langle z^{n+1}, z^n\rangle_{\tilde{B}} := re^{i(2\pi j - \psi)/2}$ with $r > 0$ and $j \in \mathbf{Z}$. If $\alpha \neq 0$, then proceeding as in the $n = 1$ case we define $\langle z^{n+1}, z^n\rangle_{\tilde{B}} = \frac{1}{2}\bar{\alpha}\gamma_{nn} + it/\alpha$ for some fixed $t \in \mathbf{R}$. Let

$$\langle z^{n+1}, \bar{z}^i z^j\rangle_{\tilde{B}} := \begin{cases} \gamma_{j,n+i+1} & i + j \leq n - 1 \\ \bar{\alpha}\gamma_{j,n+i} + \bar{\beta}\langle z^{n+1}, \bar{z}^{i-1} z^{j+1}\rangle_{\tilde{B}} & i \geq 1,\ j = n - i \end{cases} \tag{2.3}$$

and in $\mathcal{C}_{\tilde{B}}$ let

$$\bar{Z}^k Z^\ell := \alpha \bar{Z}^{k-1} Z^\ell + \beta \bar{Z}^{k-1} Z^{\ell+1} \quad (k \geq 1,\ \ell = n + 1 - k). \tag{2.4}$$

(All of these columns have length equal to the size of $M(n)$, that is, $(n+1)(n+2)/2$.)

We may write $\tilde{B}$ as a block column matrix $\tilde{B} = (\tilde{B}_0, \ldots, \tilde{B}_n)^T$, where, for each j, the columns of $\tilde{B}_j$ are indexed lexicographically by Z^{n+1}, …, $\bar{Z}^{n+1}$, and the rows by $Z^j, \ldots, \bar{Z}^j$. For $p \in \mathcal{P}_{n+1}$ and $q \in \mathcal{P}_n$ we define $\langle p, q\rangle_{\tilde{B}} := (p(Z,\bar{Z}), \hat{q})$, where $p(Z,\bar{Z})$ is defined in the usual way using the columns of $M(n)$ and of $\tilde{B}$; note that if p, $r \in \mathcal{P}_{n+1}$ and $p(Z,\bar{Z}) = r(Z,\bar{Z})$, then $\langle p, q\rangle_{\tilde{B}} = \langle r, q\rangle_{\tilde{B}}$.

Observe the following consequence of (RG): If $0 \le \ell + m \le n$, $\ell \ge 1$, then $\bar{Z}^\ell Z^m = \alpha \bar{Z}^{\ell-1} Z^m + \beta \bar{Z}^{\ell-1} Z^{m+1}$ in $\mathcal{C}_{M(n)}$; thus for $0 \le r + s \le n$,

$$\gamma_{\ell+r,m+s} = \alpha\gamma_{\ell-1+r,m+s} + \beta\gamma_{\ell-1+r,m+1+s}\,. \tag{2.5}$$

Lemma 2.3. *$\tilde{B} = M(n)W$ for some W.*

PROOF. From (2.4), it is enough to check that $Z^{n+1} \in Ran(M(n))$. Since $\{1, Z, \ldots, Z^n\}$ is independent and $M(n) \ge 0$, the Extension Principle [Fia] implies that the compression of $M(n)$ to the analytic rows and columns is invertible. Thus there exist complex numbers $a_0, \ldots, a_n$ such that

$$\sum_{i=0}^{n} a_i \left\langle z^i, z^\ell\right\rangle_{M(n)} = \left\langle z^{n+1}, z^\ell\right\rangle_{\tilde{B}} \quad (0 \le \ell \le n).$$

We shall show that the same relation holds for non-analytic rows, those determined by monomials of the form $\bar{Z}^k Z^\ell$, $k \ge 1$, $k + \ell \le n$. We use induction on $k \ge 1$. For $k = 1$ we have

$$\sum_{i=0}^{n} a_i\langle z^i, \bar{z}z^\ell\rangle_{M(n)} = \bar{\alpha}\sum_{i=0}^{n} a_i\langle z^i, z^\ell\rangle_{M(n)} + \bar{\beta}\sum_{i=0}^{n} a_i\langle z^i, z^{\ell+1}\rangle_{M(n)};$$

since Z^ℓ and $Z^{\ell+1}$ are analytic, we have

$$\begin{aligned}\sum a_i\langle z^i, \bar{z}z^\ell\rangle_{M(n)} &= \bar{\alpha}\langle z^{n+1}, z^\ell\rangle_{\tilde{B}} + \bar{\beta}\langle z^{n+1}, z^{\ell+1}\rangle_{\tilde{B}} \\ &= \bar{\alpha}\gamma_{\ell,n+1} + \bar{\beta}\langle z^{n+1}, z^{\ell+1}\rangle_{\tilde{B}}\end{aligned} \tag{2.6}$$

(by (2.3), since $\ell \le n-1$). For $\ell < n-1$, the last expression in (2.6) equals

$$\begin{aligned}\bar{\alpha}\gamma_{\ell,n+1} + \bar{\beta}\gamma_{\ell+1,n+1} &= \gamma_{\ell,n+2} && \text{(by (2.5))} \\ &= \langle z^{n+1}, \bar{z}z^\ell\rangle_{\tilde{B}} && \text{(by (2.3)).}\end{aligned}$$

For $\ell = n-1$, the final expression in (2.6) coincides with $\langle z^{n+1}, \bar{z}z^\ell\rangle_{\tilde{B}}$ by (2.3).

For $k \ge 1$, we have

$$\begin{aligned}\sum_{i=0}^{n} a_i\left\langle z^i, \bar{z}^k z^\ell\right\rangle_{M(n)} &= \sum_{i=0}^{n} a_i\left\langle z^i, (\alpha + \beta z)\bar{z}^{k-1}z^\ell\right\rangle_{M(n)} \\ &= \bar{\alpha}\sum_{i=0}^{n} a_i\left\langle z^i, \bar{z}^{k-1}z^\ell\right\rangle_{M(n)} + \bar{\beta}\sum_{i=0}^{n} a_i\left\langle z^i, \bar{z}^{k-1}z^{\ell+1}\right\rangle_{M(n)} \\ &= \bar{\alpha}\left\langle z^{n+1}, \bar{z}^{k-1}z^\ell\right\rangle_{\tilde{B}} + \bar{\beta}\left\langle z^{n+1}, \bar{z}^{k-1}z^{\ell+1}\right\rangle_{\tilde{B}} \quad \text{(by induction)} \\ &= \bar{\alpha}\gamma_{\ell,n+k} + \bar{\beta}\left\langle z^{n+1}, \bar{z}^{k-1}z^{\ell+1}\right\rangle_{\tilde{B}} = \left\langle z^{n+1}, \bar{z}^k z^\ell\right\rangle_{\tilde{B}}\end{aligned}$$

(by (2.3) if $k + \ell = n$ and by (2.3) and (2.5) if $k + \ell \le n-1$), as desired. □

The next lemma shows that for $j \le n-1$, $\tilde{B}_j = B_{j,n+1}(\gamma)$.

Lemma 2.4. *For $i+j=n+1$ and $p+q \le n-1$,*

$$\left\langle \bar{z}^i z^j, \bar{z}^p z^q \right\rangle_{\tilde{B}} = \left\langle \bar{z}^i z^{j-1}, \bar{z}^{p+1} z^q \right\rangle_{M(n)} = \gamma_{i+q,j+p}.$$

PROOF. We use induction on $i \ge 0$. For $i=0$, (2.3) implies

$$\left\langle z^{n+1}, \bar{z}^p z^q \right\rangle_{\tilde{B}} = \gamma_{q,n+p+1} = \left\langle z^n, \bar{z}^{p+1} z^q \right\rangle_{M(n)}.$$

When $i \ge 1$,

$$\begin{aligned} \left\langle \bar{z}^i z^{n+1-i}, \bar{z}^p z^q \right\rangle_{\tilde{B}} &= \alpha \left\langle \bar{z}^{i-1} z^{n+1-i}, \bar{z}^p z^q \right\rangle_{\tilde{B}} + \beta \left\langle \bar{z}^{i-1} z^{n+2-i}, \bar{z}^p z^q \right\rangle_{\tilde{B}} \\ &= \alpha \left\langle \bar{z}^{i-1} z^{n-i}, \bar{z}^{p+1} z^q \right\rangle_{M(n)} + \beta \left\langle \bar{z}^{i-1} z^{n+1-i}, \bar{z}^{p+1} z^q \right\rangle_{M(n)} \end{aligned}$$

(by Theorem 1.5 for the first term and by induction for the second term)

$$= \left\langle \bar{z}^i z^{n-i}, \bar{z}^{p+1} z^q \right\rangle_{M(n)}. \qquad \square$$

The next lemma establishes normality between columns Z^{n+1} and $Z^n \bar{Z}$ of $\tilde{B}_n$.

Lemma 2.5. *For $p+q=n$, $p \ge 1$,*

$$\left\langle z^{n+1}, \bar{z}^{p-1} z^{q+1} \right\rangle_{\tilde{B}} = \langle \bar{z} z^n, \bar{z}^p z^q \rangle_{\tilde{B}}.$$

PROOF. We use induction on $p \ge 1$. For $p=1$,

$$\begin{aligned} \left\langle \bar{z} z^n, \bar{z} z^{n-1} \right\rangle_{\tilde{B}} &= \langle \alpha z^n + \beta z^{n+1}, \bar{z} z^{n-1} \rangle_{\tilde{B}} \\ &= \alpha \langle z^n, \bar{z} z^{n-1} \rangle_{M(n)} + \beta \langle z^{n+1}, \bar{z} z^{n-1} \rangle_{\tilde{B}} \\ &= \alpha \gamma_{n-1,n+1} + \beta (\bar{\alpha} \gamma_{n-1,n+1} + \bar{\beta} \left\langle z^{n+1}, z^n \right\rangle_{\tilde{B}}) \\ &= \alpha \gamma_{n-1,n+1} - \alpha \gamma_{n-1,n+1} + \langle z^{n+1}, z^n \rangle_{\tilde{B}} = \langle z^{n+1}, z^n \rangle_{\tilde{B}}. \end{aligned}$$

The inductive step is a bit more complex: For $p \ge 2$,

$$\begin{aligned} \left\langle z^{n+1}, \bar{z}^{p-1} z^{q+1} \right\rangle_{\tilde{B}} &= \bar{\alpha} \gamma_{q+1,n+p-1} + \bar{\beta} \left\langle z^{n+1}, \bar{z}^{p-2} z^{q+2} \right\rangle_{\tilde{B}} \quad \text{(by (2.3))} \\ &= \bar{\alpha} \left\langle \bar{z} z^n, \bar{z}^{p-1} z^q \right\rangle_{\tilde{B}} + \bar{\beta} \left\langle \bar{z} z^n, \bar{z}^{p-1} z^{q+1} \right\rangle_{\tilde{B}}, \end{aligned}$$

by Lemma 2.4 and the induction hypothesis; then

$$\begin{aligned} \left\langle z^{n+1}, \bar{z}^{p-1} z^{q+1} \right\rangle_{\tilde{B}} &= \bar{\alpha} \left\langle \alpha z^n + \beta z^{n+1}, \bar{z}^{p-1} z^q \right\rangle_{\tilde{B}} \\ &\quad + \bar{\beta} \left\langle \alpha z^n + \beta z^{n+1}, \bar{z}^{p-1} z^{q+1} \right\rangle_{\tilde{B}} \quad \text{(by 2.4)} \\ &= \alpha (\bar{\alpha} \left\langle z^n, \bar{z}^{p-1} z^q \right\rangle_{M(n)} + \bar{\beta} \left\langle z^n, \bar{z}^{p-1} z^{q+1} \right\rangle_{M(n)}) \\ &\quad + \beta (\bar{\alpha} \left\langle z^{n+1}, \bar{z}^{p-1} z^q \right\rangle_{\tilde{B}} + \bar{\beta} \left\langle z^{n+1}, \bar{z}^{p-1} z^{q+1} \right\rangle_{\tilde{B}}) \\ &= \alpha \left\langle z^n, \bar{z}^p z^q \right\rangle_{M(n)} + \beta \left\langle z^{n+1}, \bar{z}^p z^q \right\rangle_{\tilde{B}} \\ &= \langle \bar{z} z^n, \bar{z}^p z^q \rangle_{\tilde{B}} \quad \text{(by (2.4))}. \end{aligned} \qquad \square$$

We next establish normality for $\tilde{B}_n$.

Lemma 2.6. *For $i+j=n+1$, $j\geq 1$, $p+q=n$, $p\geq 1$,*

$$\langle \bar{z}^i z^j, \bar{z}^{p-1} z^{q+1}\rangle_{\tilde{B}} = \langle \bar{z}^{i+1} z^{j-1}, \bar{z}^p z^q\rangle_{\tilde{B}}.$$

PROOF. The proof is by induction on $i\geq 0$; for $i=0$, Lemma 2.5 implies

$$\langle z^{n+1}, \bar{z}^{p-1} z^{q+1}\rangle_{\tilde{B}} = \langle \bar{z} z^n, \bar{z}^p z^q\rangle_{\tilde{B}}.$$

For $i\geq 1$, we have

$$\begin{aligned}
&\langle \bar{z}^i z^j, \bar{z}^{p-1} z^{q+1}\rangle_{\tilde{B}} \\
&\quad = \alpha\langle \bar{z}^{i-1} z^j, \bar{z}^{p-1} z^{q+1}\rangle_{M(n)} + \beta\langle \bar{z}^{i-1} z^{j+1}, \bar{z}^{p-1} z^{q+1}\rangle_{\tilde{B}} \quad \text{(by (2.4))} \\
&\quad = \alpha\langle \bar{z}^i z^{j-1}, \bar{z}^p z^q\rangle_{M(n)} + \beta\langle \bar{z}^i z^j, \bar{z}^p z^q\rangle_{\tilde{B}} \\
&\qquad\qquad \text{(by Theorem 1.5–(5) and by induction)} \\
&\quad = \langle \bar{z}^{i+1} z^{j-1}, \bar{z}^p z^q\rangle_{\tilde{B}} \qquad \text{(by (2.4))}.
\end{aligned}$$

□

To establish symmetry for $\tilde{B}_n$, we first show that the relationship between column Z^{n+1} and row Z^n of $\tilde{B}_n$ is compatible with the structure of a moment matrix block $B_{n,n+1}(\gamma)$.

Lemma 2.7. *For $k+\ell=n+1$, $k\geq 1$,*

$$\langle \bar{z}^k z^\ell, z^n\rangle_{\tilde{B}} = \overline{\langle z^{n+1}, \bar{z}^{k-1} z^\ell\rangle}_{\tilde{B}}.$$

PROOF. We use induction on $k\geq 1$. For $k=1$, the $\alpha=0$ case is trivial, so we assume $\alpha\neq 0$:

$$\begin{aligned}
\langle \bar{z} z^n, z^n\rangle_{\tilde{B}} - \overline{\langle z^{n+1}, z^n\rangle}_{\tilde{B}} &= \alpha\langle\, z^n, z^n\rangle_{\tilde{B}} + \beta\left\langle z^{n+1}, z^n\right\rangle_{\tilde{B}} - \overline{\langle z^{n+1}, z^n\rangle}_{\tilde{B}} \\
&= \frac{1}{\bar{\alpha}}(|\alpha|^2\gamma_{nn} - \alpha\left\langle z^{n+1}, z^n\right\rangle_{\tilde{B}} - \overline{\alpha\left\langle z^{n+1}, z^n\right\rangle_{\tilde{B}}}) \\
&= \frac{1}{\bar{\alpha}}(|\alpha|^2\gamma_{nn} - 2Re(\alpha\left\langle z^{n+1}, z^n\right\rangle_{\tilde{B}}) = 0,
\end{aligned}$$

by the definition of $\langle z^{n+1}, z^n\rangle_{\tilde{B}}$. As for the inductive step, consider $\overline{\langle z^{n+1}, \bar{z}^{k-1} z^\ell\rangle}_{\tilde{B}}$ with $k>1$. By (2.3) and the induction hypothesis, this is equal to

$$\alpha\gamma_{n+k-1,\ell} + \beta\overline{\langle z^{n+1}, \bar{z}^{k-2} z^{\ell+1}\rangle}_{\tilde{B}} = \alpha\overline{\langle z^{n+1}, \bar{z}^{k-2}\, z^\ell\rangle}_{\tilde{B}} + \beta\left\langle \bar{z}^{k-1} z^{\ell+1}, z^n\right\rangle_{\tilde{B}},$$

which in turn is equal to $\alpha\overline{\langle z^n, \bar{z}^{k-1} z^\ell\rangle}_{M(n)} + \beta\left\langle \bar{z}^{k-1} z^{\ell+1}, z^n\right\rangle_{\tilde{B}}$, by Lemma 2.4. Thus,

$$\begin{aligned}
\overline{\langle z^{n+1}, \bar{z}^{k-1} z^\ell\rangle}_{\tilde{B}} &= \alpha\overline{\langle z^n, \bar{z}^{k-1}\, z^\ell\rangle}_{M(n)} + \beta\left\langle \bar{z}^{k-1} z^{\ell+1},\ z^n\right\rangle_{\tilde{B}} \\
&= \alpha\langle \bar{z}^{k-1} z^\ell, z^n\rangle_{M(n)} + \beta\langle \bar{z}^{k-1} z^{\ell+1}, z^n\rangle_{\tilde{B}} = \left\langle \bar{z}^k z^\ell, z^n\right\rangle_{\tilde{B}},
\end{aligned}$$

(by (2.4)), as desired. □

We next establish symmetry for $\tilde{B}_n$.

Lemma 2.8. *For $i+j=n+1$, $k+\ell=n$,*

$$\overline{\langle \bar{z}^i z^j, \bar{z}^k z^\ell \rangle}_{\tilde{B}} = \langle z^i \bar{z}^j, z^k \bar{z}^\ell \rangle_{\tilde{B}}.$$

PROOF. We give the proof only for $k \geq i$ and leave the other case to the reader.

$$\begin{aligned} \overline{\langle \bar{z}^i z^j, \bar{z}^k z^\ell \rangle}_{\tilde{B}} &= \overline{\langle z^{n+1}, \bar{z}^{k-i} z^{\ell+i} \rangle}_{\tilde{B}} \quad \text{(by Lemma 2.6)} \\ &= \langle \bar{z}^{k-i+1} z^{\ell+i}, z^n \rangle_{\tilde{B}} \quad \text{(by Lemma 2.7)} \\ &= \langle z^i \bar{z}^j, z^k \bar{z}^\ell \rangle_{\tilde{B}} \quad \text{(by Lemma 2.6).} \end{aligned}$$ □

For $i+j=2n+1$, we now define γ_{ij} as follows.

$$\begin{aligned} 0 \leq i \leq n : \gamma_{ij} &= \langle z^{n+1}, \bar{z}^{n-i} z^i \rangle_{\tilde{B}} \\ n < i \leq 2n+1 : \gamma_{ij} &= \langle \bar{z}^{i-n} z^j, z^n \rangle_{\tilde{B}}. \end{aligned}$$

It follows readily from normality and symmetry in $\tilde{B}_n$ that $\tilde{B}_n$ is of the form $B_{n,n+1}(\gamma)$. Since $\tilde{B}_n$ satisfies (1.11) and (1.12), to complete the proof of Theorem 2.1 we must show that $C \equiv W^* M(n) W$ is Toeplitz. Let $M := \begin{pmatrix} M(n) & \tilde{B} \\ \tilde{B}^* & C \end{pmatrix}$ denote the unique flat extension of $M(n)$ subordinate to $\tilde{B}$, and let $\langle \cdot, \cdot \rangle_M$ be the associated form. Since $M = M^*$, if p, r, $s \in \mathcal{P}_{n+1}$, with $r(Z,\bar{Z}) = s(Z,\bar{Z})$ in $\mathcal{C}_M$, then $\langle p, r \rangle_M = \langle p, s \rangle_M$. By flatness and Lemma 1.8, the columns of M of order $n+1$ satisfy the relations of (2.4).

Lemma 2.9. *For $i+j=p+q=n+1$, $j \geq 1$, $q \geq 1$,*

$$\langle \bar{z}^i z^j, \bar{z}^p z^q \rangle_M = \langle \bar{z}^{i+1} z^{j-1}, \bar{z}^{p+1} z^{q-1} \rangle_M.$$

PROOF. We first consider $i=0$, $j=n+1$ and proceed by induction on $p \geq 0$. For $p=0$, the $\alpha=0$ case is trivial so we assume $\alpha \neq 0$.

$$\begin{aligned} &\langle \bar{z} z^n, \bar{z} z^n \rangle_M - \langle z^{n+1}, z^{n+1} \rangle_M \\ &= \langle \alpha z^n + \beta z^{n+1}, \alpha z^n + \beta z^{n+1} \rangle_M - \langle z^{n+1}, z^{n+1} \rangle_M \\ &= |\alpha|^2 \langle z^n, z^n \rangle_M + 2\operatorname{Re}(\bar{\alpha}\beta \langle z^{n+1}, z^n \rangle_M) \\ &= |\alpha|^2 \gamma_{nn} - 2\operatorname{Re}(\alpha \langle z^{n+1}, z^n \rangle_M) = 0. \end{aligned}$$

For $p \geq 1$,

$$\begin{aligned} &\langle z^{n+1}, \bar{z}^p z^q \rangle_M \\ &= \langle z^{n+1}, (\alpha + \beta z) \bar{z}^{p-1} z^q \rangle_M \\ &= \bar{\alpha} \langle z^{n+1}, \bar{z}^{p-1} z^q \rangle_M + \bar{\beta} \langle z^{n+1}, \bar{z}^{p-1} z^{q+1} \rangle_M \\ &= \bar{\alpha} \langle \bar{z} z^n, \bar{z}^p z^{q-1} \rangle_M + \bar{\beta} \langle \bar{z} z^n, \bar{z}^p z^q \rangle_M \end{aligned}$$

(by normality outside C for the first term and by induction on p for the second term)

$$= \langle \bar{z} z^n, \bar{z}^{p+1} z^{q-1} \rangle_M.$$

We now induct on $i \geq 0$. For $i \geq 1$, we use (2.4), normality in M outside C, and induction to obtain

$$\begin{aligned}\langle \bar{z}^i z^j, \bar{z}^p z^q\rangle_M &= \alpha\langle \bar{z}^{i-1} z^j, \bar{z}^p z^q\rangle_M + \beta\langle \bar{z}^{i-1} z^{j+1}, \bar{z}^p z^q\rangle_M \\ &= \alpha\langle \bar{z}^i z^{j-1}, \bar{z}^{p+1} z^{q-1}\rangle_M + \beta\langle \bar{z}^i z^j, \bar{z}^{p+1} z^{q-1}\rangle_M \\ &= \langle \bar{z}^{i+1} z^{j-1}, \bar{z}^{p+1} z^{q-1}\rangle_M. \qquad \square\end{aligned}$$

The proof of Theorem 2.1 is now complete.

3. Flat extensions for $Z^k \in \langle \bar{Z}^i Z^j\rangle_{0 \leq i+j \leq k-1}$

In this section we study flat extensions of positive, recursively generated moment matrices $M(n)$ for which there is a relation $Z^k = p(Z, \bar{Z})$ for some $p \in \mathcal{P}_{k-1}$. In the case when $k \leq [n/2] + 1$, we prove the existence of a unique flat extension $M(n+1)$. For the case $[n/2] + 1 < k \leq n$, we describe a simple algorithm which can be used to determine the existence of flat extensions in numerical examples.

Theorem 3.1. *Suppose $M(n)$ is positive and recursively generated. If $1 \leq k \leq [n/2] + 1$ and $Z^k = p(Z, \bar{Z})$ for some $p \in \mathcal{P}_{k-1}$, then $M(n)$ admits a unique flat extension $M(n+1)$.*

Remark 3.2. For n odd, or for n even and $k < [n/2] + 1$, Theorem 3.1 can be obtained as a consequence of Theorems 1.1 and 1.2, since in each of these cases $M(n)$ is actually flat. Indeed, since $Z^k = p(Z, \bar{Z})$, then $\bar{Z}^k = \bar{p}(Z, \bar{Z})$ [CF4, Lemma 3.10]; thus (RG) implies that for $i + j = n - k$, $\bar{Z}^i Z^{j+k} = (\bar{z}^i z^j p)(Z, \bar{Z})$ and $\bar{Z}^{i+k} Z^j = (\bar{z}^i \bar{p} z^j)(Z, \bar{Z})$. In the indicated cases the preceding relations imply that for $r + s = n$, $\bar{Z}^r Z^s = p_{rs}(Z, \bar{Z})$ for some $p_{rs} \in \mathcal{P}_{n-1}$. The proof of Theorem 3.1 that we present below is independent of Theorems 1.1 and 1.2 and uses a more direct argument.

Example 3.3. The case when n is even and $k = [n/2] + 1$ does not follow from Theorems 1.1 and 1.2 since $M(n)$ need not be flat. For example, with $n = 2$, consider

$$M(2) = \begin{pmatrix} 1 & 0 & 0 & 0 & 1 & 0 \\ 0 & 1 & 0 & \alpha & \bar{\alpha} & \bar{\beta} \\ 0 & 0 & 1 & \beta & \alpha & \bar{\alpha} \\ 0 & \bar{\alpha} & \bar{\beta} & |\alpha|^2 + |\beta|^2 & \bar{\alpha}^2 + \bar{\beta}\alpha & 2\bar{\alpha}\bar{\beta} \\ 1 & \alpha & \bar{\alpha} & \alpha^2 + \beta\bar{\alpha} & |\alpha|^2 + |\beta|^2 & \bar{\alpha}^2 + \bar{\beta}\alpha \\ 0 & \beta & \alpha & 2\alpha\beta & \alpha^2 + \beta\bar{\alpha} & |\alpha|^2 + |\beta|^2 \end{pmatrix}.$$

Note that $Z^2 = \alpha Z + \beta \bar{Z}$ and $\bar{Z}^2 = \bar{\alpha}\bar{Z} + \bar{\beta} Z$.

Since $\{1, Z, \bar{Z}\}$ is independent, $M(2)$ satisfies (RG). Now $M(2) \geq 0 \Leftrightarrow |\beta|^2 \geq 1 + |\alpha|^2$; moreover, if $|\beta|^2 > 1 + |\alpha|^2$, then $M(2)$ is positive and satisfies (RG), but it is *not flat* (since $\operatorname{rank} M(2) = 4 > \operatorname{rank} M(1)$). The existence of a unique flat extension for $M(2)$ follows from Theorem 3.1, whence Theorem 1.2 implies the existence of a unique 4-atomic representing measure, of the form $\mu = \sum_{i=0}^{3} \rho_i \delta_{z_i}$. Using the method of [CF4] and [Fia], we see that the atoms $\{z_i\}_{i=0}^{3}$ are the four distinct roots of

$$z^4 = 2\alpha z^3 + (\beta\bar{\alpha} - \alpha^2)z^2 + \beta(|\beta|^2 - |\alpha|^2)z,$$

and the densities $\{\rho_i\}_{i=0}^{3}$ may be obtained from the Vandermonde equation

$$V(z_0, \ldots, z_3)(\rho_0, \ldots, \rho_3)^T = (\gamma_{00}, \gamma_{01}, \gamma_{02}, \gamma_{03})^T.$$

3.1. Proof of Theorem 3.1. Our first goal is to define a block $B \equiv B[n, n+1] \in M_{n+1,n+2}(\mathbf{C})$ to serve as $B_{n,n+1}$ in the extension. Denote p by $p(z, \bar{z}) = \sum_{0 \leq r+s \leq k-1} a_{rs}\bar{z}^r z^s$. For $i + j = n + 1$, we define $p_{ij} \in \mathcal{P}_n$ as follows:

$$\begin{aligned} i \geq k,\ j < k : p_{ij}(z, \bar{z}) &= \bar{z}^{i-k} z^j \bar{p}(z, \bar{z}); \\ i \geq k,\ j \geq k : p_{ij}(z, \bar{z}) &= \bar{z}^{i-k} z^{j-k} |p|^2(z, \bar{z}); \\ i < k,\ j \geq k : p_{ij}(z, \bar{z}) &= \bar{z}^i z^{j-k} p(z, \bar{z}). \end{aligned} \tag{3.1}$$

(Note that since $k \leq [n/2] + 1$, either $i \geq k$ or $j \geq k$, so p_{ij} is well defined for all $i + j = n + 1$.) Since $p(Z, \bar{Z}) = Z^k$ in $\mathcal{C}_{M(n)}$, then $\bar{p}(Z, \bar{Z}) = \bar{Z}^k$ in $\mathcal{C}_{M(n)}$ [CF4, Lemma 3.10]; since $M(n)$ satisfies (RG), Lemma 1.7 and (3.1) imply

$$\text{For } i + j = n + 1,\ \bar{V}^i V^j = p_{ij}(V, \bar{V}) \text{ in } \mathcal{C}_S. \tag{3.2}$$

(We illustrate the case when n is odd and $i = j = k$. Since $Z^k = p(Z, \bar{Z})$, Lemma 1.7 implies $(\bar{z}^k z^k)(V, \bar{V}) = (\bar{z}^k p)(V, \bar{V})$. Also, since $\bar{p}(Z, \bar{Z}) = \bar{Z}^k$, (RG) implies $(p\bar{p})(Z, \bar{Z}) = (p\bar{z}^k)(Z, \bar{Z})$ in $\mathcal{C}_{M(n)}$; thus $\bar{V}^k V^k = (\bar{z}^k p)(V, \bar{V}) = |p|^2(V, \bar{V})$. The other cases of $i + j = n + 1$ are somewhat simpler to analyze, so we omit the details.)

We now define $B \in M_{n+1,n+2}(\mathbf{C})$ by extending (3.2) through rows corresponding to degree n. Denote the columns of

$$\tilde{B} := \begin{pmatrix} B_{0,n+1} \\ \vdots \\ B_{n-1,n+1} \\ B \end{pmatrix}$$

by $\bar{Z}^i Z^j$ $(i + j = n + 1)$. We define B implicitly via the relations

$$\bar{Z}^i Z^j = p_{ij}(Z, \bar{Z}) \qquad (i + j = n + 1). \tag{3.3}$$

Note that (3.3) uniquely determines the candidate for a flat moment matrix extension $M(n+1)$. Indeed, since $M(n)$ satisfies (RG), the relation $Z^k = p(Z, \bar{Z})$ and Lemma 1.9 imply that (3.3) must hold in any flat moment matrix extension of $M(n)$; since B uniquely determines any flat extension of $M(n)$ containing this block, it follows that there is at most one flat extension $M(n+1)$; our goal is to prove that the flat extension determined by (3.3) is indeed of the form $M(n+1)$.

Since $\deg p \le k-1$, then $\deg p_{ij} \le n$, so $\operatorname{Ran} \tilde{B} \subseteq \mathcal{C}_{M(n)}$, which establishes (1.12). We next establish that B satisfies (1.9) (symmetric property) and (1.10) (normality).

- Symmetric property for B. From (3.3) and Lemma 1.6, it suffices to show that for $i+j=n+1$, $p_{ij} = \overline{p_{ji}}$, but this is clear from (3.1).
- Normality for B. For $i+j=n+1$, $\ell+m=n$, $m \ge 1$, $j \ge 1$ we must show
$$\langle \bar{z}^i z^j, \bar{z}^\ell z^m \rangle_B = \langle \bar{z}^{i+1} z^{j-1}, \bar{z}^{\ell+1} z^{m-1} \rangle_B.$$
We divide the proof into several cases.

Case B1.

$$\begin{aligned} i \ge k,\ j < k : \langle \bar{z}^i z^j, \bar{z}^\ell z^m \rangle_B &= \langle \bar{z}^{i-k} z^j \bar{p}, \bar{z}^\ell z^m \rangle_{M(n)} && \text{(by (3.3))} \\ &= \langle \bar{z}^{i-k+1} z^{j-1} \bar{p}, \bar{z}^{\ell+1} z^{m-1} \rangle_{M(n)} && \text{(by Theorem 1.5)} \\ &= \langle \bar{z}^{i+1} z^{j-1}, \bar{z}^{\ell+1} z^{m-1} \rangle_B && \text{(by (3.3)).} \end{aligned}$$

Case B2. $j \ge k$, $i < k$: (3.3) implies

$$\langle \bar{z}^i z^j, \bar{z}^\ell z^m \rangle_B = \langle \bar{z}^i z^{j-k} p, \bar{z}^\ell z^m \rangle_{M(n)}.$$

Subcase B2a. $i+1 < k$. Then $j-1 \ge k$ (for otherwise, $n+1 = i+j = (i+1)+(j-1) \le (k-1)+(k-1) \le n$). Thus,

$$\begin{aligned} \langle \bar{z}^i z^{j-k} p, \bar{z}^\ell z^m \rangle_{M(n)} &= \langle \bar{z}^{i+1} z^{j-k-1} p, \bar{z}^{\ell+1} z^{m-1} \rangle_{M(n)} && \text{(by Theorem 1.5)} \\ &= \langle \bar{z}^{i+1} z^{j-1}, \bar{z}^{\ell+1} z^{m-1} \rangle_B && \text{(by (3.3)).} \end{aligned}$$

Subcase B2b. $i+1=k$, $j=k$. (This case occurs only when n is even, i.e., $n=2d$, $k=d+1$, $i=d$, $j=d+1$, $i+j=2d+1=n+1$.) We must show that

$$\langle \bar{z}^{k-1} z^k, \bar{z}^\ell z^m \rangle_B = \langle \bar{z}^k z^{k-1}, \bar{z}^{\ell+1} z^{m-1} \rangle_B,$$

or, equivalently,

$$\langle \bar{z}^{k-1} p, \bar{z}^\ell z^m \rangle_{M(n)} = \langle \bar{p} z^{k-1}, \bar{z}^{\ell+1} z^{m-1} \rangle_{M(n)}. \tag{3.4}$$

Since B is symmetric, it suffices to consider upper diagonals, with $m \ge k-1$ (see the remarks following (1.10)). Now

$$\langle \bar{z}^{k-1} p, \bar{z}^\ell z^m \rangle_{M(n)} = \sum_{0 \le r+s \le k-1} a_{rs} \langle \bar{z}^{r+k-1} z^s, \bar{z}^\ell z^m \rangle_{M(n)}.$$

We have $r+k-1 \leq k-1+k-1 = n$ and $s+(r+k-1) \leq 2k-2 = n$; moreover, $m \geq k-1 \geq s$ and $r+k-1+m \geq r+k-1+k-1 = r+n \geq n$. Thus,

$$\begin{aligned}
&\langle \bar{z}^{r+k-1} z^s, \bar{z}^\ell z^m \rangle_{M(n)} \\
&\quad = \langle \bar{z}^{r+s+k-1}, \bar{z}^{\ell+s} z^{m-s} \rangle_{M(n)} && \text{(by Theorem 1.5 (5))} \\
&\quad = \langle \bar{z}^n, \bar{z}^{\ell+s} z^{(m-s)-(n-(r+s+k-1))} \rangle_{M(n)} && \text{(by Theorem 1.5 (3)–(4)).}
\end{aligned}$$

Now

$$\begin{aligned}
\langle \bar{z}^{k-1} p, \bar{z}^\ell z^m \rangle_{M(n)} &= \sum a_{rs} \langle \bar{z}^n, \bar{z}^{\ell+s} z^{r+k-1+m-n} \rangle_{M(n)} \\
&= \sum a_{rs} \langle \sum \overline{a_{tq}} z^t \bar{z}^{q+n-k}, \bar{z}^{\ell+s} z^{r+k-1+m-n} \rangle_{M(n)} \quad \text{(by (RG))} \\
&= \sum a_{rs} \sum \overline{a_{tq}} \gamma_{q+r+m-1, t+s+\ell} \\
&= \sum \overline{a_{tq}} \sum a_{rs} \gamma_{q+r+m-1, t+s+\ell}.
\end{aligned}$$

Also, since $M(n)$ is recursively generated, for $0 \leq t+q \leq k-1$,

$$Z^{t+k-1} \bar{Z}^q = (z^{t-1} \bar{z}^q p)(Z, \bar{Z}) = \sum a_{rs} \bar{Z}^{r+q} Z^{s+t-1};$$

thus

$$\begin{aligned}
\langle z^{t+k-1} \bar{z}^q, \bar{z}^{\ell+1} z^{m-1} \rangle_{M(n)} &= \sum a_{rs} \langle \bar{z}^{r+q} z^{s+t-1}, \bar{z}^{\ell+1} z^{m-1} \rangle_{M(n)} \\
&= \sum a_{rs} \gamma_{r+q+m-1, t+s+\ell}.
\end{aligned}$$

Now

$$\begin{aligned}
\langle \bar{z}^{k-1} p, \bar{z}^\ell z^m \rangle_{M(n)} &= \sum \overline{a_{tq}} \langle z^{t+k-1} \bar{z}^q, \bar{z}^{\ell+1} z^{m-1} \rangle_{M(n)} \\
&= \langle z^{k-1} \bar{p}, \bar{z}^{\ell+1} z^{m-1} \rangle_{M(n)},
\end{aligned}$$

which establishes (3.4).

Subcase B2c. $i+1 = k$, $j-1 \geq k$. In $\mathcal{C}_{M(n)}$, $\bar{Z}^k = \bar{p}(Z, \bar{Z})$, so by (RG),

$$(z^{j-k-1} p \bar{z}^k)(Z, \bar{Z}) = (z^{j-k-1} |p|^2)(Z, \bar{Z});$$

thus

$$\begin{aligned}
\langle \bar{z}^i z^{j-k} p, \bar{z}^\ell z^m \rangle_{M(n)} &= \langle \bar{z}^{i+1} z^{j-k-1} p, \bar{z}^{\ell+1} z^{m-1} \rangle_{M(n)} \\
&= \langle z^{j-k-1} |p|^2, \bar{z}^{\ell+1} z^{m-1} \rangle_{M(n)} \\
&= \langle \bar{z}^{i+1} z^{j-1}, \bar{z}^{\ell+1} z^{m-1} \rangle_B.
\end{aligned}$$

Case B3. $i \geq k$, $j \geq k$. $\langle \bar{z}^i z^j, \bar{z}^\ell z^m \rangle_B = \langle \bar{z}^{i-k} z^{j-k} |p|^2, \bar{z}^\ell z^m \rangle_{M(n)}$. If $j - k > 0$, then

$$\begin{aligned} \langle \bar{z}^{i-k} z^{j-k} |p|^2, \bar{z}^\ell z^m \rangle_{M(n)} &= \langle \bar{z}^{i-k+1} z^{j-k-1} |p|^2, \bar{z}^{\ell+1} z^{m-1} \rangle_{M(n)} \\ &= \langle \bar{z}^{i+1} z^{j-1}, \bar{z}^{\ell+1} z^{m-1} \rangle_B. \end{aligned}$$

Suppose $j = k$; since $Z^k = p(Z, \bar{Z})$, (RG) implies

$$(\bar{z}^{i-k} z^k \bar{p})(Z, \bar{Z}) = (\bar{z}^{i-k} |p|^2)(Z, \bar{Z}) = (\bar{z}^{i-k} z^{j-k} |p|^2)(Z, \bar{Z}).$$

Thus

$$\begin{aligned} \langle \bar{z}^{i-k} z^{j-k} |p|^2, \bar{z}^\ell z^m \rangle_{M(n)} &= \langle \bar{z}^{i-k} z^k \bar{p}, \bar{z}^\ell z^m \rangle_{M(n)} \\ &= \langle \bar{z}^{i-k+1} z^{k-1} \bar{p}, \bar{z}^{\ell+1} z^{m-1} \rangle_{M(n)} \\ &= \langle \bar{z}^{i+1} z^{j-1}, \bar{z}^{\ell+1} z^{m-1} \rangle_B. \end{aligned}$$

Following the plan outlined in Section 1 ((1.11)–(1.13)), we now define $B_{n,n+1} := B$, $\tilde{B} := (B_{i,n+1})_{0 \leq i \leq n}$, we let

$$M := \begin{pmatrix} M(n) & \tilde{B} \\ \tilde{B}^* & C \end{pmatrix}$$

denote the unique flat extension of $M(n)$ subordinate to $\tilde{B}$, and we let $\langle \cdot, \cdot \rangle_M$ denote the associated form. If $W = (\widehat{p_{0,n+1}}, \widehat{p_{1,n}}, \ldots, \widehat{p_{n,1}}, \widehat{p_{n+1,0}})$, then $\tilde{B} = M(n)W$ and thus $C = W^* M(n) W$. Since (1.11) and (1.12) hold, to complete the proof it suffices to verify that C is constant on upper diagonals ((1.13)). By flatness and Lemma 1.8, the columns of M are defined by (3.3). Note that by the moment-matrix block structures of $M(n)$, $\tilde{B}$, and $\tilde{B}^*$, M satisfies the following properties which do not involve block C:

$$\langle \bar{z}^i z^j, \bar{z}^k z^\ell \rangle_M = \gamma_{i+\ell, j+k} \tag{3.5}$$

for $0 \leq i + j \leq n$, $0 \leq k + \ell \leq n + 1$ and for $0 \leq i + j \leq n + 1$, $0 \leq k + \ell \leq n$;

$$\langle pz, q \rangle_M = \langle p, \bar{z} q \rangle_M \quad \text{and} \quad \langle p\bar{z}, q \rangle_M = \langle p, zq \rangle_M \tag{3.6}$$

for p, $q \in \mathcal{P}_n$;

$$\langle pz, qz \rangle_M = \langle p\bar{z}, q\bar{z} \rangle_M \tag{3.7}$$

for $p \in \mathcal{P}_{n-1}$, $q \in \mathcal{P}_n$ and for $p \in \mathcal{P}_n$, $q \in \mathcal{P}_{n-1}$.

We further note the following property of M:

$$\begin{gathered} \text{If } p,\ q,\ pq \in \mathcal{P}_n \text{ and } p(Z, \bar{Z}) = 0 \text{ in } \mathcal{C}_M, \text{ then} \\ (pq)(Z, \bar{Z}) \text{ in } \mathcal{C}_M. \end{gathered} \tag{3.8}$$

Indeed, since $p(Z,\bar{Z}) = 0$ in $\mathcal{C}_{M(n)}$, (RG) implies $(pq)(Z,\bar{Z}) = 0$ in $\mathcal{C}_{M(n)}$; since M is a positive extension of $M(n)$, [Fia] implies $(pq)(Z,\bar{Z})$ in $\mathcal{C}_M$.

To prove that C is constant on upper diagonals we must verify

For $i+j = \ell+m = n+1$, $m \ge j \ge 1$, $i \ge \ell$,

$$\langle \bar{z}^i z^j, \bar{z}^\ell z^m \rangle_M = \langle \bar{z}^{i+1} z^{j-1}, \bar{z}^{\ell+1} z^{m-1} \rangle_M. \tag{3.9}$$

Case C1. $i \ge k$, $j < k$.

$$\begin{aligned}
\langle \bar{z}^i z^j, \bar{z}^\ell z^m \rangle_M &= \langle \bar{z}^{i-k} z^j \bar{p}, \bar{z}^\ell z^m \rangle_M \\
&= \sum \overline{a_{rs}} \langle \bar{z}^{i-k+s} z^{j+r}, \bar{z}^\ell z^m \rangle_M \\
&= \sum \overline{a_{rs}} \langle \bar{z}^{i-k+s+1} z^{j+r-1}, \bar{z}^{\ell+1} z^{m-1} \rangle_M
\end{aligned}$$

(by (3.7), since $i-k+s+j+r-1 = (i+j)+(s+r)-k-1 \le n+1+k-1-k-1 = n-1$)

$$\begin{aligned}
&= \langle \bar{z}^{i+1-k} z^{j-1} \bar{p}, \bar{z}^{\ell+1} z^{m-1} \rangle_M \\
&= \langle \bar{z}^{i+1} z^{j-1}, \bar{z}^{\ell+1} z^m \rangle_M.
\end{aligned}$$

Case C2. $i < k$, $j \ge k$.

Subcase C2a. $i < k-1$. As in Case B2a, $j-1 \ge k$, so

$$\begin{aligned}
\langle \bar{z}^i z^j, \bar{z}^\ell z^m \rangle_M &= \langle \bar{z}^i z^{j-k} p, \bar{z}^\ell z^m \rangle_M \\
&= \sum a_{rs} \langle \bar{z}^{i+r} z^{j-k+s}, \bar{z}^\ell z^m \rangle_M \\
&= \sum a_{rs} \langle \bar{z}^{i+r+1} z^{j-k+s-1}, \bar{z}^{\ell+1} z^{m-1} \rangle_M
\end{aligned}$$

(by (3.7), since $i+r+j-k+s-1 \le n-1$)

$$\begin{aligned}
&= \langle \bar{z}^{i+1} z^{j-k-1} p, \bar{z}^{\ell+1} z^{m-1} \rangle_M \\
&= \langle \bar{z}^{i+1} z^{j-1}, \bar{z}^{\ell+1} z^{m-1} \rangle_M.
\end{aligned}$$

Subcase C2b. $i+1 = k$, $j = k$. (n even). Note that $m \ge j = k$.

$$\begin{aligned}
\langle \bar{z}^i z^j, \bar{z}^\ell z^m \rangle_M &= \langle \bar{z}^{k-1} p, \bar{z}^\ell z^m \rangle_M \\
&= \sum a_{rs} \langle \bar{z}^{r+k-1} z^s, \bar{z}^\ell z^m \rangle_M \\
&= \sum a_{rs} \langle \bar{z}^{r+s+k-1}, \bar{z}^{\ell+s} z^{m-s} \rangle_M
\end{aligned}$$

(by (3.7), since $m-s \ge m-(k-1) > m-k \ge 0$ and $r+s+k-1 \le 2k-2 = n$)

$$= \sum a_{rs} \langle \bar{z}^n, \bar{z}^{\ell+s} z^{m-(n-(r+k-1))} \rangle_M$$

(by (3.6), since $n - (r + k - 1) \geq s$ and since $m \geq k \Rightarrow r + k - 1 + m \geq 2k - 1 > n$)

$$= \langle \bar{z}^{i+1} z^{j-1}, \bar{z}^{\ell+1} z^{m-1} \rangle_M$$

(exactly as in the proof of Case B2b, but replacing $M(n)$ by M; note only that if $r + s,\ t + q \leq k - 1$, then $r + q + s + t - 1 \leq n - 1$, so (3.5) implies $\langle \bar{z}^{r+q} z^{s+t-1}, \bar{z}^{\ell+1} z^{m-1} \rangle_M = \gamma_{r+q+m-1,t+s+\ell}$).

Subcase C2c. $i + 1 = k$, $j - 1 \geq k$. In $\mathcal{C}_M$ we have $\bar{Z}^k = \bar{p}(Z, \bar{Z})$ and, from (3.8), $(z^{j-k-1} p \bar{z}^k)(Z, \bar{Z}) = (z^{j-k-1} |p|^2)(Z, \bar{Z})$. Thus

$$\begin{aligned} \langle \bar{z}^i z^j, \bar{z}^\ell z^m \rangle_M &= \langle \bar{z}^i z^{j-k} p, \bar{z}^\ell z^m \rangle_M \\ &= \langle \bar{z}^{i+1} z^{j-k-1} p, \bar{z}^{\ell+1} z^{m-1} \rangle_M \qquad \text{(by (3.7))} \\ &= \langle \bar{z}^{i+1} z^{j-1}, \bar{z}^{\ell+1} z^{m-1} \rangle_M. \end{aligned}$$

Case C3. $i \geq k, j \geq k$. The proof is identical to that of Case B3 (replacing $M(n)$ by M and using (3.7)).

The proof of Theorem 3.1 is now complete.

We conclude this section by considering the case when $Z^k = p(Z, \bar{Z})$ for $p \in \mathcal{P}_{k-1}$ and $[n/2]+1 < k \leq n$. Let $p(z, \bar{z}) \equiv \sum_{0 \leq i+j \leq k-1} a_{ij} \bar{z}^i z^j$. In constructing a flat extension $M(n+1)$, $B_{n,n+1}$ is uniquely determined by the relation $Z^{n+1} = \sum a_{ij} Z^{n+1-k+j} \bar{Z}^i$; indeed this relation uniquely determines $\gamma_{n,n+1}, \ldots, \gamma_{0,2n+1}$, and thus also $\gamma_{n+1,n} = \overline{\gamma_{n,n+1}}, \ldots, \gamma_{2n+1,0} = \overline{\gamma_{0,2n+1}}$. If the resulting block $\tilde{B} \equiv (B_{i,n+1})_{0 \leq i \leq n}$ satisfies $\operatorname{Ran} \tilde{B} \not\subseteq \mathcal{C}_{M(n)}$, then there is no flat extension $M(n+1)$. If $\operatorname{Ran} \tilde{B} \subseteq \mathcal{C}_{M(n)}$, let W be such that $\tilde{B} = M(n)W$; then $M(n)$ admits a flat extension $M(n+1)$ if and only if $W^* M(n) W$ is Toeplitz.

4. A Positive Invertible Moment Matrix Admitting No Representing Measure

Using results from algebraic geometry, D. Hilbert established in [Hil] the existence of a polynomial $q \in \mathbf{R}[x, y]$ of total degree 6 which is nonnegative on the real plane $\mathbf{R}^2$, but which cannot be expressed as a finite sum of squares of polynomials; an explicit such polynomial was later found by K. Schmüdgen [Sch1]. (Another concrete example is given in [BCJ].) We will use Schmüdgen's example to construct data $\gamma \equiv \{\gamma_{ij}\}_{0 \leq i+j \leq 6}$ whose associated moment matrix $M(3)(\gamma)$ is positive and invertible but admits no representing measure; in particular, $M(3)(\gamma)$ does not have a flat extension $M(4)$. This will disprove Conjecture 1.1 in [CF4], since invertible moment matrices satisfy property (RG) vacuously. We begin by recalling Schmüdgen's result. Let $\mathbf{C}[x, y]$ denote the polynomials in Hermitian variables x

and y with complex coefficients ($x = \bar{x}$, $y = \bar{y}$). Let $\mathcal{P}$ denote the cone in $\mathbf{C}[x, y]$ consisting of "sums of squares" $\sum \bar{p}_i p_i$. Let $\mathcal{C}_+ = \{p \in \mathbf{C}[x, y] : p(x, y) \geq 0$ for all real x, $y\}$. A linear functional $F : \mathbf{C}[x, y] \to \mathbf{C}$ is *positive* if $F|\mathcal{P}$ is non-negative; F is *strongly positive* if $F|\mathcal{C}_+$ is non-negative; F has a positive Borel representing measure if and only if F is strongly positive [ShT], [Sch1] (cf. [Hav1], [Hav2]).

Theorem 4.1. ([Sch1, Theorem])

(1) *The polynomial*

$$q(x, y) := 200(x^3 - 4x)^2 + 200(y^3 - 4y)^2 + (y^2 - x^2)x(x + 2)[x(x - 2) + 2(y^2 - 4)]$$

is nonnegative on $\mathbf{R}^2$, *but cannot be written as a sum of squares.*

(2) *There exists a positive linear functional F on $\mathbf{C}[x, y]$ with $F(q) < 0$. Thus, F cannot be represented as integration with respect to a positive Borel measure with support in $\mathbf{R}^2$.*

F is defined first on the space $\mathbf{C}_6[x, y]$, the complex polynomials of total degree at most 6, as a linear combination of evaluation functionals (and is then extended to all of $\mathbf{C}[x, y]$):

$$F(p) := 32 \sum_{i=1}^{8} p(A_i) + p(B_1) + p(B_2) - p(A_9) \quad (p \in \mathbf{C}_6[x, y]),$$

where

$$\begin{array}{llll} A_1 := (-2, -2), & A_2 := (0, -2), & A_3 := (2, -2), & A_4 := (-2, 0), \\ A_5 := (0, 0), & A_6 := (-2, 2), & A_7 := (0, 2), & A_8 := (2, 2), \\ A_9 := (2, 0), & B_1 := (\frac{1}{100}, 0), & B_2 := (0, \frac{1}{100}). & \end{array}$$

We define $\gamma_{k\ell} := F((x - iy)^k (x + iy)^\ell)$ $(0 \leq k + \ell \leq 6)$. Observe that

$$\gamma_{k\ell} = \sum_{r=0}^{k} \sum_{s=0}^{\ell} (-1)^{k-r} i^{k+\ell-r-s} \binom{k}{r} \binom{\ell}{s} F(x^{r+s} y^{k+\ell-r-s}),$$

and that

$$F(x^r y^s) = \begin{cases} 257 & r = 0 \text{ and } s = 0 \\ 96[(-2)^s + 2^s] + 10^{-2s} & r = 0 \text{ and } s \geq 1 \\ 32[3(-2)^r + 2^{r+1}] + 10^{-2r} - 2^r & r \geq 1 \text{ and } s = 0 \\ [1 + (-1)^r][1 + (-1)^s] 2^{r+s+5} & r \geq 1 \text{ and } s \geq 1 \end{cases}.$$

The associated matrix $M(3)$ is built using the following values:

$$\begin{aligned}
\gamma_{00} &= 257 \\
\gamma_{01} &= 10^{-2}(1-6599i) \\
\gamma_{02} &= 132 \\
\gamma_{03} &= 10^{-6}(1+263999999i) \\
\gamma_{04} &= \frac{333599999999}{50000000} \\
\gamma_{05} &= 10^{-10}(1-10559999999999i) \\
\gamma_{06} &= 2112
\end{aligned}
\qquad
\begin{aligned}
\gamma_{11} &= \frac{7020001}{5000} \\
\gamma_{12} &= \bar{\gamma}_{03} \\
\gamma_{13} &= 528 \\
\gamma_{14} &= \bar{\gamma}_{05} \\
\gamma_{15} &= -\frac{29727999999999999}{500000000000}
\end{aligned}
\quad \ldots$$

$$\ldots \quad
\begin{aligned}
\gamma_{22} &= \frac{485600000001}{50000000} \\
\gamma_{23} &= \gamma_{05} \\
\gamma_{24} &= \gamma_{06}
\end{aligned}
\qquad
\gamma_{33} = \frac{35808000000000001}{500000000000}.$$

A straightforward calculation using the Nested Determinants Test now shows that $M(3) \geq 0$ and that $\det M(3) > 0$. Since the presence of a representing measure for γ would immediately give a corresponding measure for $F|\mathbf{C}_6[x,y]$, it follows from Theorem 4.1(2) that $M(3)$ cannot admit a representing measure.

In view of the preceding example we modify [CF4, Conjecture 1.1] as follows.

Conjecture 4.2. *The following are equivalent for a truncated moment sequence* $\gamma \equiv \gamma^{(2n)}$:

(i) γ *has a representing measure;*
(ii) γ *has a representing measure with moments of all orders;*
(iii) γ *has a compactly supported representing measure;*
(iv) γ *has a finitely atomic representing measure;*
(v) γ *has a* rank $M(n)$*-atomic representing measure;*
(vi) $M(n) \geq 0$ *admits a flat extension* $M(n+1)$.

Added in Proof. In recent work [CF5], we have adapted results of V. Tchakaloff [Tch] and I.P. Mysovskikh [Mys] to prove (i) $\Rightarrow$ (iv) in Conjecture 4.2; thus, conditions (i), (ii), (iii) and (iv) are all equivalent. Independently, M. Putinar [P5] has found a different proof of (i) $\Rightarrow$(iv), also based on extending results of [Tch]. (Somewhat earlier, we had obtained (iii) $\Rightarrow$(iv) by adapting [Tch], and J. McCarthy had communicated to us another proof of the same implication, using convexity theory.)

Theorem 1.2 shows that (v) and (vi) of Conjecture 4.2 are equivalent, and clearly (v) $\Rightarrow$(iv); however, J. McCarthy [McC], in response to Conjecture 4.2, has recently proved that there exist truncated moment sequences γ having representing measures, but such that $M(n)(\gamma)$ does *not* have a flat extension $M(n+1)$. Thus (i) $\not\Rightarrow$(v) and Conjecture 4.2 is false as stated. McCarthy's dimension-theoretic result actually shows that moment sequences γ admitting no flat extensions are in a sense generic: among moment sequences γ with representing measures, those with rank $M(n)(\gamma)$-atomic representing measures are rare. On the other hand, it follows from the equivalence of (i) and (iv) and from the equivalence of (v) and (vi) that a truncated moment sequence γ has a representing measure if and only if for some $k \geq 0$, $M(n)(\gamma)$ admits a positive extension $M(n+k)$ which in turn has a flat extension $M(n+k+1)$.

In [CF5] we continue to study concrete necessary or sufficient conditions for the existence of flat extensions. In particular, we exhibit several examples of positive, recursively generated moment matrices which do not admit representing measures and which are much easier to construct and analyze than the example of Theorem 4.1.

REFERENCES

[AK] N. I. AHIEZER AND M. G. KREIN, *Some Questions in the Theory of Moments*, Transl. Math. Monographs **2** (1962), Amer. Math. Soc., Providence.

[Akh] N. I. AKHIEZER, *The Classical Moment Problem*, Hafner Publ. Co., New York, 1965.

[Atz] A. ATZMON, *A moment problem for positive measures on the unit disc*, Pacific J. Math. **59** (1975), 317–325.

[Ber] C. BERG, *The multidimensional moment problem and semigroups*, Moments in Mathematics, Proc. Symposia Appl. Math. **37** (1987), 110–124.

[BCJ] C. BERG, J. P. R. CHRISTENSEN AND C. U. JENSEN, *A remark on the multidimensional moment problem*, Math. Ann. **223** (1979), 163–169.

[BM] C. BERG AND P. H. MASERICK, *Polynomially positive definite sequences*, Math. Ann. **259** (1982), 487–495.

[Cas] G. CASSIER, *Problème des moments sur un compact de* $\mathbf{R}^n$ *et décomposition des polynômes à plusieurs variables*, J. Funct. Anal. **58** (1984), 254–266.

[CF1] R. CURTO AND L. FIALKOW, *Recursively generated weighted shifts and the subnormal completion problem*, Integral Equations and Operator Theory **17** (1993), 202–246.

[CF2] R. CURTO AND L. FIALKOW, *Recursively generated weighted shifts and the subnormal completion problem, II*, Integral Equations and Operator Theory **18** (1994).

[CF3] R. CURTO AND L. FIALKOW, *Recursiveness, positivity, and truncated moment problems*, Houston J. Math. **17** (1991), 603–635.

[CF4] R. CURTO AND L. FIALKOW, *Solution of the truncated complex moment problem for flat data*, Memoirs Amer. Math. Soc. **568** (1996), x + 52 pp.

[CF5] R. CURTO AND L. FIALKOW, *Flat extensions of positive moment matrices, II: Recursively generated relations*, preprint 1996.

[CP] R. CURTO AND M. PUTINAR, *Nearly subnormal operators and moment problems*, J. Funct. Anal. **115** (1993), 480–497.

[Fia] L. FIALKOW, *Positivity, extensions and the truncated complex moment problem*, Contemporary Math. **185** (1995), 133–150.

[Fug] B. FUGLEDE, *The multidimensional moment problem*, Expo. Math. **1** (1983), 47–65.

[Hau] F. HAUSDORFF, *Momentprobleme für ein endliches Intervall*, Math. Zeit. **16** (1923), 220–248.

[Hav1] E. K. HAVILAND, *On the momentum problem for distributions in more than one dimension*, Amer. J. Math. **57** (1935), 562–568.

[Hav2] E. K. HAVILAND, *On the momentum problem for distributions in more than one dimension, Part II*, Amer. J. Math. **58** (1936), 164–168.

[Hil] D. HILBERT, *Über die Darstellung definiter Formen als Summen von Formenquadraten*, Math. Ann. **32** (1888), 342–350.

[KrN] M.G. KREIN AND A. NUDEL'MAN, *The Markov Moment Problem and Extremal Problems*, Transl. Math. Monographs **50** (1977), Amer. Math. Soc., Providence.

[Lan] H. LANDAU, *Classical background of the moment problem*, Moments in Mathematics, Proc. Symposia Appl. Math. **37** (1987), 1–15.

[McC] J. MCCARTHY, *Private Communication*, 1995.

[McG] J. L. MCGREGOR, *Solvability criteria for certain N-dimensional moment problems*, J. Approx. Theory **30** (1980), 315–333.

[Mys] I. P. MYSOVSKIKH, *On Chakalov's Theorem*, USSR Comp. Math. **15** (1975), 221–227.

[P1] M. PUTINAR, *A two-dimensional moment problem*, J. Funct. Anal. **80** (1988), 1–8.

[P2] M. PUTINAR, *The L problem of moments in two dimensions*, J. Funct. Anal. **94** (1990), 288–307.

[P3] M. PUTINAR, *Positive polynomials on compact semi-algebraic sets*, Indiana Univ. Math. J. **42** (1993), 969–984.

[P4] M. PUTINAR, *Extremal solutions of the two-dimensional L-problem of moments*, preprint 1994.

[P5] M. PUTINAR, *On Tchakaloff's Theorem*, preprint 1995.

[Rez] B. REZNICK, *E-mail Communication*, 1995.

[Sar] D. SARASON, *Moment problems and operators on Hilbert space*, Moments in Mathematics, Proc. Symposia Appl. Math. **37** (1987), 54–70.

[Sch1] K. SCHMÜDGEN, *An example of a positive polynomial which is not a sum of squares of polynomials. A positive, but not strongly positive functional*, Math. Nachr. **88** (1979), 385–390.

[Sch2] K. SCHMÜDGEN, *The K-moment problem for semi-algebraic sets*, Math. Ann. **289** (1991), 203–206.

[ShT] J. SHOHAT AND J. TAMARKIN, *The Problem of Moments*, Math. Surveys I, Amer. Math. Soc., Providence, 1943.

[Smu] J. L. SMUL'JAN, *An operator Hellinger integral (Russian)*, Mat. Sb. **91** (1959), 381–430.

[Sto] J. STOCHEL, *Private Correspondence*, 1994.

[StSz] J. STOCHEL AND F. SZAFRANIEC, *Algebraic operators and moments on algebraic sets*, Portugaliae Math. **51** (1994), 1–21.

[Sza] F. SZAFRANIEC, *Moments on compact sets*, in "Prediction Theory and Harmonic Analysis", V. Mandrekar and H. Salehi, eds., North-Holland, Amsterdam, 1983; pp. 379–385.

[Tch] V. TCHAKALOFF, *Formules de cubatures mécaniques à coefficients non négatifs*, Bull. Sc. Math. **81** (1957), 123–134.

[Wol] WOLFRAM RESEARCH, INC., *Mathematica*, Version 2.1, Wolfram Research, Inc., Champaign, Illinois, 1992.

RAÚL E. CURTO
Department of Mathematics
The University of Iowa
Iowa City, Iowa 52242
E-MAIL: curto@math.uiowa.edu

L. A. FIALKOW
Dept. of Mathematics and Computer Science
SUNY at New Paltz
New Paltz, NY 12561
E-MAIL: fialkow@mcs.newpaltz.edu

Received: August 23rd, 1995.

Operator Theory:
Advances and Applications, Vol. 104

Geometric Invariants for Resolutions of Hilbert Modules*

RONALD G. DOUGLAS AND GADADHAR MISRA

Dedicated to Carl M. Pearcy on his sixtieth birthday

0. INTRODUCTION

The development of complex function theory beyond the one-variable case required new techniques and approaches, not just an extension of what had worked already. The same is proving true of multi–variate operator theory. Still it is reasonable to start by seeking to understand in the larger context results that have proved important and useful in the study of single operator theory. For that reason the first author showed [6] how to frame the canonical model theory of Sz.-Nagy and Foias for contraction operators in the language of Hilbert module resolutions over the disk algebra. This point of view made clear why a straightforward extension of model theory failed in the multi–variate case. Since the appropriate algebra in this case would be higher dimensional, one would expect module resolutions, if they existed, to be of longer length and hence not expressible as a canonical model. While work on this topic has shed light on what one might expect to be true (cf. [7]), useful results are still scarce.

Now one can raise two kinds of questions about module resolutions. The first concerns their existence, that is, under what hypotheses on a given Hilbert module does a resolution by "nice" Hilbert modules exist. Here little progress has been made except to understand better what being a nice module should mean. The current working definition [7] requires the spectral sheaf to be locally free or a vector bundle with acyclic Koszul cohomology. (We are assuming all modules are finitely generated.) We do not intend to explore this issue here.

The second question asks how, given such a resolution, does one extract information from it about the module one is trying to study. In the case considered by Sz.-Nagy and Foias, this problem was solved [6] by localizing the connecting homomorphism in the resolution to recover the characteristic operator function. However, here one can impose conditions effectively making the resolution unique. In the multi-variate case, this is not possible.

One of the important impetuses for the development of homological algebra was the problem of extracting information from module resolutions in the context

*Research supported in part by grants from the National Science Foundation. The second author was also supported by a travel grant from the CDE.

of pure algebra. The techniques developed there, however, will not carry over without some adaptation since the algebraic hypotheses required are seldom satisfied in the context of Hilbert modules. While developing an adaptation of the algebraic theory to the multi–variate operator theory context is important, here we are exploring another approach. We seek to build on the complex geometric structure inherent in the situation and attempt to relate curvatures and other geometric quantities. Such an approach can be viewed as generalizing that which leads to the characteristic operator function in the one-dimensional case.

At present, we do not understand the situation well enough to formulate theorems, let alone prove them. One of the difficulties is the scarcity of concrete examples with calculations of the associated geometric objects such as curvature. The statement of Theorem 1.4 which relates the curvature in a quotient module under very special circumstances to that of the module, became apparent only after the calculation of particular cases outlined in Sections 2.4 and 2.5. Moreover, these calculations themselves suggest the possibility that the difference of the curvatures, corresponding to a moudle and a submodule defined by a zero set, may depend only on the geometry of the zero set and its embedding. Some of the other examples support such a conjecture but the family discussed in Section 2.1 shows it can't be true without further restrictions. What the nature of such hypotheses might be or, indeed, if there are any remains a mystery.

There is another way to arrive at the existence of such a conjecture. In the single-variable case, the alternating difference of the curvatures and a curvature–like contribution from the connecting map, yields an atomic measure with support equal to the spectrum of the quotient module. One seeks an analogue of such a formula in the multi–variable case assuming that the other connecting map contributes nothing since it is co–isometric. In various examples, such a formula is shown to hold. One can also consider these expressions restricted to the zero set. A chief purpose of the calculations in Section 2 is to determine if such formulae hold and in what generality.

Although Theorem 1.4 provides a good method for calculating the curvature of a quotient module, the approach requires the submodule to be prime-like; it does not work in the presence of nilpotents. Coming to grips with such cases requires first that one generalize the notion of functional Hilbert space since the quotient module will not be one in the sense of Aronszajn if the submodule is not prime-like. Further, one will need to consider geometric invariants that arise from higher-order localization, that is, localizations by modules with one point spectrum but which are not one-dimensional. We take some preliminary steps in this direction in the last section.

Here we consider some natural examples of short exact sequences of Hilbert modules that are built from a family of modules, which contains the Hardy and Bergman modules, and a submodule arising from the closure of a principal ideal. The quotient module is supported on the zero set of the ideal and we seek to relate its curvature to that of the curvatures of the other two modules and an analogous quantity defined from the connecting homomorphism. All are considered as currents. Although we attempt to suggest more general conclusions and to formulate conjectures, this paper records work in progress.

1. Restrictions

Let Ω be a bounded domain in $\mathbb{C}^m$ and $K : \Omega \times \Omega \to \mathbb{C}$ be a function which is holomorphic in the first variable and anti - holomorphic in the second variable. If K is also a non negative (nnd) function in the sense that the matrix $((K(z_i, z_j)))$ is nnd for all finite subsets $\{z_1, \ldots, z_n\} \subseteq \Omega$, then K determines a unique Hilbert space $\mathcal{M}$ of holomorphic functions on Ω such that

$$\langle f, K(\cdot, w)\rangle = f(w) \text{ for all } f \in \mathcal{M},\ w \in \Omega. \tag{1.1}$$

In the usual terminology (cf. [1]), we say that $\mathcal{M}$ is a functional Hilbert space and K is said to be the reproducing kernel for $\mathcal{M}$.

Let $\mathcal{A}(\Omega)$ denote the closure of the polynomials in the supremum norm on Ω. We assume throughout that the algebra $\mathcal{A}(\Omega)$ acts boundedly on the Hilbert space $\mathcal{M}$. This means that the pointwise product $f \cdot h$ is in $\mathcal{M}$ for each $f \in \mathcal{A}(\Omega)$ and each $h \in \mathcal{M}$. Note that the closed graph theorem ensures the boundedness of the operator $(f, h) \to f \cdot h$ so that $\mathcal{M}$ is a Hilbert module in the sense of [8]. Let $\mathcal{M}_0$ denote the closure in $\mathcal{M}$ of an ideal $\mathcal{I} \subseteq \mathcal{A}(\Omega)$. If $\mathcal{Z}$ denotes the set of common zeros of the functions in $\mathcal{I}$, then the case when $\mathcal{Z}$ is discrete and finite can be analyzed since the quotient module $\mathcal{M}_{\mathrm{q}}$ is finite-dimensional. However, if $\mathcal{M}_0$ is taken to be the subspace of all functions vanishing on an infinite subset $\mathcal{Z}$ of Ω, then the situation is considerably more complex. We assume in what follows that $\mathcal{M}_0$ is the submodule of $\mathcal{M}$ consisting of all functions h that vanish on an analytic hypersurface $\mathcal{Z}$ of Ω. In general, it is not enough for $\mathcal{M}_0$ to be the closure of an ideal $\mathcal{I}$ with zero set $\mathcal{Z}$ unless we assume, among other assumptions, that $\mathcal{I}$ is the intersection of prime ideals. Some further complications arise in case $\mathcal{M}_0$ is the closure of an ideal $\mathcal{I}$ that is not but for the present we restrict attention to the largest submodule with zero set $\mathcal{Z}$. However, the approach developed here can be applied to analyze the case in which $\mathcal{M}_0$ is the closure of a principal ideal and we plan to take up this and related questions in a future paper.

We can view $\mathcal{M}_{\mathrm{q}}$ as a module over $\mathcal{A}(\Omega)$ by compressing the module action on $\mathcal{M}$ to $\mathcal{M}_{\mathrm{q}}$ (cf. [8, p. 41]). It follows from [1] that both $\mathcal{M}_0$ and $\mathcal{M}_{\mathrm{q}}$ are functional Hilbert spaces over Ω and $\mathcal{Z}$ respectively. We let $K(\cdot, w)$, $K_0(\cdot, w)$ and $K_{\mathrm{q}}(\cdot, w)$ denote the reproducing kernels for $\mathcal{M}$, $\mathcal{M}_0$ and $\mathcal{M}_{\mathrm{q}}$, respectively.

It is easy to verify that

$$K(\cdot, w) = K_{\mathrm{q}}(\cdot, w) + K_0(\cdot, w).$$

We consider the Hilbert space $\mathcal{M}_{\mathrm{res}}$ obtained by restricting the functions in $\mathcal{M}$ to the set $\mathcal{Z}$, that is,

$$\mathcal{M}_{\mathrm{res}} = \left\{ h_0 \colon \mathcal{Z} \to \mathbb{C}, \text{holomorphic} \;\middle|\; h_0 = h|_{\mathcal{Z}} \text{ for some } h \in \mathcal{M} \right\}.$$

The norm of $h_0 \in \mathcal{M}_{\text{res}}$ is

$$\|h_0\| = \inf\{\|h\| : h|_{\mathcal{Z}} = h_0 \text{ for } h \in \mathcal{M}\}.$$

Aronszajn [1, p. 351] shows that the restriction map $R : \mathcal{M}_{\text{q}} \to \mathcal{M}_{\text{res}}$ is unitary as follows.

Let $P : \mathcal{M} \to \mathcal{M}_{\text{q}}$ be the projection. For $h \in \mathcal{M}$, we have

$$\begin{aligned} \|Ph\| &= \inf\{\|h + h_1\| : h_1 \in \mathcal{M}_0\} \\ &= \inf\{\|\tilde{h}\| : \tilde{h}\,|_{\mathcal{Z}} = h|_{\mathcal{Z}}\} \end{aligned}$$

Since the functions Ph and h have the same restriction to $\mathcal{Z}$, it follows that the map $R : Ph \to h|_{\mathcal{Z}}$ is an isometry. If h_0 is in $\mathcal{M}_{\text{res}}$, then there exists $h \in \mathcal{M}$ such that $h|_{\mathcal{Z}} = h_0$ and it follows that $R^* h_0 = Ph$. Thus R is unitary. This can be used to show that the reproducing kernel $K_{\text{res}}(\cdot, w)$ for $\mathcal{M}_{\text{res}}$ is $K_{\text{res}}(\cdot, w) = K(\cdot, w)|_{\mathcal{Z}}$, $w \in \mathcal{Z}$ [1, p. 351].

Restricting the module map $(f, h) \to f \cdot h$ in both arguments to $\mathcal{Z}$, we see that $\mathcal{M}_{\text{res}}$ is a module over the algebra

$$\mathcal{A}_{\text{res}}(\Omega) \stackrel{\text{def}}{=} \{f|_{\mathcal{Z}} : f \in \mathcal{A}(\Omega)\}.$$

Although $\mathcal{A}_{\text{res}}(\Omega)$ need not be a function algebra, its completion is and $\mathcal{M}_{\text{res}}$ is a Hilbert module over it. Let $i : \mathcal{Z} \to \Omega$ be the inclusion map and $i^* : \mathcal{A}(\Omega) \to \mathcal{A}_{\text{res}}(\Omega)$ be the map $i^* f = f \circ i$. We can push forward the module defined over $\mathcal{A}_{\text{res}}(\Omega)$ to a module over $\mathcal{A}(\Omega)$ via the map

$$(f, h) \to (i^* f) \cdot h, \quad f \in \mathcal{A}(\Omega), \ h \in \mathcal{M}_{\text{res}}. \tag{1.2}$$

Recall that two Hilbert modules $\mathcal{M}_1$ and $\mathcal{M}_2$ over the algebra $\mathcal{A}(\Omega)$ are said to be isomorphic if there is an unitary operator $T : \mathcal{M}_1 \to \mathcal{M}_2$ intertwining the two module actions, that is, $f \cdot Th = Tf \cdot h$ for $f \in \mathcal{A}(\Omega)$ and $h \in \mathcal{M}_1$. Any operator satisfying the latter condition is said to be a module map.

Theorem 1.1. *Let $\mathcal{M}$ be a Hilbert module over the algebra $\mathcal{A}(\Omega)$ and $\mathcal{M}_0$ be the submodule of functions vanishing on a fixed subset $\mathcal{Z} \subseteq \Omega$. The push forward $i_* \mathcal{M}_{\text{res}}$ is isomorphic to the quotient module $\mathcal{M}_{\text{q}}$.*

PROOF. First, we reprove the fact that the restriction map is unitary in a way that gells well with the spirit of this paper. We have $K_{\text{q}}(\cdot, w) = K(\cdot, w)$ for $w \in \mathcal{Z}$. The map

$$h \stackrel{R}{\to} \langle h, K_{\text{q}}(\cdot, w)\rangle = \langle h, K(\cdot, w)\rangle = h|_{\mathcal{Z}}, \quad h \in \mathcal{M}_{\text{q}}, \ w \in \mathcal{Z}$$

is the restriction map on $\mathcal{M}_{\text{q}}$. If we define an inner product on $R(\mathcal{M}_{\text{q}})$ as $\langle Rh, Rh'\rangle \stackrel{\text{def}}{=} \langle h, h'\rangle$, then it follows that

$$\langle Rh, K|_{\text{res}}(\cdot, w)\rangle = \langle h, K(\cdot, w)\rangle = \langle h, K_{\text{q}}(\cdot, h)\rangle = h(w), \quad w \in \mathcal{Z}.$$

Thus the reproducing kernel for the space $R(\mathcal{M}_q)$ is $K|_{\rm res}(\cdot, w)$, $w \in \mathcal{Z}$. Therefore, $R(\mathcal{M}_q)$ is the Hilbert space $\mathcal{M}_{\rm res}$. By our construction R is an onto isometry.

We only need to verify that $R : \mathcal{M}_q \to \mathcal{M}_{\rm res}$ is a module map, that is, $(f \circ i) \cdot (Rh) = RP(f \cdot h)$ for all $h \in \mathcal{M}_q$. Note that for $w \in \mathcal{Z}$, we have $\langle h, M_f^* P K_q(\cdot, w)\rangle = \langle h, M_f^* K(\cdot, w)\rangle$. This implies that $\langle PM_f h, K_q(\cdot, w)\rangle = f(w)\langle h, K_q(\cdot, w)\rangle$ for $h \in \mathcal{M}_q$, $w \in \mathcal{Z}$. Further,

$$\begin{aligned}\langle f \circ i \cdot Rh, K_{\rm res}(\cdot, w)\rangle &= \langle Rh, M_{f\circ i}^* K_{\rm res}(\cdot, w)\rangle \\ &= \langle h, \overline{f(w)} R^* K_{\rm res}(\cdot, w)\rangle \\ &= \langle h, \overline{f(w)} K_q(\cdot, w)\rangle \\ &= \langle PM_f h, K_q(\cdot, w)\rangle \\ &= \langle P(f \cdot h), R^* K_{\rm res}(\cdot, w)\rangle \\ &= \langle RP(f \cdot h), K_{\rm res}(\cdot, w)\rangle.\end{aligned}$$

This calculation verifies that R is a module map and the proof is complete. □

We point out that if $\mathcal{Z}$ happens to be an open subset of Ω, then $\mathcal{M}_q$ equals $\mathcal{M}$. In this case, $i_* \mathcal{M}_{\rm res}$ and $\mathcal{M}$ are isomorphic. Thus, we don't distinguish the modules $\mathcal{M}$ and $\mathcal{M}_{\rm res}$.

For our analysis, we will assume that $\mathcal{Z}$ is an analytic hypersurface in Ω in the sense of [11, Definition 8, p. 17]. Let $U \subseteq \Omega$ be a fixed open set containing a given point $z_0 \in \mathcal{Z}$. We may choose local co–ordinates ([11, Theorem 9, p.17]), $\phi \stackrel{\rm def}{=} (\phi^1, \ldots, \phi^m) : U \subseteq \Omega \to \mathbb{C}^m$ such that

$$\mathcal{Z} \cap U = \{z \in U : \phi^1(z) = 0\}. \tag{1.3}$$

In view of the remark preceeding Corollary 3 in [11, p. 34], if the second Cousin problem is solvable for Ω, then there exists a global defining function, which we will again denote by ϕ^1, for the hypersurface $\mathcal{Z}$. It is easy to see that, even though the function ϕ need not define global co-ordinates for Ω, it extends to a holomorphic function on Ω. We will assume throughout that the function ϕ is bounded and that the neighbourhood U of $z_0 \in \mathcal{Z}$ has been chosen such that ϕ is bi-holomorphic on U. If $i : U \to \Omega$ is the inclusion map, then $\phi \circ i : U \to V \subseteq \mathbb{C}^m$, V open in $\mathbb{C}^m$, is a bi-holomorphic map.

Let $\Gamma x(v) = \langle x, K(\cdot, \phi^{-1}(v))\rangle$, $x \in \mathcal{M}$ and $v \in V$. Let $\mathcal{N}$ be the set of holomorphic functions $\{\Gamma x : x \in \mathcal{M}\}$ on V with the inner product $\langle \Gamma x, \Gamma y\rangle_{\mathcal{N}} \stackrel{\rm def}{=} \langle x, y\rangle_{\mathcal{M}}$. Thus Γ is an onto isometry. Obviously, the kernel function for $\mathcal{N}$ is

$$K_\phi(v_1, v_2) = K(\phi^{-1}(v_1), \phi^{-1}(v_2)), \quad v_1, v_2 \in V. \tag{1.4}$$

As remarked immediately after Theorem 1.1, the module $\mathcal{M}_{\rm res\ U}$ is isomorphic to $\mathcal{M}$ and we are merely looking at a different realization of the functional Hilbert space, this time as holomorphic functions on U. We push forward the module $\mathcal{M}_{\rm res\ U}$ under the map ϕ so that the module action for $\phi_* \mathcal{M}_{\rm res\ U}$ is given by $(f, h) \to (f \circ \phi) \cdot h$ for $f \in \mathcal{A}(V)$ and $h \in \mathcal{M}_{\rm res\ U}$.

Lemma 1.2. *The modules $\mathcal{N}$ and the push forward $\phi_*\mathcal{M}_{\text{res U}}$ are isomorphic.*

PROOF. For the proof we need only to verify that Γ is a module map. This amounts to verifying $\Gamma(f\circ\phi)\cdot h = f\cdot\Gamma h$. However,

$$\begin{aligned}\langle f\cdot(\Gamma h), K_\phi(\cdot,v)\rangle &= \langle \Gamma h, M_f^* K_\phi(\cdot,v)\rangle\\ &= \overline{f(v)}\langle \Gamma h, K_\phi(\cdot,v)\rangle\\ &= \overline{f\circ\phi\circ\phi^{-1}(v)}\Gamma h(v)\\ &= \langle h, M_{f\circ\phi}^* K(\cdot,\phi^{-1}(v))\rangle\\ &= \big(\Gamma(f\circ\phi)\cdot h\big)(v)\\ &= \langle \Gamma(f\circ\phi)\cdot h, K_\phi(\cdot,v)\rangle\,.\end{aligned}$$

This verification completes the proof. □

Since $(\partial\phi^1/\partial z_1)(z_0)\neq 0$, $z_0\in\mathcal{Z}$, we may choose $\phi=(\phi^1,\dots,\phi^m)$, where ϕ^ℓ $(\ell\neq 1)$ is the projection to z_ℓ, to be our new co-ordinate system.

We now relate this description of the various modules to complex geometry (cf. [5]). Let

$$0\longleftarrow \mathcal{M}_{\mathrm{q}} \overset{\mathrm{Q}}{\longleftarrow} \mathcal{M} \overset{X}{\longleftarrow} \mathcal{M}_0 \longleftarrow 0 \tag{1.5}$$

be an exact sequence of Hilbert modules, where X is the inclusion map. We obtain a localisation by tensoring with the one-dimensional module $\mathbb{C}_w$. Then it is not hard to see that while $\dim \mathcal{M}\otimes_{\mathcal{A}(\Omega)}\mathbb{C}_w$ and $\dim\mathcal{M}_0\otimes_{\mathcal{A}(\Omega)}\mathbb{C}_w$ equal 1, the dimension of $\mathcal{M}_{\mathrm{q}}\otimes_{\mathcal{A}(\Omega)}\mathbb{C}_w$ is one or zero according as w is in $\mathcal{Z}$ or not. These localisations give rise to hermitian anti–holomorphic vector bundles. An anti–holomorphic frame determines such a bundle. In the following description s is an anti–holomorphic frame, which is described in terms of the reproducing kernel. We obtain two hermitian anti-holomorphic vector bundles

$$E \overset{\text{def}}{=} \Big\{\big(\mathcal{M}\otimes_{\mathcal{A}(\Omega)}\mathbb{C}_w, w\big)\to w,\ s(w)=K(\cdot,w)\in\mathcal{M}\otimes_{\mathcal{A}(\Omega)}\mathbb{C}_w\Big\}$$

and

$$E_0 \overset{\text{def}}{=} \Big\{\big(\mathcal{M}_0\otimes_{\mathcal{A}(\Omega)}\mathbb{C}_w, w\big)\to w,\ s(w)=K_0(\cdot,w)\in\mathcal{M}_0\otimes_{\mathcal{A}(\Omega)}\mathbb{C}_w, w\notin\mathcal{Z}\Big\},$$

which live on Ω, while

$$E_{\mathrm{q}} \overset{\text{def}}{=} \Big\{\big(\mathcal{M}_{\mathrm{q}}\otimes_{\mathcal{A}(\Omega)}\mathbb{C}_w, w\big)\to w,\ s(w)=K_{\mathrm{q}}(\cdot,w)\in\mathcal{M}_{\mathrm{q}}\otimes_{\mathcal{A}(\Omega)}\mathbb{C}_w,\ w\in\mathcal{Z}\Big\}$$

lives only on $\mathcal{Z}$.

From Theorem 1.1, it follows that this last bundle is equivalent to

$$\begin{aligned}E_{\text{res}} \overset{\text{def}}{=} &\Big\{\big(\mathcal{M}_{\text{res}}\otimes_{\mathcal{A}_{\text{res}}(\Omega)}\mathbb{C}_w, w\big)\to w,\\ &\quad s(w)=K_{\text{res}}(\cdot,w)\in\mathcal{M}_{\text{res}}\otimes_{\mathcal{A}_{\text{res}}(\Omega)}\mathbb{C}_w,\ w\in\mathcal{Z}\Big\}.\end{aligned}$$

Further, the metric for the bundle is obtained by restriction, that is, $K_{\text{res}}(w,w) = K(w,w), \quad w \in \mathcal{Z}$. The following diagram captures our situation.

$$\begin{array}{ccccc} \mathcal{M}_{\text{res}} \otimes_{\mathcal{A}_{\text{res}}(\Omega)} \mathbb{C}_w \cong & i^*\big(\mathcal{M} \otimes_{\mathcal{A}(\Omega)} \mathbb{C}_w\big) & \xleftarrow{i^*} & \mathcal{M} \otimes_{\mathcal{A}(\Omega)} \mathbb{C}_w & \\ & \tilde{s}\big\uparrow\!\!\downarrow & & \uparrow\!\!\downarrow s & , \\ & \mathcal{Z} & \xrightarrow{i} & \Omega & \end{array}$$

$$\begin{array}{rcl} s(w) & = & K(\cdot, w) \\ \tilde{s}(w) & = & K_{\text{res}}(\,\cdot\,, w) \\ \langle \tilde{s}(w), \tilde{s}(w)\rangle & = & \langle s(w), s(w)\rangle|_{\mathcal{Z}} \,. \end{array}$$

Every anti–hermitian holomorphic vector bundle has a unique connection compatible with both the complex structure and the metric. (This construction is usually stated for hermitian holomorphic vector bundles but only formal changes are required to handle the anti–holomorphic case.) The curvature matrix of this connection is a hermitian matrix of $(1,1)$–forms. In the case of a line bundle, there is a simple expression for the curvature [12, p. 184, (14)]

$$\mathcal{K}(w) = -\sum_{i,j=1}^{m} \frac{\partial^2}{\partial w_i \partial \bar{w}_j} \log \|s(w)\|^2 dw_i \wedge d\bar{w}_j .$$

There is a natural way in which we may restrict the curvature of a bundle on an open set V to a submanifold of the form $\{u \in V : u_1 = 0\}$, which is a submanifold with co-ordinates $\{u_2, \dots, u_m\}$. The differential operators ∂ and $\bar{\partial}$ restrict naturally to this submanifold. Thus the restriction of the curvature is

$$\begin{aligned} \mathcal{K}|_{\text{res}\{\text{u}_1=0\}}(u) &\stackrel{\text{def}}{=} \sum_{i,j=2}^{m} \Big(\frac{\partial^2}{\partial u_i\, \partial \bar{u}_j} \log \gamma\Big)(0, u_2, \dots, u_m)\, du_i \wedge d\bar{u}_j \\ &= \sum_{i,j=2}^{m} \Big(\frac{\partial^2}{\partial u_i\, \partial \bar{u}_j} \log \gamma(0, u_2, \dots, u_m)\Big)\, du_i \wedge d\bar{u}_j \\ &= \sum_{i,j=2}^{m} \Big(\frac{\partial^2}{\partial u_i\, \partial \bar{u}_j} \log \gamma|_{\text{res}\{\text{u}_1=0\}}\Big)\, du_i \wedge d\bar{u}_j , \end{aligned}$$

where γ is the hermitian metric for the bundle on V.

The bundles E_{q} and E_{res} are equivalent by Theorem 1.1. This implies the corresponding curvatures are equal. However, since the hypersurface $\mathcal{Z}$ is an arbitrary analytic hypersurface, there is no obvious way to relate the curvature for E_{res} to that of E. For this, we must realise the hypersurface $\mathcal{Z}$ as $\{u \in \phi(\Omega) : u_1 = 0\}$. The pull–back of the bundle on $V = \phi(U)$ obtained from localising the module $\mathcal{N}$ is equivalent to the bundle E. This follows from Lemma 1.2, which says that the bundles obtained from localisations of $\phi_* \mathcal{M}_{\text{res U}}$ and $\mathcal{N}$ are equivalent. This in turn means that the pull – back of the bundle E under the map ϕ^{-1} is equivalent to the bundle $E_{\mathcal{N}}$ obtained from the localisation of $\mathcal{N}$. Applying the pull-back operation once more, this time under the map ϕ, we see that $\phi^* E_{\mathcal{N}} \simeq \phi^*\big((\phi^{-1})^* E\big) \simeq E$. As it was pointed out earlier, the curvature for bundle $E_{\mathcal{N}}$ restricts naturally to the submanifold $\phi(\mathcal{Z} \cap U) = \{0, u_2, \dots, u_m\}$; we will pull-back this restriction under the map $\phi|_{\text{res } \mathcal{Z}}$ to obtain the curvature of the bundle E_{res}. Indeed, we have

$$
\begin{aligned}
& \phi|^*_{\text{res}}\Big(\mathcal{K}_{\mathcal{N}}(u)\Big) && (1.6)\\
\stackrel{\text{def}}{=} \;& \phi^*\Big(\sum_{i,j=2}^{n}\big(\frac{\partial^2}{\partial u_i\,\partial \bar{u}_j}\log(K_\phi|_{\text{res}})\big)(u,u)\,du_i\wedge d\bar{u}_j\Big)\\
= \;& \sum_{i,j=2}^{m}\Big(\frac{\partial^2}{\partial u_i\,\partial \bar{u}_j}\log(K_\phi|_{\text{res}})\Big)(\phi|_{\text{res}}(z),\phi|_{\text{res}}(z))\;\phi|^*_{\text{res}}(du_i\wedge d\bar{u}_j)\\
= \;& \sum_{i,j=2}^{m}\frac{\partial^2}{\partial u_i\,\partial \bar{u}_j}\log K(\phi^{-1}(0,u_2,\dots,u_m),\phi^{-1}(0,u_2,\dots,u_m))\;\phi|^*_{\text{res}}(du_i\wedge d\bar{u}_j).
\end{aligned}
$$

With the choice of co-ordinates we have made, $D\phi$ acts as the identity on the normal subspace to the zero set $\mathcal{Z} \subseteq \Omega$. Thus, if we first restrict the curvature $\mathcal{K}_{\mathcal{N}}$ to the zero set $\{u_1 = 0\}$, then the pull - back operation is redundant except for identifying the basis $\{(\partial/\partial u_2), \dots, (\partial/\partial u_m)\}$ with the basis $\{(\partial/\partial z_2), \dots, (\partial/\partial z_m)\}$ in the normal subspace of $T\Omega$ corresponding to $\mathcal{Z}$. If we take K_{res} to be the metric restricted to the zero set with respect to this particular co-ordinate system, then the calculation above proves:

Theorem 1.3. *The curvature of the bundle E_{res} is the restriction*

$$\sum_{i,j=2}^{m}\Big(\frac{\partial^2}{\partial z_i \partial \bar{z}_j}\log K|_{\text{res}}\Big)\,dz_i\wedge d\bar{z}_j$$

for a suitable choice of co-ordinates.

The statement of this theorem is perhaps related to the Adjunction formula I [10, p. 146]. It is possible to state Theorem 1.3 in a co-ordinate free manner, that is, in a form that doesn't require $\mathcal{Z}$ to be expressed in special co-ordinates. The tangent bundle for $\mathcal{Z}$ is a subbundle of the tangent bundle on Ω with the induced hermitian metrics agreeing. This fact induces an orthogonal projection from the bundle with two-form sections over Ω to the corresponding bundle over $\mathcal{Z}$ which acts to restrict the curvature of E viewed only on $\mathcal{Z}$ to yield the curvature of E_{res}. In the co-ordinates introduced above for $\mathcal{Z}$, this is the action described in the statement of the theorem.

The previous discussion identifies the quotient module as $\mathcal{M}_{\text{res}}$. If $\mathcal{M}_{\text{q}}$ is in $B_1(\mathcal{Z})$ (cf. [5]), then the curvature of the bundle E_{q} is a complete unitary invariant for the quotient module. Further, Theorem 1.3 shows how to calculate this curvature. In spite of this we look for simpler invariants for the quotient module. The reasons for this are twofold. First, the curvature of the bundle E_{res} is not always easy to calculate. Secondly, we suspect that the curvature calculation for the quotient module has some analogy with the earlier work of Bott and Chern [3]. They start out with a short exact sequence of complex hermitian vector bundles. One of their results relates the Chern classes associated with these bundles in a very simple manner. Even though our starting point is a short exact sequence of Hilbert modules and we can associate complex hermitian vector bundles via the localisation technique, the resulting vector bundles do not form a short exact sequence. We proceed somewhat differently to look for purely topological or geometric invariants in our situation.

In addition to the bundles E and E_0 which we have discussed, we will need to consider the bundle $\text{Hom}(E_0, E)$. The localisation $X(w)$ of the inclusion map $X : \mathcal{M}_0 \to \mathcal{M}$ from (1.5) provides a section for this bundle which is non zero off the set $\mathcal{Z}$. The maps $w \to X(w)^*X(w)$ and $w \to X(w)X(w)^*$ define metrics for E_0 and E respectively. The map $X(w)$ is in $\text{Hom}(E_0, E)$ and mediates between these metrics and the ones defined by the hermitian structure on E.

The following theorem attempts to expand on the discussion on page 119 in [8]. Let $\phi^1 = f_1^{r_1} \cdots f_k^{r_k}$ be the factorisation of ϕ^1 into irreducibles $f_1, \ldots, f_k$. Then one may drop the multiplicities $r_1, \ldots, r_k$ and take the product of $f_1, \ldots, f_k$ to be the defining function, in the sense of [11, p. 33], for the hypersurface $\mathcal{Z}$. As pointed out earlier, we may choose a global defining function for the hypersurface $\mathcal{Z}$ as long as the second Cousin problem can be solved on Ω. We refer the reader to [10, chapter 3] for a discussion of currents.

Theorem 1.4. *Let Ω be a domain in $\mathbb{C}^m$ for which the second Cousin problem is solvable. If ϕ^1 is the defining function for the hypersurface $\mathcal{Z}$, then*

$$\begin{aligned}&\sum_{i,j=1}^{m} \frac{\partial^2}{\partial w_i \partial \bar{w}_j} \log X(w)^*X(w)\, dw_i \wedge d\bar{w}_j \\ &= \sum_{i,j=1}^{m} \frac{\partial^2}{\partial w_i \partial \bar{w}_j} |\phi^1|^2 dw_i \wedge d\bar{w}_j - \mathcal{K}_0(w) + \mathcal{K}(w).\end{aligned}$$

is valid as an equation for currents on Ω.

PROOF. Let U be an open subset of Ω containing a point $z_0 \in \mathcal{Z}$. Recall that $\mathcal{M}_0$ is the space of all functions in $\mathcal{M}$ which vanish on $\mathcal{Z}$. If ϕ^1 is a defining function for $\mathcal{Z}$, then ϕ^1 and all $h \in \mathcal{M}_0$ vanish on $\mathcal{Z}$. Hence $h = \phi^1 \cdot g$, for some holomorphic function g defined on the open set U with $g \neq 0$ on U. Let e_n be an orthonormal basis for $\mathcal{M}_0$. The reproducing kernel has the expansion $K_0(z, w) = \sum_{n=0}^{\infty} e_n(z)\overline{e_n(w)}$. Since $e_n(z) = \phi^1(z) g_n(z)$ on the set U for each n, it follows that $K_0(z, w) = \phi^1(z)\overline{\phi^1(w)}\chi_U(z, w)$ on U, where $\chi_U(z, w) = \sum_{n=0}^{\infty} g_n(z)\overline{g_n(w)}$. The reproducing property of $K_0(\cdot, w)$ implies that $K_0(w, w)$ does not vanish on $\Omega \backslash \mathcal{Z}$. Since ϕ^1 is a defining function, it follows that $\chi_U(w, w) \neq 0$ off the set $\mathcal{Z} \cap U$.

We point out that, in fact $\chi_U(w,w)$ is never zero on U. If $\chi_U(w,w)=0$ for some $w\in U$, then $\sum_{n=0}^{\infty}|g_n(w)|^2=0$. It follows that $g_n(w)=0$ for each n. This in turn would mean the order of the zero at w for each $f\in\mathcal{M}_0$ is strictly greater than 1. This contradiction proves our assertion.

Note that $K_0(w,w)$ differs from $\chi_U(w,w)$ by the absolute value of a nonvanishing holomorphic function on an open set which does not intersect $\mathcal{Z}$. Therefore,

$$\sum_{i,j=1}^{m}\frac{\partial^2}{\partial w_i\partial\bar{w}_j}\log|\chi_U(w,w)|^2 dw_i\wedge d\bar{w}_j=-\mathcal{K}_0(w), \tag{1.7}$$

for w in any open subset of U disjoint from $\mathcal{Z}$. The real analytic nature of the curvature determines it everywhere once we know it on any open set. Since χ_U is not zero on U, it follows that (1.7) is valid on all of U

It is easy to see that $X(w)^*X(w)=K_0(w,w)/K(w,w)$. The following calculation is valid on the open set $U\subseteq\Omega$ in the distributional sense.

$$\begin{aligned}
&\sum_{i,j=1}^{m}\frac{\partial^2}{\partial w_i\partial\bar{w}_j}\log X(w)^*X(w)dw_i\wedge d\bar{w}_j \qquad (1.8)\\
&=\sum_{i,j=1}^{m}\frac{\partial^2}{\partial w_i\partial\bar{w}_j}\log\Big((|\phi^1(w)|^2\chi_U(w,w))/K(w,w)\Big)dw_i\wedge d\bar{w}_j\\
&=\sum_{i,j=1}^{m}\frac{\partial^2}{\partial w_i\partial\bar{w}_j}\Big(\log|\phi^1(w)|^2+\log|\chi_U(w,w)|\Big)dw_i\wedge d\bar{w}_j+\mathcal{K}(w)\\
&=\sum_{i,j=1}^{m}\frac{\partial^2}{\partial w_i\partial\bar{w}_j}\log|\phi^1(w)|^2dw_i\wedge d\bar{w}_j-\mathcal{K}_0(w)+\mathcal{K}(w)
\end{aligned}$$

This calculation for an arbitrary $z_0\in U\cap\mathcal{Z}$ together with the fact that ϕ^1 is a global defining function completes the proof. □

The Poincaré-Lelong equation [10, p. 388] relates the current

$$c(f)\stackrel{\text{def}}{=}\sum_{i,j=1}^{m}\frac{\partial^2}{\partial w_i\partial\bar{w}_j}\log|f(w)|^2dw_i\wedge d\bar{w}_j \tag{1.9}$$

to the fundamental class of the divisor associated with f.

Any two defining functions for the hypersurface $\mathcal{Z}$ differ by a nonvanishing holomorphic function. The expression in (1.9) depends only on the hypersurface $\mathcal{Z}$ if $f=\phi^1$ is a defining function for the hypersurface $\mathcal{Z}$.

2. EXAMPLES

We have identified the curvature of the quotient module in Theorem 1.3. Further, the three term alternating sum $\sum_{i,j=1}^{m} \frac{\partial^2}{\partial w_i \partial \bar{w}_j} \log \left(X(w)^* X(w) \right) dw_i \wedge d\bar{w}_j - \mathcal{K}_0 + \mathcal{K}$ is the current $c(\phi^1)$. This current depends only on $\mathcal{Z}$ since ϕ^1 is a defining function for $\mathcal{Z}$. In the following subsections we calculate the difference $\mathcal{K}_0|_{\mathcal{Z}} - \mathcal{K}|_{\mathcal{Z}}$.

2.1. EXAMPLES (n, ε). Let $\mathcal{M}$ denote the functional Hilbert space on the bi–disk with the reproducing kernel

$$K(z,w) = (1 - z_1 \bar{w}_1)^{-n} (1 - z_2 \bar{w}_2)^{-n}, \quad z, w \in \mathbb{D}^2, \text{ for a fixed } n \in \mathbb{N}. \tag{2.10}$$

Let $p_\varepsilon = z_1 - \varepsilon z_2$ and $\mathcal{Z}_\varepsilon$ be the associated zero set.

We know that the vectors $\{z_1^k z_2^\ell : k, \ell \geq 0\}$ form a complete orthogonal set in $\mathcal{H}$. The norm of these vectors is obtained from the power series expansion

$$\begin{aligned} K(z,z) &= \frac{1}{(1-|z_1|^2)^n} \frac{1}{(1-|z_2|^2)^n} \\ &= 1 + n(|z_1|^2 + |z_2|^2) + \frac{n(n+1)}{2}|z_1|^4 + n^2 |z_1|^2 \; |z_2|^2 + \frac{n(n+1)}{2}|z_2|^4 + \cdots, \end{aligned}$$

where the co–efficient of $|z_1|^{2\ell}|z_2|^{2k}$ is $\|z_1^\ell z_2^k\|^{-2}$. In the following calculation, we have fixed an arbitrary pair n and ε. To describe the subspace $\mathcal{M}_q$, consider the homogeneous polynomial $\sum_{\ell=0}^{k} a_\ell z_1^{k-\ell} z_2^\ell$, and note that

$$\begin{aligned} \left\langle \sum_{\ell=0}^{k} a_\ell z_1^{k-\ell} z_2^\ell, z_1^i z_2^j (z_1 - \varepsilon z_2) \right\rangle &= \left\langle \sum_{\ell=0}^{k} a_\ell z_1^{k-\ell} z_2^\ell, \left(z_1^{i+1} z_2^j - \varepsilon z_1^i z_2^{j+1} \right) \right\rangle \\ &= \sum_{\ell=0}^{k} a_\ell \Big(\langle z_1^{k-\ell} z_2^\ell, z_1^{i+1} z_2^j \rangle - \varepsilon \langle z_1^{k-\ell} z_2^\ell, z_2^i z_2^{j+1} \rangle \Big). \end{aligned}$$

Since

$$\langle z_1^{k-\ell} z_2^\ell, z_1^{i+1} z_2^j \rangle = \begin{cases} 0, & i+1 \neq k-\ell \text{ or } j \neq \ell \\ 1, & i+1 = k-\ell \text{ and } j = \ell \end{cases},$$

it follows that if we put $a_0 = 1/\|z_1^k\|^2$, then $\sum_{\ell=0}^{k} a_\ell z_1^{k-\ell} z_2^\ell$ is orthogonal to $\mathcal{M}_0$ if and only if

$$a_1 = \frac{1}{\varepsilon \|z_1^{k-1} z_2\|^2}, \; a_2 = \frac{1}{\varepsilon^2 \|z_1^{k-2} z_2^2\|^2}, \; \ldots, \; a_k = \frac{1}{\varepsilon^k \|z_2^k\|^2}. \tag{2.11}$$

Let $\mathcal{P}_k$ denote the space of homogeneous polynomials of degree k on $\mathbb{D}^2$. Then $\mathcal{M}_q = \bigoplus_{k=0}^{\infty} (\mathcal{P}_k \cap \mathcal{M}_q)$. It is easily seen that the dimension of $\mathcal{M}_q \cap \mathcal{P}_k = 1$. Hence

$\{e_k : k \geq 0\}$ is an orthogonal spanning set for $\mathcal{M}_q$, where

$$\begin{aligned} e_k(z_1, z_2) &= \sum_{\ell=0}^{k} \frac{z_1^{k-\ell} z_2^{\ell}}{\varepsilon^{\ell} \| z_1^{k-\ell} z_2^{\ell} \|^2} \\ \| e_k(z_1, z_2) \|^2 &= \sum_{\ell=0}^{k} \frac{\| z_1^{k-\ell} z_2^{\ell} \|^2}{\varepsilon^{2\ell} \| z_1^{k-\ell} z_2^{\ell} \|^4} = \sum_{\ell=0}^{k} \| z_1^{k-\ell} z_2^{\ell} \|^{-2} \varepsilon^{-2\ell}. \end{aligned}$$

We calculate the module action for the Hardy space. Note that, in this case, $\| z_1^{k-\ell} z_2^{\ell} \| = 1$. Hence $\| e_k \|^2 = \sum_{\ell=0}^{k} \varepsilon^{-2\ell}$.

$$\begin{aligned} PM_1 \frac{e_k}{\|e_k\|} &= P \frac{\left(z_1^{k+1} + \cdots + z_2^k z_1 \varepsilon^{-k} + z_2^{k+1} \varepsilon^{-(k+1)} - z_2^{k+1} \varepsilon^{-(k+1)} \right)}{\|e_{k+1}\| \, \|e_k\|} \|e_{k+1}\| \\ &= \frac{1}{\|e_k\|} \left\{ \|e_{k+1}\| \frac{e_{k+1}}{\|e_{k+1}\|} - \left\langle \frac{e_{k+1}}{\|e_{k+1}\|}, \frac{z_2^{k+1}}{\varepsilon^{k+1}} \right\rangle \frac{e_{k+1}}{\|e_{k+1}\|} \right\} \\ &= \frac{1}{\|e_k\|} \frac{e_{k+1}}{\|e_{k+1}\|} \left\{ \|e_{k+1}\| - \frac{1}{\|e_{k+1}\| \varepsilon^{2k+2}} \right\}, \end{aligned}$$

$$\frac{\varepsilon^{2k+2} \|e_{k+1}\|^2 - 1}{\varepsilon^{2k+2} \|e_{k+1}\|} = \frac{\varepsilon^{2k+2} + \cdots + \varepsilon^2}{\varepsilon^{2k+2} \|e_{k+1}\|} = \frac{\left(1 + \cdots + \varepsilon^{-2k}\right)}{\|e_{k+1}\|} = \frac{\|e_k\|^2}{\|e_{k+1}\|},$$

and it follows that

$$\begin{aligned} PM_1 \frac{e_k}{\|e_k\|} &= \frac{e_{k+1}}{\|e_k\| \, \|e_{k+1}\|} \frac{\varepsilon^{2k+2} \|e_{k+1}\|^2 - 1}{\varepsilon^{2k+2} \|e_{k+1}\|} = \frac{\|e_k\|^2}{\|e_{k+1}\|} \frac{e_{k+1}}{\|e_k\| \, \|e_{k+1}\|} \\ &= \frac{\|e_k\|}{\|e_{k+1}\|} \frac{e_{k+1}}{\|e_{k+1}\|}. \end{aligned}$$

Similarly,

$$\begin{aligned} P(\varepsilon^{-1}) M_2 \frac{e_k}{\|e_k\|} &= P \frac{\frac{z_1^k}{\varepsilon} z_2 + \cdots + \frac{z_2^{k+1}}{\epsilon^{k+1}} + z_1^{k+1} - z_1^{k+1}}{\|e_k(z)\|} \\ &= \frac{1}{\|e_k(z)\|} \left\{ \frac{\|e_{k+1}\| e_{k+1}}{\|e_{k+1}\|} - \left\langle \frac{e_{k+1}}{\|e_{k+1}}, z_1^{k+1} \right\rangle \frac{e_{k+1}}{\|e_{k+1}\|} \right\} \\ &= \frac{e_{k+1}}{\|e_{k+1}\| \|e_k\|} \left\{ \|e_{k+1}\| - \frac{1}{\|e_{k+1}\|} \right\} \\ &= \frac{1}{\varepsilon^2} \frac{\left(\frac{1}{\varepsilon^{2k}} + \cdots + 1 \right)}{\|e_{k+1}\|} \frac{e_{k+1}}{\|e_k\| \, \|e_{k+1}\|} \\ &= \frac{1}{\varepsilon^2} \frac{\|e_k\|}{\|e_{k+1}\|} \frac{e_{k+1}}{\|e_{k+1}\|}. \end{aligned}$$

Thus, we have

$$PM_1 \frac{e_k}{\|e_k\|} = \varepsilon(PM_2)\frac{e_k}{\|e_k\|} = \frac{\|e_k\|}{\|e_{k+1}\|} e_{k+1}.$$

We are now ready to calculate the kernel function for $\mathcal{M}_0$. We know that

$$\begin{aligned}
e_0\|e_0\|^{-1} &= 1 \\
e_1\|e_1\|^{-1} &= \left(\frac{z_1}{\|z_1\|^2} + \frac{z_2}{\varepsilon\|z_2\|^2}\right)\|e_1\|^{-1} = \frac{n(z_1 + \frac{z_2}{\varepsilon})}{\sqrt{n(1+\varepsilon^{-2})}} \\
e_2\|e_2\|^{-1} &= \left(\frac{z_1^2}{\|z_1^2\|^2} + \frac{z_1 z_2}{\varepsilon\|z_1 z_2\|^2} + \frac{z_1^2}{\varepsilon^2\|z_2\|^2}\right)\left(\frac{n(n+1)}{2} + \frac{n^2}{\varepsilon^2} + \frac{n(n+1)}{2\varepsilon^4}\right)^{-1/2} \\
&= \left(\varepsilon^2 \frac{n(n+1)}{2} z_1^2 + \varepsilon n^2 z_1 z_2 + \frac{n(n+1)}{2} z_2^2\right) \\
&\qquad \times \left(\frac{n(n+1)}{2} + \frac{n^2}{\varepsilon^2} + \frac{n(n+1)}{2\varepsilon^4}\right)^{-1/2}
\end{aligned}$$

The kernel function $K_q(z,z)$ admits an expansion of the form $\sum_{k=0}^{\infty} |e_k|^2 \, \|e_k\|^{-2}$. We will write

$$K(z,z) = \sum_{\ell=0}^{\infty} f_\ell(z_1, z_2),$$

where each f_ℓ is a homogeneous real analytic polynomial of degree ℓ. Of course,

$$K_0 = K - K_q = \sum_{k=0}^{\infty} (f_k - |e_k|^2 \, \|e_k\|^{-2}).$$

It is not hard to see that $|p_\varepsilon|^2$ is a factor of $f_k - |e_k|^2 \, \|e_k\|^{-2}$ for $k \geq 0$. We will carry out this factorisation for $k \leq 2$ since it will carry all the information relevant to the curvature calculation. Since $f_0 - |e_0|^2 \, \|e_0\|^{-2} = 0$, we start with $k = 1$:

$$\begin{aligned}
&n(|z_1|^2 + |z_2|^2) - |e_1(z_1,z_2)|^2 \, \|e_1\|^{-2} \\
&\quad = n(|z_1|^2 + |z_2|^2) - n\frac{\left|z_1 + \frac{z_2}{\varepsilon}\right|^2}{1+\varepsilon^{-2}} \\
&\quad = n\{|z_1|^2 + |z_2|^2 - \frac{|z_1\varepsilon + z_2|^2}{1+\varepsilon^2}\} \\
&\quad = n\{(1+\varepsilon^2)(|z_1|^2 + |z_2|^2) - (|z_1|^2\varepsilon^2 + |z_2|^2 + \varepsilon z_1\bar{z}_2 + \varepsilon\bar{z}_1 z_2)\} \\
&\quad = \frac{n}{1+\varepsilon^2}\{\varepsilon^2|z_2|^2 + |z_1|^2 + \varepsilon z_1\bar{z}_2 + \varepsilon\bar{z}_1 z_2\} \\
&\quad = \frac{n}{1+\varepsilon^2}|z_1 - \varepsilon z_2|^2.
\end{aligned}$$

Now we calculate the case with $k = 2$:

$$\left(\frac{n(n+1)}{2}|z_1|^4 + n^2|z_1|^2|z_2|^2 + \frac{n(n+1)}{2}|z_2|^4\right) - |e_2(z_1, z_2)|^2 \, \|e_2\|^{-2}$$

$$\begin{aligned}
&= \frac{1}{2}\left(n((1+\varepsilon^2)n + (1+\varepsilon^4))\right)\left(\left(\frac{n(n+1)}{2}|z_1|^4 + n^2|z_1|^2|z_2|^2 + \frac{n(n+1)}{2}|z_2|^4\right)\right. \\
&\qquad \left. - \left|\varepsilon^2\frac{n(n+1)}{2}z_1^2 + \varepsilon n^2 z_1 z_2 + \frac{n(n+1)}{2}z_2^2\right|^2\right)\frac{2}{n((1+\varepsilon^2)^2n + (1+\varepsilon^4))} \\
&= \left[\frac{n^2(n+1)}{4}((1+\varepsilon^2)^2 n + (1+\varepsilon^4))|z_1|^4 + \frac{n^3((1+\varepsilon^2)^2n + (1+\varepsilon^4))}{2}|z_1|^2\,|z_2|^2\right. \\
&\qquad + \frac{n^2(n+1)}{4}((1+\varepsilon^2)^2n + (1+\varepsilon^4))|z_2|^4 \\
&\qquad - \left\{\frac{\varepsilon^4 n^2(n+1)^2}{4}|z_1|^4 + \frac{\varepsilon^3 n^3(n+1)}{2}z_1 z_2 \overline{z}_1^2 + \frac{\varepsilon^2 n^2(n+1)^2}{4}z_2^2\overline{z}_1\overline{z}_1^2\right. \\
&\qquad\qquad + \frac{\varepsilon^3 n^3(n+1)}{2}z_1^2\overline{z}_1\overline{z}_2 + \varepsilon^2 n^4|z_1|^2|z_2|^2 + \frac{\varepsilon n^3(n+1)}{2}z_2^2\overline{z}_1\overline{z}_2 \\
&\qquad\qquad \left.\left. + \frac{\varepsilon^2 n^2(n+1)^2}{4}z_1^2\overline{z}_2^2 + \frac{\varepsilon n^3(n+1)}{2}z_1 z_2\overline{z}_2^2 + \frac{n^2(n+1)^2}{4}|z_2|^4\right\}\right] \\
&\qquad\qquad \times \frac{2}{n((1+\varepsilon^2)^2n + (1+\varepsilon^4))} \\
&= (z_1 - \varepsilon z_2)\left\{\frac{\varepsilon^2 n^3(n+1)}{2}|z_1|^2(\overline{z}_1 - \varepsilon\overline{z}_2) + \frac{n^3(n+1)}{2}|z_2|^2(\overline{z}_1 - \varepsilon\overline{z}_2)\right. \\
&\qquad \left. + \frac{n^2(n+1)^2}{4}(\overline{z}_1^2 - \varepsilon^2\overline{z}_1^2)(z_1 + \varepsilon z_2)\right\}\frac{2}{n((1+\varepsilon^2)^2n + (1+\varepsilon^4))} \\
&= |z_1 - \varepsilon z_2|^2\left\{\frac{\varepsilon^2 n^3(n+1)}{2}|z_1|^2 + \frac{n^3(n+1)}{2}|z_2|^2 + \frac{n^2(n+1)^2}{4}|z_1 + \varepsilon z_2|^2\right\} \\
&\qquad\qquad \times \frac{2}{n((1+\varepsilon^2)^2n + (1+\varepsilon^4))}.
\end{aligned}$$

Consequently, we have

$$\begin{aligned}
K_0(z,z) = |z_1 - \varepsilon z_2|^2 &\left\{\frac{n}{1+\varepsilon^2} + \frac{2}{n((1+\varepsilon^2)^2n + (1+\varepsilon^4))}\left\{\frac{\varepsilon^2 n^3(n+1)}{2}|z_1|^2\right.\right. \\
&\qquad \left.\left. + \frac{n^3(n+1)}{2}|z_2|^2 + \frac{n^2(n+1)^2}{4}|z_1 + \varepsilon z_2|^2\right\} + \cdots\right\}.
\end{aligned}$$

Thus we have determined the norm $K_0(w,w) = |w_1 - w_2|^2 \chi(w,w)$. As pointed out earlier, we may use $\chi(w,w)$ to calculate the curvature $\mathcal{K}_0$ for the bundle E_0. In view of Lemma 2.3 [14], at least we can read off the curvature of E_0 at 0 from this calculation. To compute the curvature of the restriction we use Theorem 1.3 and find that

$$\mathcal{K}_0\big|_{\mathcal{Z}}(0) = -\frac{2\left(\frac{n^3(n+1)}{2}(1+\varepsilon^4) + n^2(n+1)^2\varepsilon^2\right)}{n((1+\varepsilon^2)^2 n + (1+\varepsilon^4))}\,\frac{1+\varepsilon^2}{n} \tag{2.12}$$

A similar application of [14, Lemma 2.3] to the expansion for the kernel function K shows that $\mathcal{K}|_{\mathcal{Z}}(0) = -n(1+\varepsilon^2)$. This proves the following result:

Proposition 2.1. *Let $\mathcal{M}$ denote the functional Hilbert space on the bi–disk with the reproducing kernel given in (2.10). Then $\mathcal{M}$ is a module over the algebra $\mathcal{A}(\mathbb{D}^2)$. Let $\mathcal{M}_0$ be the submodule of functions vanishing on $\mathcal{Z} = \{(z_1,z_2) \in \mathbb{D}^2 : z_1 - \varepsilon z_2 = 0\}$. Then*

$$\mathcal{K}_0\big|_{\mathcal{Z}}(0) - \mathcal{K}\big|_{\mathcal{Z}}(0) = -\frac{2(2n+1)\varepsilon^2}{((1+\varepsilon^2)^2 n + (1+\varepsilon^4))}(1+\varepsilon^2).$$

We find that the difference $\mathcal{K}_0\big|_{\mathcal{Z}}(0) - \mathcal{K}\big|_{\mathcal{Z}}(0)$ is not independent of n for arbitrary ε. This shows that the difference of the curvatures for E and E_0 does not depend on just the geometry of $\mathcal{Z}$ and its embedding in Ω.

2.2. Examples $(n,1)$. We discuss the case $\varepsilon = 1$ at some length. In this case, $\mathcal{K}_0|_{\mathcal{Z}}(0) = 2n+2$. For $w \in \mathbb{D}$, let $\varphi(z) = (z-w)/(1-\bar{w}z)$ be the Möbius map of the unit disk.

Proposition 2.2. *For $\varepsilon = 1$ and any n, we have $\mathcal{K}_0|_{\mathcal{Z}}(w) = |\varphi'(w)|^2 \mathcal{K}_0(0)$. Similarly, $\mathcal{K}|_{\mathcal{Z}}(w) = |\varphi'(w)|^2 \mathcal{K}(0)$.*

Proof. Clearly, a theorem similar to the one in [13, p. 457] is valid for the module $\mathcal{M}$. Explicitly, the module maps $h \to f \cdot h$ and $h \to (f \circ (\phi,\phi)) \cdot h$ for $f \in \mathcal{A}(\mathbb{D}^2)$ are unitarily equivalent. In other words, $(\varphi,\varphi)_*\mathcal{M}$ and $\mathcal{M}$ are isomorphic. The unitary inducing the unitary equivalence is given by the formula

$$U_\varphi : h \to \Big(\det D(\varphi,\varphi)\Big)^{n/2}\Big(h \circ (\varphi,\varphi)\Big), \quad h \in \mathcal{M}. \tag{2.13}$$

It is clear that each of these unitary operators leaves $\mathcal{M}_0$ invariant. As pointed out in [4, p. 115], in this case, the submodule $\mathcal{M}_0 \simeq (\varphi,\varphi)_*\mathcal{M}_0$ and the quotient module $\mathcal{M}_q \simeq (\varphi,\varphi)_*\mathcal{M}_q$. We can now calculate the curvature for these modules at any point $(\varphi(0),\varphi(0))$ using the equivalence under the Möbius map and a change of variable formula (cf. [13]). This calculation applied to the module $\mathcal{M}$ will give us the second part of the theorem. In view of Theorem 1.3, the restriction of the curvature $\mathcal{K}_0$ corresponds to the curvature of the module $\mathcal{M}_0|_{\text{res }\mathcal{Z}}$. But the unitary U_φ leaves the module $\mathcal{M}_0|_{\text{res }\mathcal{Z}}$ invariant. Thus the previous discussion applies to this last module and completes the proof of the first part. □

This proposition enables us to calculate the difference of the restrictions of the two curvature at an arbitrary point if the zero set is of the form $\{(z_1, z_2) \in \mathbb{D}^2 : z_1 = z_2\}$.

Proposition 2.3. *Let $\mathcal{M}$ denote the functional Hilbert space on the bi–disk with the reproducing kernel given in (2.10). Then $\mathcal{M}$ is a module over the algebra $\mathcal{A}(\mathbb{D}^2)$. Let $\mathcal{M}_0$ be the submodule of functions vanishing on $\mathcal{Z} = \{(z_1, z_2) \in \mathbb{D}^2 : z_1 - z_2 = 0\}$. Then*

$$\left(\mathcal{K}_0 - \mathcal{K}\right)\big|_{\mathcal{Z}}(z_2) = -2(1 - |z_2|^2)^{-2} dz_2 \wedge d\bar{z}_2.$$

PROOF. It follows from [14, Lemma 2.3] that

$$\mathcal{K}_0|_{\mathcal{Z}}(0) = -(2n + 2) dz_2 \wedge d\bar{z}_2.$$

Proposition 2.2 implies that at an arbitrary point z in the unit disk we must have

$$\mathcal{K}_0|_{\mathcal{Z}}(z_2) = -(2n + 2)(1 - |z_2|^2)^{-2}\, dz_2 \wedge d\bar{z}_2.$$

Since $\mathcal{K}|_{\mathcal{Z}}(z_2) = -2n(1 - |z_2|^2)^{-2} dz_2 \wedge d\bar{z}_2$, it follows that the difference of the curvatures restricted to the zero set $\mathcal{Z}$ is given by

$$\left(\mathcal{K}_0 - \mathcal{K}\right)\big|_{\mathcal{Z}}(z_2) = -2(1 - |z_2|^2)^{-2} dz_2 \wedge d\bar{z}_2.$$

This completes the proof. □

All our calculations so far are for an arbitrary n. However, the final expression for the differnece of the curvatures is independent of n. As we pointed out at the end of the previous section, this is not always the case. However, if $\varepsilon = 0$, then the difference of the two curvatures is again independent of n. We will discuss the case $\varepsilon = 0$ in the next section in slightly greater generality.

2.3. THE CASE OF A PRODUCT DOMAIN.

Proposition 2.4. *Let $\Omega = \Omega_1 \times \Omega_2$ be a product domain in $\mathbb{C}^2$. Let $\mathcal{M}_1$ and $\mathcal{M}_2$ be functional Hilbert spaces over Ω_1 and Ω_2, respectively. Suppose that $\mathcal{M}_i$ is a Hilbert module over the algebra $\mathcal{A}(\Omega_i)$, $i = 1, 2$. Then the Hilbert space tensor product $\mathcal{M} = \mathcal{M}_1 \otimes \mathcal{M}_2$ is a Hilbert module over the algebra $\mathcal{A}(\Omega_1 \times \Omega_2)$. Let $\mathcal{M}_0$ be the submodule of functions in $\mathcal{M}$ which vanish on $\mathcal{Z} = \{(w_1, w_2) \in \Omega : w_1 = 0\}$. Then*

$$\Big(\mathcal{K}_0(w) - \mathcal{K}(w)\Big)\Big|_{\mathcal{Z}} = \mathcal{K}_{\mathrm{q}}(w) - \mathcal{K}|_{\mathcal{Z}}(w) = 0.$$

PROOF. If $\mathcal{M}_1$ and $\mathcal{M}_2$ are functional Hilbert spaces over Ω_1 and Ω_2, respectively, then the Hilbert space tensor product $\mathcal{M} = \mathcal{M}_1 \otimes \mathcal{M}_2$ is a functional Hilbert space over Ω. The reproducing kernel $K(\cdot, w)$ for $\mathcal{M}$ is simply the product of the two reproducing kernels for $\mathcal{M}_1$ and $\mathcal{M}_2$ (cf. [1]). Since $\mathcal{M} = \mathcal{M}_1 \otimes \mathcal{M}_2$, it follows that $\mathcal{M}_0 = z_1\mathcal{M}_1 \otimes \mathcal{M}_2$. Let $w = (w_1, w_2) \in \Omega$. We denote the kernel functions for $\mathcal{M}_1$

and $\mathcal{M}_2$ by $K_1(\cdot, w)$ and $K_2(\cdot, w)$, respectively. To calculate the kernel function for the module $z_1\mathcal{M}_1$, we assume, without loss of generality, that $K_1(\,\cdot\,, 0) = 1$, that is,

$$K_1(z_1, w_1) = \sum_{n,m=0}^{\infty} a_{nm} z_1^n \bar{w}_1^m, \quad a_{00} = 1,\ a_{10} = 0 = a_{01}.$$

The subspace $z_1\mathcal{M}_1$ is the ortho-complement of the 1–dimensional space of scalars in $\mathcal{M}_1$. Consequently, the reproducing kernel for this subspace is

$$\begin{aligned} K_1(z_1, w_1) - 1 &= z_1\bar{w}_1 \left(\sum_{n,m=1}^{\infty} a_{nm} z_1^{n-1} \bar{w}_1^{m-1} \right) \\ &= z_1 \bar{w}_1 f(z_1, w_1), \end{aligned}$$

where $f(z_1, w_1) = \sum_{n,m=1}^{\infty} a_n z_1^{n-1} \bar{w}_1^{m-1}$. Thus, the reproducing kernel $K_0(z, w)$ for the subspace $\mathcal{M}_0$ is

$$K_0(z, w) = (z_1\bar{w}_1) f(z_1, w_1) K_2(z_2, w_2).$$

The reproducing kernel $K_{\mathrm{q}}(\cdot, w)$ is the difference

$$\begin{aligned} K_{\mathrm{q}}(z, w) &= K(z, w) - K_0(z, w) \\ &= K(z, w) - (z_1\bar{w}_1) f(z_1, w_1) K_2(z_2, w_2) \\ &= \Big(K_1(z_1, w_1) - (z_1\bar{w}_1) f(z_1, w_1)\Big) K_2(z_2, w_2) \\ &= K_2(z_2, w_2). \end{aligned} \tag{2.14}$$

We can now calculate the curvatures of these three modules.

$$\begin{aligned} \mathcal{K}(w) &= -\sum_{i,j=1}^{2} \frac{\partial^2}{\partial w_i\, \partial \bar{w}_j} \log K_1(w_1, w_1) K_2(w_2, w_2)\, dw_i \wedge d\bar{w}_j \\ &= -\sum_{i=1}^{2} \frac{\partial^2}{\partial w_i\, \partial \bar{w}_i} \log K_i(w_i, w_i)\, dw_i \wedge d\bar{w}_j. \\ \mathcal{K}_{\mathrm{q}}(w) &= -\frac{\partial^2}{\partial w_2\, \partial \bar{w}_2} \log K_2(w_2, w_2)\, dw_2 \wedge d\bar{w}_2. \\ \mathcal{K}_0(w) &= -\sum_{i,j=1}^{2} \frac{\partial^2}{\partial w_i\, \partial \bar{w}_j} \log(w_1\bar{w}_1) f(w_1, w_1) K_2(w_2, w_2)\, dw_i \wedge d\bar{w}_j \\ &= -\Big(1 + \frac{\partial^2}{\partial w_1\, \partial \bar{w}_1} \log f(w_1, w_1)\Big)\, dw_1 \wedge d\bar{w}_1 \\ &\qquad - \frac{\partial^2}{\partial w_2\, \partial \bar{w}_2} \log K_2(w_2, w_2) dw_2 \wedge d\bar{w}_2. \end{aligned}$$

Now restricting the curvatures, we find that

$$(\mathcal{K}_0(w) - \mathcal{K}(w))\big|_{\mathcal{Z}} = \mathcal{K}_{\mathrm{q}}(w) - \mathcal{K}\big|_{\mathcal{Z}}(w) = 0.$$

This completes the proof. □

2.4. EXAMPLE (1, 1). In this subsection, we make more detailed calculations for the special case of the Hardy module. The reproducing kernel has an expansion of the form

$$\begin{aligned} K_{\mathrm{q}}(w,w) &= \frac{\sum_{n=0}^{\infty} |e_n(w_1,w_2)|^2}{n+1} \\ &= \frac{\sum_{n=0}^{\infty}\left(\sum_{j=0}^{n} |w_1^{n-j} w_2^{j}|^2\right)}{n+1} \\ &= \sum_{n=0}^{\infty} \frac{|w_1^{n+1} - w_2^{n+1}|^2}{(|w_1 - w_2|^2)(n+1)}. \end{aligned}$$

Since this series is absolutely convergent, we can sum term by term to obtain

$$\begin{aligned} &K_{\mathrm{q}}(w,w) \\ &= \frac{1}{|w_1-w_2|^2}\left(-\log(1-|w_1|^2) + \log(1-\overline{w}_1 w_2) + \log(1 - w_2\overline{w}_1) - \log(1-|w_2|^2)\right) \\ &= \frac{1}{|w_1-w_2|^2}\log\left\{\frac{|1-w_1\overline{w}_2|^2}{(1-|w_1|^2)(1-|w_2|^2)}\right\} \\ &= \frac{1}{|w_1-w_2|^2}\log\left(1 + \frac{|(w_1-w_2)|^2}{(1-|w_1|^2)(1-|w_2|^2)}\right) \\ &= \frac{1}{|w_1-w_2|^2}\left(\frac{|(w_1-w_2)|^2}{(1-|w_1|^2)(1-|w_2|^2)} - \frac{1}{2}\frac{|w_1-w_2|^4}{(1-|w_1|^2)^2(1-|w_2|^2)^2} + \cdots\right) \\ &= K(w,w)\left(1 - \frac{1}{2}K(w,w)|w_1-w_2|^2 + \cdots\right). \end{aligned}$$

The reproducing kernel $K_0(\cdot, w)$ is obtained by taking the difference

$$\begin{aligned} K_0(w,w) &= K(w,w) - K_{\mathrm{q}}(w,w) \\ &= K(w,w) - K(w,w)\left(1 - \frac{1}{2}K(w,w)|w_1-w_2|^2 + \cdots\right) \\ &= |w_1-w_2|^2 (K(w,w))^2 \left(\frac{1}{2} - \frac{1}{3}K(w,w)|w_1-w_2|^2 + \cdots\right). \end{aligned}$$

Recalling (1.8), we calculate

$$\begin{aligned}
&\mathcal{K}(X(w)X(w)^*)\\
&\stackrel{\text{def}}{=} -\sum_{i,j=1}^{2} \frac{\partial^2}{\partial w_i\,\partial \bar{w}_j} \log \frac{K_0(w,w)}{K(w,w)} dw_i \wedge d\bar{w}_j\\
&= -\sum_{i,j=1}^{2} \frac{\partial^2}{\partial w_i\,\partial \bar{w}_j} \log\left\{|w_1-w_2|^2 K(w,w)\left(\frac{1}{2}-\frac{1}{3}|w_1-w_2|^2K(w,w)+\cdots\right)\right\}\\
&\qquad\qquad dw_i \wedge d\bar{w}_j\\
&= -\sum_{i,j=1}^{2} \frac{\partial^2}{\partial w_i\,\partial \bar{w}_j} \log|w_1-w_2|^2\, dw_i\wedge d\bar{w}_j - \sum_{i,j=1}^{2} \frac{\partial^2}{\partial w_i\,\partial \bar{w}_j} \log K(w,w)\, dw_i \wedge d\bar{w}_j\\
&\qquad - \sum_{i,j=1}^{2} \frac{\partial^2}{\partial w_i\,\partial \bar{w}_j} \log\left(\frac{1}{2}-\frac{1}{3}|w_1-w_2|^2K(w,w)+\cdots\right) dw_i \wedge d\bar{w}_j.
\end{aligned}$$

The following calculation occurs frequently in what follows.

$$\begin{aligned}
&\frac{\partial^2}{\partial w_i\,\partial \bar{w}_j}|w_1-w_2|^{2n}K(w,w)\,dw_i\wedge d\bar{w}_j\\
&= \frac{\partial}{\partial w_i}\left((w_1-w_2)^n\left((-1)^{j+1}n\big(\bar{w}_1-\bar{w}_2\big)^{n-1}+\Big(\frac{\partial}{\partial \bar{w}_j}K(w,w)^n\Big)\big(\bar{w}_1-\bar{w}_2\big)^n\right)\right)\\
&\qquad dw_i\wedge d\bar{w}_j \quad =\\
&= (-1)^{j+1}n\big(\bar{w}_1-\bar{w}_2\big)^{n-1}\Big\{\ (-1)^{i+1}(w_1-w_2)^{n-1}K(w,w)\\
&\qquad \times\ \Big(\frac{\partial}{\partial w_i}K(w,w)^n\Big)(w_1-w_2)^n\Big\}\ dw_i\wedge d\bar{w}_j\\
&\qquad +\ \big(\bar{w}_1-\bar{w}_j\big)^n\Big\{(-1)^i n(w_1-w_2)^{n-1}K(w,w)\\
&\qquad \times\ \left(\frac{\partial^2}{\partial w_i\,\partial \bar{w}_j}K(w,w)^n\right)(w_1-w_2)^n\Big\}\ dw_i\wedge d\bar{w}_j.
\end{aligned}$$

Thus,

$$\left.\frac{\partial^2}{\partial w_i\,\partial \bar{w}_j}|w_1-w_2|^{2n}K(w,w)\,dw_i\wedge d\bar{w}_j\right|_{\mathcal{Z}} = \begin{cases} 0 & \text{if } n>1\\ (-1)^{i+j}K(w,w)dw_i\wedge d\bar{w}_j & \text{otherwise}\ .\end{cases} \tag{2.15}$$

We now restrict these curvatures to the set $\mathcal{Z}$. By the calculation (2.15),

$$\sum_{i,j=1}^{n} \frac{\partial^2}{\partial w_i \, \partial \bar{w}_j} \log \left(\frac{1}{2} - \frac{1}{3} |w_1 - w_2|^2 K(w,w) + \cdots \right) \, dw_i \wedge d\bar{w}_j = 0.$$

Also the singular support of the current

$$T_{\mathcal{Z}} = \sum_{i,j=1}^{n} \frac{\partial^2}{\partial w_i \, \partial \bar{w}_j} \log |w_1 - w_2|^2 \, dw_i \wedge d\bar{w}_j$$

is $\mathcal{Z}$. We may restrict $T_{\mathcal{Z}}$ to its singular support. It now follows that

$$\begin{aligned} \mathcal{K}(X(w)X(w)^*)|_{\mathcal{Z}} &= -\sum_{i,j=1}^{2} \frac{\partial^2}{\partial w_i \, \partial \bar{w}_j} \log K(w,w) \, dw_i \wedge d\bar{w}_j \Big|_{\mathcal{Z}} - T_{\mathcal{Z}}|_{\mathcal{Z}} \\ &= -2\big(1 - |w_2|^2\big)^{-2} \, dw_2 \wedge d\bar{w}_2 - T_{\mathcal{Z}}|_{\mathcal{Z}}. \end{aligned} \tag{2.16}$$

Thus we have proved the following:

Proposition 2.5. *Let $\mathcal{M}$ be the Hardy module and $\mathcal{M}_0$ be the submodule of functions vanishing on the diagonal set $\mathcal{Z} = \{(w_1, w_2) \in \mathbb{D}^2 : w_1 = w_2\}$. Then*

$$\mathcal{K}(X(w)X(w)^*)|_{\mathcal{Z}} = -2\big(1 - |w_2|^2\big)^{-2} \, dw_2 \wedge d\bar{w}_2 - T_{\mathcal{Z}}\Big|_{\mathcal{Z}}.$$

REMARK: In [2], the second author together with B. Bagchi obtained the Sz.-Nagy and Foias characteristic operator function of the contractive module $\mathcal{M}_{\mathrm{q}}$ in an explicit and usable form as a product involving the extended discrete series representations D_1 and D_3 of the group of bi-holomorphic automorphisms of the disk. Using this formula, it is shown that the compression of $\mathrm{Id} \otimes M_z$ on the Hardy module is precisely the Sz.-Nagy and Foias model for $\mathcal{M}_{\mathrm{q}}$. The results in [2] are more general and cover all the homogeneous operators discussed in [13].

2.5. EXAMPLE $(2,1)$. Throughout this subsection $\mathcal{M}$ will stand for the Bergman module $\mathrm{B}^2(\mathbb{D}^2)$. Calculations similar to the ones in the previous subsection show that the kernel function $K_{\mathrm{q}}(\cdot, w)$ is obtained from the expansion

$$\begin{aligned} &K_{\mathrm{q}}(w,w) \qquad \qquad (2.17) \\ &= \sum_{k=0}^{\infty} |e_k(w)|^2 \, \|e_k\|^{-2} \\ &= \sum_{k=0}^{\infty} \frac{6}{(k+1)(k+2)(k+3)} |w_1 - w_2|^{-6} \Big\{ \big| (k+1)\big(w_1^{k+3} - w_2^{k+3}\big) \\ &\qquad - (k+3) w_1 w_2 \big(w_1^{k+1} - w_2^{k+1}\big) \big|^2 \Big\} \end{aligned}$$

$$
\begin{aligned}
&= 6|w_1 - w_2|^{-6}\Big\{-2|w_1|^2 + 2\bar{w}_1 w_2 + 2w_1\bar{w}_2 - 2|w_2|^2 \\
&\qquad + \big(2 - (|w_1|^2 + \bar{w}_1 w_2 + w_1\bar{w}_2 + |w_2|^2) + 2|w_1\bar{w}_2|^2\big) \\
&\qquad\qquad \times \log\frac{|1 - w_1\bar{w}_2|^2}{(1-|w_1|^2)(1-|w_2|^2)}\Big\} \\
&= 6|w_1 - w_2|^{-6}\Bigg\{ -2|w_1 - w_2|^2 + \big(|w_1 - w_2|^2 + 2(1-|w_1|^2)(1-|w_2|^2)\big) \\
&\qquad\qquad \times \log\Big(1 + \frac{|w_1 - w_2|^2}{(1-|w_1|^2)(1-|w_2|^2)}\Big)\Bigg\} \\
&= \frac{1}{(1-|w_1|^2)^2(1-|w_2|^2)^2} - \frac{|w_1 - w_2|^2}{(1-|w_1|^2)^3(1-|w_2|^2)^3} \\
&\qquad + \frac{9}{10}\frac{|w_1 - w_2|^4}{(1-|w_1|^2)^4(1-|w_2|^2)^4} - \cdots \\
&= K(w,w)\left(1 - \frac{|w_1 - w_2|^2}{(1-|w_1|^2)(1-|w_2|^2)} + \frac{9}{10}\frac{|w_1 - w_2|^4}{(1-|w_1|^2)^2(1-|w_2|^2)^2} - \cdots\right).
\end{aligned}
$$

The kernel function for the submodule $\mathcal{M}_0$ is now calculated as the difference

$$
\begin{aligned}
&K_0(w,w) \qquad\qquad (2.18)\\
&= K(w,w) - K_{\mathrm{q}}(w,w) \\
&= K(w,w)\frac{|w_1 - w_2|^2}{(1-|w_1|^2)(1-|w_2|^2)}\Big(1 - \frac{9}{10}\frac{|w_1 - w_2|^2}{(1-|w_1|^2)(1-|w_2|^2)} + \cdots\Big).
\end{aligned}
$$

Now we can calculate the curvature

$$
\begin{aligned}
&\mathcal{K}(X(w)X(w)^*) \\
&\stackrel{\text{def}}{=} -\sum_{i,j=1}^{2}\frac{\partial^2}{\partial w_i\,\partial\bar{w}_j}\log\frac{K_0(w,w)}{K(w,w)}\,dw_i\wedge d\bar{w}_j \\
&= -\sum_{i,j=1}^{2}\frac{\partial^2}{\partial w_i\,\partial\bar{w}_j}\log\Bigg(\frac{|w_1 - w_2|^2}{(1-|w_1|^2)(1-|w_2|^2)} \\
&\qquad\qquad \Big(1 - \frac{9}{10}\frac{|w_1 - w_2|^2}{(1-|w_1|^2)(1-|w_2|^2)} + \cdots\Big)\Bigg)\,dw_i\wedge d\bar{w}_j
\end{aligned}
$$

$$= -\sum_{i,j=1}^{2} \frac{\partial^2}{\partial w_i \, \partial \bar{w}_j} \log |w_1 - w_2|^2 \, dw_i \wedge d\bar{w}_j$$
$$- \sum_{i,j=1}^{2} \frac{\partial^2}{\partial w_i \, \partial \bar{w}_j} \log \frac{1}{(1-|w_1|^2)(1-|w_2|^2)} \, dw_i \wedge d\bar{w}_j$$
$$- \sum_{i,j=1}^{2} \frac{\partial^2}{\partial w_i \, \partial \bar{w}_j} \log \left(1 - \frac{9}{10} \frac{|w_1 - w_2|^2}{(1-|w_1|^2)(1-|w_2|^2)} + \cdots \right) dw_i \wedge d\bar{w}_j.$$

Using the calculation from (2.15), we have proved:

Proposition 2.6. *Let $\mathcal{M}$ be the Bergman module and $\mathcal{M}_0$ be the submodule of functions vanishing on the diagonal set $\mathcal{Z} = \{(w_1, w_2) \in \mathbb{D}^2 : w_1 = w_2\}$. Then*

$$\mathcal{K}(X(w)X(w)^*)|_{\mathcal{Z}} = -2\left(1 - |w_2|^2\right)^{-2} dw_2 \wedge d\bar{w}_2 - T_{\mathcal{Z}}\Big|_{\mathcal{Z}}.$$

3. Higher Multiplicity Localisations

Let $\Omega \subsetneq \mathbb{C}^m$ be a bounded domain and $\mathcal{A}(\Omega)$ be the algebra consisting of all functions that are holomorphic in some open set U containing the closure $\bar{\Omega}$ of the set Ω. Let $\mathcal{M}$ be a functional Hilbert space consisting of holomorphic functions on Ω. The map $(f, h) \to f \cdot h$, $f \in \mathcal{A}(\Omega), h \in \mathcal{M}$, where $f \cdot h$ is the pointwise product of complex functions, turns $\mathcal{M}$ into a module over the algebra $\mathcal{A}(\Omega)$. One possibility is that $\mathcal{M}$ is the closure of an ideal in $\mathcal{A}(\Omega)$. The problem of characterizing such modules and, in particular, deciding when two are equivalent, was solved in [9]. A rigidity phenomenon intervenes and is detected by the use of higher multiplicity localization. We recall now the definition of second-order localization and its calculation for modules that are functional Hilbert spaces.

For $w \in \Omega$ and $\alpha \in \mathbb{C}$ possibly depending on w, let us fix a two dimensional module $\mathbb{C}^2_{\omega,\alpha}$ over the algebra $\mathcal{A}(\Omega)$ via the action:

$$\left(f, \begin{pmatrix} \lambda \\ \mu \end{pmatrix}\right) \to \begin{pmatrix} f(w) & (\partial_\alpha f)(w) \\ 0 & f(w) \end{pmatrix} \begin{pmatrix} \lambda \\ \mu \end{pmatrix},$$

where $\partial_\alpha f(w) = \alpha_1 \dfrac{\partial f}{\partial z_1}(w) + \cdots + \alpha_m \dfrac{\partial f}{\partial z_m}(w)$.
The module tensor product $\mathcal{M} \otimes_{\mathcal{A}(\Omega)} \mathbb{C}^2_{\omega,\alpha}$ is the orthogonal complement of the following subspace $\mathcal{N}$ in the Hilbert space $\mathcal{M} \otimes \mathbb{C}^2_{\omega,\alpha}$.

$$\mathcal{N} = \Big\{ f \cdot h \otimes \begin{pmatrix} a \\ 0 \end{pmatrix} + f \cdot k \otimes \begin{pmatrix} 0 \\ b \end{pmatrix} - h \otimes f \cdot \begin{pmatrix} a \\ 0 \end{pmatrix} - k \otimes f \cdot \begin{pmatrix} 0 \\ b \end{pmatrix} :$$
$$h,\, k \in \mathcal{M},\ \begin{pmatrix} a \\ b \end{pmatrix} \in \mathbb{C}^2 \text{ and } f \in \mathcal{A}(\Omega) \Big\}$$

$$= \left\{a(f - f(w)) \cdot h \otimes \begin{pmatrix}1\\0\end{pmatrix} + b((f - f(w)) \cdot k) \otimes \begin{pmatrix}0\\1\end{pmatrix} + b\partial_\alpha f(w)k \otimes \begin{pmatrix}1\\0\end{pmatrix} : \right.$$

$$\left. h,\ k \in \mathcal{M},\ \begin{pmatrix}a\\b\end{pmatrix} \in \mathbb{C}^2 \text{ and } f \in \mathcal{A}(\Omega)\right\}.$$

Since $\mathcal{M}$ is a functional Hilbert space, it admits a kernel function $K(\cdot, w)$; that is,

(i) the function $K(\cdot, w)\colon \Omega \to \mathbb{C}$ is anti-holomorphic for each $w \in \Omega$,
(ii) $\overline{K(z, w)} = K(w, z)$, and
(iii) $\langle h, K(\cdot, w)\rangle = h(w)$ for $h \in \mathcal{M}$.

It is easy to see, as a consequence of the reproducing property, that

$$\langle h, \partial_\alpha K(\cdot, w)\rangle = \partial_\alpha h(w),$$

where

$$\partial_\alpha K(\cdot, w) = \alpha_1 \frac{\partial}{\partial w_1} K(\cdot, w) + \cdots + \alpha_m \frac{\partial}{\partial w_m} K(\cdot, w).$$

Using these properties of the reproducing kernel, we easily verify that

(i) $u(w) = K(\cdot, w) \otimes \binom{0}{1} \perp \mathcal{N}$,
(ii) $v(w) = (K(\cdot, w) \otimes \binom{1}{0} + \partial_\alpha K(\cdot, w) \otimes \binom{0}{1}) \perp \mathcal{N}$, and
(iii) $\{u(w), v(w)\}$ span $\mathcal{N}^\perp$.

We infer that the set $\{u(w), v(w)\}$ is a basis for $\mathcal{M} \otimes_{\mathcal{A}(\Omega)} \mathbb{C}^2_{\omega,\alpha}$, and

$$\dim \mathcal{M} \otimes_{\mathcal{A}(\Omega)} \mathbb{C}^2_{\omega,\alpha} = 2.$$

We will now obtain an orthonormal basis for the localisation $\mathcal{M} \otimes_{\mathcal{A}(\Omega)} \mathbb{C}^2_{\omega,\alpha}$.

The vector $\mu(w) = u(w)/\|u(w)\|$ is a unit vector in $\mathcal{M} \otimes_{\mathcal{A}(\Omega)} \mathbb{C}^2_{\omega,\alpha}$. To obtain another unit vector orthogonal to $\mu(w)$ in $\mathcal{M} \otimes_{\mathcal{A}(\Omega)} \mathbb{C}^2_{\omega,\alpha}$, we set

$$\begin{aligned} e(w) &= v(w) - \langle v(w), \mu(w)\rangle \mu(w) \\ &= K(\cdot, w) \otimes \begin{pmatrix}1\\0\end{pmatrix} + \left(\partial_\alpha K(\cdot, w) - \frac{\langle \partial_\alpha K(\cdot, w), K(\cdot, w)\rangle}{\|K(\cdot, w)\|^2} K(\cdot, w)\right) \otimes \begin{pmatrix}0\\1\end{pmatrix}. \end{aligned}$$

The unit vector $\eta(w) = e(w)/\|e(w)\| \in \mathcal{M} \otimes_{\mathcal{A}(\Omega)} \mathbb{C}^2_{\omega,\alpha}$ is orthogonal to $\mu(w)$. Thus $\{\mu(w), \eta(w)\}$ is an orthonormal basis for $\mathcal{M} \otimes_{\mathcal{A}(\Omega)} \mathbb{C}^2_{\omega,\alpha}$. We will need the following expression for the norm $\|e(w)\|$.

$$\begin{aligned}
\|e(w)\|^2 &= \|K(\cdot,w)\|^2 + \left(\|\partial_\alpha K(\cdot,w)\|^2 - \frac{|\langle \partial_\alpha K(\cdot,w), K(\cdot,w)\rangle|^2}{\|K(\cdot,w)\|^2}\right)\\
&= \|K(\cdot,w)\|^2 - \|K(\cdot,w)\|^2\\
&\qquad \times \left(\frac{|\langle \partial_\alpha K(\cdot,w), K(\cdot,w)\rangle|^2 - \|\partial_\alpha K(\cdot,w)\|^2\|K(\cdot,w)\|^2}{\|K(\cdot,w)\|^4}\right)\\
&= \|K(\cdot,w)\|^2 \left\{1 - \left(\frac{|\langle \partial_\alpha K(\cdot,w), K(\cdot,w)\rangle|^2 - \|\partial_\alpha K(\cdot,w)\|^2\|K(\cdot,w)\|^2}{\|K(\cdot,w)\|^4}\right)\right\}.
\end{aligned}$$

Let $\pi \colon E \stackrel{\text{def}}{=} \{\mathcal{M} \otimes_{\mathcal{A}(\Omega)} \mathbb{C}^2_{\omega,\alpha} : w \in \Omega\} \to \Omega$, $\pi(\mathcal{M} \otimes_{\mathcal{A}(\Omega)} \mathbb{C}^2_{\omega,\alpha}, w) = w$. The map $w \to \{u(w), v(w)\}$ provides an anti–holomorphic frame for the bundle E over Ω. Further, each fibre is an inner product space and we see that the Grammian matrix $H(w)$ is

$$H(w) = \begin{pmatrix} \|u(w)\|^2 & \langle u(w), v(w)\rangle \\ \overline{\langle u(w), v(w)\rangle} & \|v(w)\|^2 \end{pmatrix},$$

where

$$\begin{aligned}
\|u(w)\|^2 &= \|K(\cdot,w)\|^2,\\
\langle u(w), v(w)\rangle &= \langle K(\cdot,w), \partial_\alpha K(\cdot,w)\rangle,\\
\|v(w)\|^2 &= \|K(\cdot,w)\|^2 + \|\partial_\alpha K(\cdot,w)\|^2.
\end{aligned}$$

The determinant of the Grammian

$$\begin{aligned}
\det H(w) &= \|K(\cdot,w)\|^2 \left(\|K(\cdot,w)\|^2 + \|\partial_\alpha K(\cdot,w)\|^2\right) + |\langle K(\cdot,w), \partial_\alpha K(\cdot,w)\rangle|\\
&= \|K(\cdot,w)\|^4 + \|K(\cdot,w)\|^2 \; \|\partial_\alpha K(\cdot,w)\|^2 - |\langle K(\cdot,w), \partial_\alpha K(\cdot,w)\rangle|\\
&= \|K(\cdot,w)\|^2 \; \|e(w)\|^2.
\end{aligned}$$

A factorisation for the Gram matrix $H(w)$ is obtained via the matrix $\Gamma(w)$ as follows. Let

$$\Gamma(w) = \begin{pmatrix} \|K(\cdot,w)\| & \dfrac{\langle K(\cdot,w), \partial_\alpha K(\cdot,w)\rangle}{\|K(\cdot,w)\|} \\ 0 & \|e(w)\| \end{pmatrix}.$$

The factorisation

$$\Gamma(w)^*\Gamma(w) = H(w), \tag{3.19}$$

follows from

$$\frac{|\langle K(\cdot,w), \partial_\alpha K(\cdot,w)\rangle|^2}{\|K(\cdot,w)\|} + \|e(w)\|^2 = \|K(\cdot,w))\|^2 + \|\partial_\alpha K(\cdot,w)\|^2.$$

We now obtain a projection formula for $\mathrm{pr} : \mathcal{M} \otimes \mathbb{C}^2_{\omega,\alpha} \longrightarrow \mathcal{M} \otimes_{\mathcal{A}(\Omega)} \mathbb{C}^2_{\omega,\alpha}$,

$$\begin{aligned}
\mathrm{pr} : h \otimes \begin{pmatrix} a \\ 0 \end{pmatrix} + k \otimes \begin{pmatrix} 0 \\ b \end{pmatrix} &\longrightarrow \left(\left\langle h \otimes \begin{pmatrix} a \\ 0 \end{pmatrix} + k \otimes \begin{pmatrix} 0 \\ b \end{pmatrix}, \mu(w) \right\rangle \right) \mu(w) \\
&\quad + \left(\left\langle h \otimes \begin{pmatrix} a \\ 0 \end{pmatrix} + k \otimes \begin{pmatrix} 0 \\ b \end{pmatrix}, \eta(w) \right\rangle \right) \eta(w) \\
= \; &\left\{ a\langle h, K(\cdot, w)\rangle + \left(\langle k, \partial_\alpha K(\cdot, w)\rangle - \frac{\langle K(\cdot, w), \partial_\alpha K(\cdot, w)\rangle}{\|K(\cdot, w)\|^2} \langle k, K(\cdot, w)\rangle \right) b \right\} \frac{\eta(w)}{\|e(w)\|} \\
&\quad + \frac{b}{\|K(\cdot, w)\|} \langle k, K(\cdot, w)\rangle \mu(w).
\end{aligned}$$

Set

$$a = \frac{\|e(w)\|}{\langle h, K(\cdot, w)\rangle}$$

and $b = 0$. From the preceding calculation, it follows that

$$\mathrm{pr}\big(h \otimes \begin{pmatrix} a \\ 0 \end{pmatrix}\big) = a\big(\frac{\langle h, K(\cdot, w)\rangle}{\|e(w)\|}\big) = \eta(w). \tag{3.20}$$

Similarly, set

$$\begin{aligned}
a &= -\langle h, K(\cdot, w)\rangle^{-1} \left(\langle k, \partial_\alpha K(\cdot, w)\rangle - \frac{\langle K(\cdot, w), \partial_\alpha K(\cdot, w)\rangle}{\|K(\cdot, w)\|^2} \langle k, K(\cdot, w)\rangle \right) \\
&\qquad \times \frac{\|K(\cdot, w)\|}{\langle k, K(\cdot, w)\rangle}, \\
b &= \frac{\|K(\cdot, w)\|}{\langle k, K(\cdot, w)\rangle}.
\end{aligned}$$

It now follows that

$$\mathrm{pr}\big(h \otimes \begin{pmatrix} a \\ 0 \end{pmatrix} + k \otimes \begin{pmatrix} 0 \\ b \end{pmatrix}\big) = \mu(w). \tag{3.21}$$

Let $\mathcal{M}_0$ be a submodule of $\mathcal{M}$. We let $K_0(\cdot, w)$ denote the kernel function for $\mathcal{M}_0$. Similarly, let $\{\mu_0(w), \eta_0(w)\}$ be the orthonormal basis for $\mathcal{M}_0 \otimes_{\mathcal{A}(\Omega)} \mathbb{C}^2_{\omega,\alpha}$. Let $X : \mathcal{M}_0 \to \mathcal{M}$ be the inclusion map. Then $X^* : \mathcal{M} \to \mathcal{M}_0$ is the projection and $X^* K(\cdot, w) = K_0(\cdot, w)$. Again, there is a short exact sequence of modules as in previous sections with the initial module defined as the "quotient" $\mathcal{M}_q$ of $\mathcal{M}$ by the range of X. If $\mathcal{M}_0$ is not prime-like, then ordinary localization will not determine sufficiently many geometric invariants for $\mathcal{M}_q$ and we must allow higher multiplicity localizations. The following calculations show some of the invariants obtained and their relation to those obtained from ordinary localization.

We now see that

$$\mu_0(w) \xrightarrow{\mathrm{pr}^{-1}} h\otimes\begin{pmatrix}a\\0\end{pmatrix} + k\otimes\begin{pmatrix}0\\b\end{pmatrix} \xrightarrow{X\otimes\mathrm{Id}} Xh\otimes\begin{pmatrix}a\\0\end{pmatrix} + Xk\otimes\begin{pmatrix}0\\b\end{pmatrix}$$
$$\xrightarrow{\mathrm{pr}} b\frac{\langle Xk, K(\cdot,w)\rangle}{\|K(\cdot,w)\|}\mu(w) + \Big\{a\langle Xh, K(\cdot,w)\rangle + b\Big(\langle Xk, \partial_\alpha K(\cdot,w)\rangle$$
$$-\frac{\langle K(\cdot,w), \partial_\alpha K(\cdot,w)\rangle}{\|K(\cdot,w)\|^2}\langle Xk, K(\cdot,w)\rangle\Big)\Big\}\frac{\eta(w)}{\|e(w)\|}.$$

In view of the calculations

$$\begin{aligned} b\langle Xk, K(\cdot,w)\rangle\frac{\mu(w)}{\|K(\cdot,w)\|} &= b\langle Xk, K(\cdot,w)\rangle\frac{\mu(w)}{\|K(\cdot,w)\|} \\ &= \left\langle k, \frac{K_0(\cdot,w)}{\|K_0(\cdot,w)\|}\right\rangle^{-1}\langle Xk, K(\cdot,w)\rangle\frac{\mu(w)}{\|K(\cdot,w)\|} \\ &= \frac{\|K_0(\cdot,w)\|}{\|K(\cdot,w)\|}\mu(w) \end{aligned}$$

and

$$a\langle Xh, K(\cdot,w)\rangle + b\left(\langle Xk, \partial_\alpha K(\cdot,w)\rangle - \frac{\langle K(\cdot,w), \partial_\alpha K(\cdot,w)\rangle}{\|K(\cdot,w)\|^2}\langle Xk, K(\cdot,w)\rangle\right)$$

$$= b\left\{\left(-\langle k, \partial_\alpha K_0(\cdot,w)\rangle - \frac{\langle K_0(\cdot,w), \partial_\alpha K_0(\cdot,w)\rangle}{\|K_0(\cdot,w)\|^2}\langle k, K_0(\cdot,w)\rangle\right) + \right.$$
$$\left. + \left(\langle Xk, \partial_\alpha K(\cdot,w)\rangle - \frac{\langle K(\cdot,w), \partial_\alpha K(\cdot,w)\rangle}{\|K(\cdot,w)\|^2}\langle Xk, K(\cdot,w)\rangle\right)\right\} =$$
$$= \|K_0(\cdot,w)\|\left(\frac{\langle K_0(\cdot,w), \partial_\alpha K_0(\cdot,w)\rangle}{\|K_0(\cdot,w)\|^2} - \frac{\langle K(\cdot,w), \partial_\alpha K(\cdot,w)\rangle}{\|K(\cdot,w)\|^2}\right),$$

we obtain the projection formula

$$\mu_0(w) \xrightarrow{X\otimes_{\mathcal{A}(\Omega)}\mathrm{Id}} \frac{K_0(w,w)}{K(w,w)}\mu(w) \tag{3.22}$$
$$+ \frac{\|K_0(\cdot,w)\|}{\|e(w)\|}\Big(\frac{\langle K_0(\cdot,w), \partial_\alpha K_0(\cdot,w)\rangle}{\|K_0(\cdot,w)\|^2} - \frac{\langle K(\cdot,w), \partial_\alpha K(\cdot,w)\rangle}{\|K(\cdot,w)\|^2}\Big)\eta(w).$$

Now we obtain the other projection formula

$$\begin{aligned}\eta_0(w) &\xrightarrow{\text{pr}^{-1}} h \otimes \begin{pmatrix} a \\ 0 \end{pmatrix} \xrightarrow{X\otimes \text{Id}} Xh \otimes \begin{pmatrix} a \\ 0 \end{pmatrix} \\ &\xrightarrow{\text{pr}} a\frac{\langle Xh, K(\cdot, w)\rangle}{\|e(w)\|}\eta(w) \\ &= \frac{\|e_0(w)\|}{\|e(w)\|}\eta(w).\end{aligned} \tag{3.23}$$

Now we can calculate the matrix for the operator

$$X(w) \stackrel{\text{def}}{=} X \otimes_{\mathcal{A}(\Omega)} \text{Id}\colon \mathcal{M}_0 \otimes_{\mathcal{A}(\Omega)} \mathbb{C}^2_{\omega,\alpha} \to \mathcal{M} \otimes_{\mathcal{A}(\Omega)} \mathbb{C}^2_{\omega,\alpha} \tag{3.24}$$

with respect to the two orthonormal bases $\{\mu(w), \eta(w)\}$ and $\{\mu_0(w), \eta_0(w)\}$. From the equations (3.22) and (3.23), we see that

$$X(w) = \begin{pmatrix} \frac{\|K_0(\cdot,w)\|}{\|K(\cdot,w)\|} & \frac{\|K_0(\cdot,w)\|}{\|e(w)\|}\left(\frac{\langle K_0(\cdot,w), \partial_\alpha K_0(\cdot,w)\rangle}{\|K_0(\cdot,w)\|^2} - \frac{\langle K(\cdot,w), \partial_\alpha K(\cdot,w)\rangle}{\|K(\cdot,w)\|^2}\right) \\ 0 & \frac{\|e_0(w)\|}{\|e(w)\|} \end{pmatrix}.$$

Let $H_0(w)$ be the Grammian for the localisation $\mathcal{M}_0 \otimes_{\mathcal{A}(.)} \mathbb{C}^2_{\omega,\alpha}$ We obtain a factorisation $H_0(w) = \Gamma_0(w)^*\Gamma_0(w)$ similar to (3.19). Calculating the inverse of Γ,

$$\Gamma(w)^{-1} = \begin{pmatrix} \|K(\cdot,w)\|^{-1} & \frac{\langle K(\cdot,w), \partial_\alpha K(\cdot,w)\rangle}{\|K(\cdot,w)\|^2\|e(w)\|} \\ 0 & \|e(w)\|^{-1} \end{pmatrix}$$

verifies that $X(w) = \Gamma_0(w)\Gamma(w)^{-1}$. Consequently,

$$\begin{aligned} X(w)X(w)^* &= \Gamma_0(w)\Gamma(w)^{-1}\Gamma(w)^{*\,-1}\Gamma_0(w)^* \\ &= \Gamma_0(w)\Big(\Gamma(w)^*\Gamma(w)\Big)^{-1}\Gamma_0(w)^* \\ &= \Gamma_0(w)H(w)^{-1}\Gamma_0(w)^* \end{aligned} \tag{3.25}$$

Similarly,

$$X(w)^*X(w) = \Gamma(w)^{-1*}H_0(w)\Gamma(w)^{-1}. \tag{3.26}$$

With this explicit calculation of the operator $X^*(w)X(w)$, we can calculate some curvatures.

Proposition 3.1. *Let $\mathcal{M}$ be a functional Hilbert space on the domain Ω. Assume that $\mathcal{M}$ is also a Hilbert module over the algebra $\mathcal{A}(\Omega)$. Let $\mathcal{M}_0$ be a submodule of $\mathcal{M}$. Let $X : \mathcal{M}_0 \to \mathcal{M}$ be the inclusion map and $X(w)$ be the map in (3.24). Then:*

$$\text{(1)} \sum_{i,j=1}^{m} \frac{\partial^2}{\partial w_i\, \partial \bar{w}_j} \log\det\big(X(w)^* X(w)\big) = \mathcal{K}_0(w) - \mathcal{K}(w)$$
$$- \sum_{i,j=1}^{m} \frac{\partial^2}{\partial w_i\, \partial \bar{w}_j} \log \frac{\|e_0(w)\|}{\|e(w)\|}, \text{ and}$$
$$\text{(2)} \sum_{i,j=1}^{m} \frac{\partial^2}{\partial w_i\, \partial \bar{w}_j} \log\operatorname{tr}\big(X(w)^* X(w)\big) = \sum_{i,j=1}^{m} \frac{\partial^2}{\partial w_i\, \partial \bar{w}_j} \log\operatorname{tr}\big(H_0(w) H(w)^{-1}\big).$$

PROOF. Since $\det\big(X(w)X(w)^*\big) = \det H_0(w)/\det H(w)$ by (3.25), the first statement in the proposition follows from:

$$\begin{aligned}
&\sum_{i,j=1}^{m} \frac{\partial^2}{\partial w_i\, \partial \bar{w}_j} \log\det\big(X(w)^* X(w)\big) \\
&\quad = \sum_{i,j=1}^{m} \frac{\partial^2}{\partial w_i\, \partial \bar{w}_j} \log\Big(\frac{\det H_0(w)}{\det H(w)}\Big) \\
&\quad = -\sum_{i,j=1}^{m} \frac{\partial^2}{\partial w_i\, \partial \bar{w}_j} \log \frac{\|K_0(\cdot,w)\|^2\, \|e_0(w)\|^2}{\|K(\cdot,w)\|^2 \|e(w)\|^2} \\
&\quad = -\sum_{i,j=1}^{m} \frac{\partial^2}{\partial w_i\, \partial \bar{w}_j} \log \|K_0(\cdot,w)\|^2 + \bar{\partial}\partial \log \|K(\cdot,w)\|^2 \\
&\qquad - \sum_{i,j=1}^{m} \frac{\partial^2}{\partial w_i\, \partial \bar{w}_j} \log \|e_0(w)\|^2 + \sum_{i,j=1}^{m} \frac{\partial^2}{\partial w_i\, \partial \bar{w}_j} \log \|e(w)\|^2.
\end{aligned}$$

Since

$$\begin{aligned}
\operatorname{tr}\Big(X(w)^* X(w)\Big) &= \operatorname{tr}\Big(\Gamma(w)^{-1*} H_0(w) \Gamma(w)^{-1}\Big) \\
&= \operatorname{tr}\Big(H_0(w) \Gamma(w)^{-1} (\Gamma(w))^{-1*}\Big) \\
&= \operatorname{tr}\Big(H_0(w) H(w)^{-1}\Big),
\end{aligned}$$

it follows that

$$\sum_{i,j=1}^{m} \frac{\partial^2}{\partial w_i\, \partial \bar{w}_j} \log\operatorname{tr}\big(X(w)^* X(w)\big) = \sum_{i,j=1}^{m} \frac{\partial^2}{\partial w_i\, \partial \bar{w}_j} \log\big(\operatorname{tr}\big(H_0(w) H(w)^{-1}\big)\big).$$

This completes the proof of the second statement. □

The first two terms in the first statement of the Proposition would have been obtained from the rank one localisation. So, the new information, if any, is contained in the last two terms, i.e., $-\partial\bar{\partial}\log(\|e_0(w)\|/\|e(w)\|)$.

Here is another expression for $X(w)^*X(w)$:

$$\begin{aligned} X(w)^*X(w) &= \Gamma(w)^{-1*}H_0(w)\Gamma(w)^{-1} \\ &= \Gamma(w)\Big(\Gamma(w)^*\Gamma(w)\Big)^{-1}H_0(w)\Gamma(w)^{-1} \\ &= \Gamma(w)H(w)^{-1}H_0(w)\Gamma(w)^{-1} \\ &= \Big(\Gamma(w)H(w)^{-1/2}\Big)H(w)^{-1/2}H_0(w)H(w)^{-1/2}\Big(\Gamma(w)H(w)^{-1/2}\Big)^{-1}. \end{aligned}$$

This last expression may relate well to the following definition of curvature for $X(w)^*X(w)$ thought of as a metric on a rank 2 bundle on Ω.

$$\mathcal{K}(X(w)^*X(w)) \stackrel{\text{def}}{=} \bar\partial\Big(\big(X(w)^*X(w)\big)^{-1}\partial\big(X(w)^*X(w)\big)\Big).$$

However, this formula is valid only if the curvature is calculated with respect to the unique connection compatible with the complex structure as well as the hermitian metric. Since we did not use a holomorphic frame in computing the matrix for $X(w)^*X(w)$, the consideration above does not apply. It is not clear how we should get back to our complex geometric set up.

REFERENCES

[1] N. ARONSZAJN, *Theory of reproducing kernels*, Trans. Amer. Math. Soc., **68**(1950), 337–404.

[2] B. BAGCHI AND G. MISRA, *Nagy-Foias characteristic functions and homogeneous Cowen-Douglas operators*, in preparation.

[3] R. BOTT AND S. S. CHERN, *Hermitian vector bundles and the equidistribution of the zeros of their holomorphic sections*, Acta Math., **114** (1968), 71–112.

[4] D. N. CLARK AND G. MISRA, *On homogeneous contractions and unitary representations of $SU(1,1)$*, J. Operator Th., **30** (1993), 109–122.

[5] M. J. COWEN AND R. G. DOUGLAS, *Complex geometry and operator theory*, Acta Math., **141** (1978), 187–261.

[6] R. G. DOUGLAS, *On Šilov resolutions of Hilbert modules*, in Operator Theory: Advances and Applications, V. 28, Birkhäuser Verlag, Basel (1988) 51–60.

[7] R. G. DOUGLAS, *Models and resolutions for Hilbert modules*, Contemporay Mathematics, to appear.

[8] R. G. DOUGLAS AND V. I. PAULSEN, *Hilbert Modules over Function Algebras*, Longman Research Notes, 217, 1989.

[9] R. G. DOUGLAS, V. I. PAULSEN, H. SAH AND K. YAN, *Algebraic reduction and rigidity for Hilbert modules*, Amer. J. Math, **117** (1995), 75–92.

[10] P. GRIFFITHS AND J. HARRIS, *Principles of Algebraic Geometry*, John Wiley & Sons, 1978.

[11] R. C. GUNNING AND H. ROSSI, *Analytic functions of Several Complex Variables*, Prentice Hall, 1965.

[12] S. KOBAYASHI AND K. NOMIZU, *Foundations of Differential Geometry*, Volume II, John Wiley & Sons, 1969.

[13] G. MISRA, *Curvature and discrete series representations of $SL_2(R)$*, J. Int. Eqn. Operator Th., **9** (1986), 452–459.
[14] R. O. WELLS, JR., *Differential Analysis on Complex Manifolds*, Springer Verlag, 1973.

R. G. DOUGLAS
Department of Mathematics
Texas A & M University
College Station, TX 77843

G. MISRA
Indian Statistical Institute
R. V. College Post
Bangalore 560 059

Received: August 23rd, 1995.

Operator Theory:
Advances and Applications, Vol. 104

Some Multplicities for Contractions with Hilbert-Schmidt Defect

GEORGE R. EXNER* AND IL BONG JUNG

Dedicated to Carl Pearcy

ABSTRACT. We show in this paper that for absolutely continuous contractions on Hilbert space with Hilbert-Schmidt defect in the classes C_{10}, $C_{1\cdot}$, or $C_{\cdot 0}$, a number of multiplicity measures coincide. These include some involving dual operator algebra class, the size of a zero operator dilated, and the multiplicity of the unitary piece of the minimal coisometric extension. The main ingredient in these results, and another equivalent multiplicity measure, is the n in an "n-fold analytic co-kernel" defined on the unit disk for such an operator.

1. INTRODUCTION AND PRELIMINARIES

Among bounded linear operators on Hilbert space the unilateral shift is arguably the most studied and best understood. One of its good properties is that for the shift S of multiplicity one, and shifts $S^{(n)}$ of higher multiplicity which are n-fold direct sums of S, the n of counting multiplicity coincides with a number of other possible measures of multiplicity, such as $\dim(\ker(S^{(n)*}))$, the maximal size of zero operator $S^{(n)}$ dilates to a semi-invariant subspace (the 2-2 position of a 3×3 operator matrix for $S^{(n)}$), and the multiplicity of an "analytic co-kernel on the unit disk $\mathbb{D}$" for $S^{(n)}$ (definitions reviewed below). As well, the place of $S^{(n)}$ in the array of dual operator algebra classes $\mathbb{A}_{i,j}$, $1 \leq i, j \leq \aleph_0$ determines and is determined by n: for n finite, $S^{(n)} \in \mathbb{A}_{n,\aleph_0} \setminus \mathbb{A}_{n+1,1}$.

In this paper we give some classes of operators, namely contractions in the classes $C_{1\cdot}$, $C_{\cdot 0}$, and C_{10} with Hilbert-Schmidt defect operator, for which there are similar coincidences of multiplicity measures. Our results constitute in part generalizations of previous work in [13] and [14] from the case in which the defect operator is of finite rank. Perhaps more importantly, we incorporate in the present paper the notion of "n-fold analytic co-kernel" related to ideas in [11] and subsequent papers and a generalization of the (multiplicity 1) analytic invariant subspaces crucial to [17], [5], and [9]; we show that, for our special case, the multiplicity measure "maximal size of analytic co-kernel" coincides with the rest.

*The authors were partially supported by KOSEF 94-1400-02-01-3 and a travel grant from Bucknell University.

This paper is dedicated to Carl M. Pearcy, on the occasion of his 60th birthday, with warm affection and gratitude for his encouragement, support, and profound influence upon our growth as mathematicians.

We begin with some notation and preliminaries. We denote by $\mathbb{N}$ and $\mathbb{C}$ the natural and complex numbers, and by $\mathbb{T}$ and $\mathbb{D}$ the unit circle and unit disk in the complex plain, respectively. $\mathcal{H}$ shall always denote a separable, complex, infinite dimensional Hilbert space, with $\mathcal{L}(\mathcal{H})$ the algebra of bounded linear operators on $\mathcal{H}$. We use $P_{\mathcal{M}}$ to denote the orthogonal projection on a (closed) subspace $\mathcal{M}$ of $\mathcal{H}$, and $\bigvee$ to denote closed linear span. We reserve I_n and 0_n, $1 \leq n \leq \aleph_0$, to denote the identity and zero operators on the spaces of the given dimension n .

We will use without further comment standard facts about the Hilbert-Schmidt, trace-class, Fredholm, and semi-Fredholm operators, as well as the Fredholm index ind$(\cdot)$ (see [19] for a discussion). Denote by $\sigma(T)$, $\sigma_e(T)$, $\sigma_{le}(T)$, and $\sigma_{re}(T)$ respectively the spectrum, essential spectrum, left essential spectrum, and right essential spectrum of $T \in \mathcal{L}(\mathcal{H})$. Recall (e.g., from [19]) that there are holes in the spectrum and "pseudo-holes" in the essential spectrum.

Recall that for a contraction T the defect operator D_T is defined by $D_T = (I - T^*T)^{\frac{1}{2}}$; we shall say that a contraction T has "Hilbert-Schmidt defect" if D_T is in the Hilbert-Schmidt class. We need as well some classes of contractions defined in [20]: a contraction T in $\mathcal{L}(\mathcal{H})$ is in the class $C_{0\cdot}$ if $\|T^n x\| \to 0$ for all $x \in \mathcal{H}$, and is in the class $C_{\cdot 0}$ if T^* is in $C_{0\cdot}$. As well, T is in $C_{1\cdot}$ if $\|T^n x\| \to 0$ implies $x = 0$, and $C_{\cdot 1}$ is again defined by duality. The classes $C_{\alpha\beta}$, α and β each 0 or 1, are defined by $C_{\alpha\beta} = C_{\alpha\cdot} \cap C_{\cdot\beta}$.

A (closed) subspace $\mathcal{M}$ is invariant for an operator T if $T\mathcal{M} \subseteq \mathcal{M}$; for any subspace $\mathcal{M}$, we denote by $T|\mathcal{M}$ the restriction of T to $\mathcal{M}$. We denote by Lat(T) the lattice of subspaces invariant for T. A subspace $\mathcal{K}$ is *semi-invariant* for T if there exist subspaces $\mathcal{M}$ and $\mathcal{N}$, each invariant for T, and with $\mathcal{N} \subseteq \mathcal{M}$, such that $\mathcal{K} = \mathcal{M} \ominus \mathcal{N} \triangleq \mathcal{M} \cap \mathcal{N}^{\perp}$. We denote by $T_{\mathcal{K}}$ the compression $P_{\mathcal{K}}T|\mathcal{K}$ of T to $\mathcal{K}$; if A is some operator on a Hilbert space (of finite or infinite dimension) such that T has some compression unitarily equivalent to A we say T *dilates* or is a dilation of A; recall that in such a case, T may be written up to unitary equivalence as a three by three operator matrix

$$T = \begin{pmatrix} * & * & * \\ 0 & A & * \\ 0 & 0 & * \end{pmatrix}.$$

It is well known that any contraction T may be written $T = T_c \oplus U$, where U is a unitary operator and T_c has no reducing subspace on which it is unitary (of course, either T_c or U may be absent). If U is absent we say T is a completely non-unitary contraction. If U is either absent or has spectral measure absolutely continuous with respect to Lebesgue measure on $\mathbb{T}$, we say T is an absolutely continuous contraction.

We sketch very briefly the background for the dual operator algebra or Scott Brown's approach to the study of contractions; see [3] for a thorough introduction, as well as a history of the subject through about 1985. Denote by $L^\infty = L^\infty(\mathbb{T})$ the Banach algebra of all complex valued Lebesgue measurable, essentially bounded functions on $\mathbb{T}$ and by H^∞ the usual Hardy subspace of L^∞. It is well known, and fundamental to the study of contractions, that there is the Sz.-Nagy–Foiaş Functional Calculus ([3, Theorem 4.1]) from $H^\infty(\mathbb{T})$ to $\mathcal{A}_T$ (the ultraweakly closed unital algebra generated by T) for T an absolutely continuous contraction. We denote by $\mathbb{A}$ the class of absolutely continuous contractions for which the functional calculus is an isometry (and then, as it turns out, has many more good properties). Further, each of $H^\infty(\mathbb{T})$ and $\mathcal{A}_T$ is the Banach dual of another space; in particular, $\mathcal{A}_T$ is the dual of Q_T, a quotient of the trace class operators. Denote cosets in Q_T by $[L]$. If for vectors x and y in $\mathcal{H}$ we let $x \otimes y$ denote the usual rank one operator, each $x \otimes y$ induces a coset $[x \otimes y]$ in Q_T. It is known from [1] and [6] that, for an operator T in $\mathbb{A}$, *each* coset $[L]$ is induced by some rank one operator (that is, for each $[L]$ we may "solve the equation" $[L] = [x \otimes y]$). But this was a matter of considerable study for some time, and the classes $\mathbb{A}_{m,n}$, $1 \leq m, n \leq \aleph_0$, were defined as part of that study: an operator T in $\mathbb{A}$ is in the class $\mathbb{A}_{m,n}$ if every $m{\times}n$ system of simultaneous equations of the form

$$[x_i \otimes y_j] = [L_{ij}], \quad 0 \leq i < m, \quad 0 \leq j < n,$$

where $\{[L_{ij}]\}_{\substack{0\leq i<m \\ 0\leq j<n}}$ is an arbitrary $m \times n$ array from $\mathcal{Q}_{\mathcal{A}}$, has a solution $\{x_i\}_{0\leq i<m}$, $\{y_j\}_{0\leq j<n}$ consisting of a pair of sequences of vectors from $\mathcal{H}$. As is customary we simplify $\mathbb{A}_{n,n}$ to $\mathbb{A}_n$. Finally, in terms of the functional calculus, we may define the class C_0 to consist of those contractions for which there exists $u \in H^\infty$, $u \not\equiv 0$, such that $u(T) = 0$ (see [20] for the definition and [2] for a definitive study).

Another notion from [20] vital for the study of contractions is that of various dilation and restriction models for a contraction. If T is a contraction, T has a unitary dilation which is, in a natural sense, minimal. Various pieces of this dilation yield a minimal isometric dilation U_T of T and a minimal co-isometric extension B_T of T. Via the Wold decomposition, U_T has a maximal unitary direct summand denoted $R(T)$ and B_T has a maximal unitary direct summand denoted $R_*(T)$ (of course, either may be absent). We shall reserve B for the bilateral shift of multiplicity one, since the count of its appearance in $R_*(T)$ will turn out to be one of the equivalent multiplicity measures for some classes of contractions.

Finally, we assemble some standard notions, weaker than but related to similarity, useful for the study of operators. An operator $X : \mathcal{H} \longrightarrow \mathcal{H}'$ is an *injection* if it is one to one, and a *quasi-affinity* if it is one to one and has dense range. Let $\mathcal{H}$ and $\mathcal{H}'$ be Hilbert spaces with $T \in \mathcal{L}(\mathcal{H})$ and $T' \in \mathcal{L}(\mathcal{H}')$. We say the operator T is *injected* into T' if there exists an injection $X : \mathcal{H} \longrightarrow \mathcal{H}'$ such that $T'X = XT$, and we write $T' \succ^i T$, (or $T \prec^i T'$). A family $\{X_\alpha\}$ of injections in $\mathcal{L}(\mathcal{H}, \mathcal{H}')$ is called *complete* if $\bigvee_\alpha X_\alpha \mathcal{H} = \mathcal{H}'$. The operator T is *completely injected* into T' if there is a complete family $\{X_\alpha\}$ of injections in $\mathcal{L}(\mathcal{H}, \mathcal{H}')$ so that $T'X_\alpha = X_\alpha T$ for each α, and we write $T' \succ^{c.i.} T$ (or $T \prec^{c.i.} T'$). The operator T is a *quasi-affine transform* of T' if there exists a quasi-affinity $X : \mathcal{H} \to \mathcal{H}'$ such that $T'X = XT$, and we denote this by $T' \succ T$ (or $T \prec T'$).

2. C_{10} OPERATORS WITH HILBERT-SCHMIDT DEFECT OPERATOR

In this section we consider operators T in the class C_{10} and with defect operator D_T in the Hilbert-Schmidt class and show that for them a number of "multiplicities" coincide. We assemble first some definitions and lemmas; the first definition is central to this investigation.

Definition 2.1. *Let T be an operator on $\mathcal{H}$. We say that T has an* n-fold analytic co-kernel, *n a positive integer, if there are $\mathcal{H}$-valued functions $f_1, f_2, \dots, f_n$ holomorphic on the disk $\mathbb{D}$ and satisfying*

(i) $(T^* - \lambda)f_i(\lambda) = 0, \quad 1 \le i \le n, \ \lambda \in \mathbb{D}$, *and*
(ii) *the vectors $f_1(\lambda), f_2(\lambda), \dots, f_n(\lambda)$ are independent for each $\lambda \in \mathbb{D}$.*

We say that T has a full *n-fold analytic co-kernel if in addition*

$$\bigvee_{\lambda \in \mathbb{D}} \{f_1(\lambda), \dots, f_n(\lambda)\} = \mathcal{H}.$$

We say that T has an $\aleph_0$–fold analytic co-kernel *if there are $\mathcal{H}$-valued functions $f_1, f_2, \dots, f_n, \dots,$ holomorphic on the disk $\mathbb{D}$ and satisfying*

(i) $(T^* - \lambda)f_i(\lambda) = 0, \quad 1 \le i < \aleph_0, \ \lambda \in \mathbb{D}$, *and*
(ii) *the vectors $f_1(\lambda), f_2(\lambda), \dots, f_n(\lambda)$ are independent for each $\lambda \in \mathbb{D}$ and $n \in \mathbb{N}$.*

We say that T has a full *$\aleph_0$-fold analytic co-kernel if in addition*

$$\bigvee_{\substack{1 \le n < \aleph_0, \\ \lambda \in \mathbb{D}}} f_n(\lambda) = \mathcal{H}.$$

We hasten to point out that these notions are not new. In the specific context of C_{10} contractions with Hilbert-Schmidt defect operator Uchiyama constructs an n-fold analytic co-kernel for a T with $\dim(\ker(T^*)) = n$ in [25], and we will recall this construction shortly. Earlier, and in a more general (even non-contractive) setting, Cowen and Douglas studied this property in combination with others in [11] and subsequent papers. Finally, the $n = 1$ case of this definition yields the notion of "analytic invariant subspace" studied in [17], [5], and [9, Theorem 6.2], as is shown by a trivial calculation using the following definition:

Definition 2.2. *Let T be a contraction and $\mathcal{M} \in \mathrm{Lat}(T)$. We say $\mathcal{M}$ is an* analytic invariant subspace *for T if there exists a non-zero conjugate analytic function $e : \lambda \to e_\lambda$ from $\mathbb{D}$ into $\mathcal{M}$ such that*

$$(T|\mathcal{M} - \lambda)^* e_\lambda = 0, \quad \lambda \in \mathbb{D}.$$

From [9, Theorem 6.2] we know that an absolutely continuous contraction with an analytic invariant subspace is in the class $\mathbb{A}_{1,\aleph_0}$; one aim of this paper is to improve this to a version including multiplicity in our special case.

Turn now to the case where T is an absolutely continuous contraction in C_{10} with Hilbert-Schmidt defect D_T and suppose $\dim(\ker(T^*)) = n$. For ease of exposition, we assume $n < \aleph_0$ and omit the minor modifications to deal with the case $n = \aleph_0$. We sketch the construction from [25] showing that T has an n-fold analytic co-kernel and fix some notation along the way. Let $\mathcal{F}$ be an n dimensional Hilbert space with orthonormal basis $\{e_1, \ldots, e_n\}$, let $\ell^2_+(\mathcal{F})$ be the usual space of sequences from $\mathcal{F}$ which are (norm) square summable, and let $S_{\mathcal{F}}$ denote the unilateral shift on $\ell^2_+(\mathcal{F})$; clearly $S_{\mathcal{F}}$ is of multiplicity n.

From [24, Theorem 2] there exists a quasi-affinity $Y : \mathcal{H} \to \ell^2_+(\mathcal{F})$ so that

$$YT = S_{\mathcal{F}}Y.$$

Defining the f_i by

$$f_i(\lambda) \triangleq Y^*(e_i, \lambda e_i, \lambda^2 e_i, \ldots), \quad 1 \le i \le n, \ \lambda \in \mathbb{D}, \tag{1}$$

and using

$$T^*Y^* = Y^*S^*_{\mathcal{F}} \tag{2}$$

the desired properties of the f_i follow easily.

We will use the notation above in what follows, and must add to it. Let

$$e_i^j \triangleq (\underbrace{0, \ldots, 0}_{j-1}, e_i, 0, \ldots), \quad 1 \le i \le n, \ 1 \le j < \aleph_0,$$

i.e., e_i occurs as the $j^{\underline{th}}$ and only nonzero component of e_i^j. For convenience, set $e_i^0 = (0, 0, \ldots)$ for each i, $1 \le i \le n$. Note that

$$e_i^j = S_{\mathcal{F}}^{j-1}(e_i, 0, 0, \ldots), \quad 1 \le i \le n, \ 1 \le j < \aleph_0.$$

Observe also that

$$S_{\mathcal{F}}(e_i^j) = e_i^{j+1}, \quad 1 \le i \le n, \ 1 \le j < \aleph_0,$$

and

$$S^*_{\mathcal{F}}(e_i^j) = e_i^{j-1}, \quad 1 \le i \le n, \ 1 \le j < \aleph_0. \tag{3}$$

Finally,

$$\bigvee_{\lambda \in \mathbb{D}} f_i(\lambda) \subseteq \bigvee_{1 \le j < \aleph_0} Y^*(e_i^j), \quad 1 \le i \le n. \tag{4}$$

(In fact, this containment may easily be shown to be an equality.)

The following lemma reflects the independence of the analytic co-kernels $\{f_i\}$.

Lemma 2.3. *Suppose T is an absolutely continuous contraction in C_{10} with Hilbert-Schmidt defect D_T and adopt the notation above. If $\dim(\ker(T^*)) = n < \aleph_0$, then*

$$\bigvee_{1 \le j < \aleph_0} Y^*(e_k^j) \not\subseteq \bigvee_{\substack{1 \le j < \aleph_0, \\ 1 \le i \le n, \\ i \ne k}} Y^*(e_i^j), \quad 1 \le k \le n. \tag{5}$$

If $\dim(\ker(T^*)) = \aleph_0$, then

$$\bigvee_{1 \le j < \aleph_0} Y^*(e_k^j) \not\subseteq \bigvee_{\substack{1 \le j < \aleph_0, \\ 1 \le i \le n, \\ i \ne k}} Y^*(e_i^j), \quad k,\ n \in \mathbb{N}. \tag{6}$$

PROOF. Suppose the containment occurs, w.l.o.g. for $k = 1$. Let $\mathcal{F}' \subseteq \mathcal{F}$ be the span of $\{e_2, \dots, e_n\}$, consider $\ell_+^2(\mathcal{F}')$ as a subset of $\ell_+^2(\mathcal{F})$ in the natural way, and let $S_{\mathcal{F}'}$ be the restriction of $S_{\mathcal{F}}$ to $\ell_+^2(\mathcal{F}')$. Let $Y_1 : \mathcal{H} \to \ell_+^2(\mathcal{F}')$ be defined by

$$Y_1^* = Y^* | \ell_+^2(\mathcal{F}').$$

Since Y^* is a quasi-affinity Y_1^* is one to one, and assuming the containment negated in (5) we have

$$\overline{\mathrm{ran}(Y_1^*)} = \overline{\mathrm{ran}(Y^*)} = \mathcal{H},$$

so Y_1 is a quasi-affinity as well. Clearly

$$T^* Y_1^* = Y_1^* S_{\mathcal{F}'}^* \qquad \text{so} \qquad Y_1 T = S_{\mathcal{F}'} Y_1.$$

But by [22, Theorem 1] we deduce

$$\mathrm{ind}(S_{\mathcal{F}}) = \mathrm{ind}(T) = \mathrm{ind}(S_{\mathcal{F}'})$$

which is a contradiction. □

We need also the following; the first assertion is from the proof of [25, Lemma 1.2] and the second is a computation.

Lemma 2.4. *Suppose T is an absolutely continuous contraction in C_{10} with Hilbert-Schmidt defect D_T, and suppose $\mathcal{M}$ is some subspace invariant for T. Let $T_1 = T|\mathcal{M}$; then D_{T_1} is Hilbert-Schmidt and $T_1 \in C_{10}$.*

The next result on "splitting" an n-fold analytic co-kernel is essential to what follows; it relies on the fact that for T an absolutely continuous contraction in C_{10} with Hilbert-Schmidt defect D_T, one knows from [25] that $\dim(\ker(T^*)) \ge n$ implies that T has an n-fold analytic co-kernel (the result is implicit in our sketch, after Definition 2.2, of a key construction from that paper).

Proposition 2.5. *Let T be an absolutely continuous contraction acting on $\mathcal{H}$, in C_{10}, with Hilbert-Schmidt defect D_T, and satisfying $\dim(\ker(T^*)) \geq n$. Then there exists a subspace $\mathcal{M}$ invariant for T so that, with $T_1 = T|\mathcal{M}$ and $T_2 = T_{\mathcal{M}^\perp}$ and thus*

$$T = \begin{pmatrix} T_1 & * \\ 0 & T_2 \end{pmatrix},$$

we have

$$\begin{aligned} &\text{(i) } T_1 \in C_{10}, \\ &\text{(ii) } D_{T_1} \text{ is Hilbert-Schmidt}, \\ &\text{(iii) } T_1 \text{ has an } (n-1)\text{-fold analytic co-kernel, and} \\ &\text{(iv) } T_2 \text{ has a 1-fold analytic co-kernel,} \end{aligned} \tag{7}$$

where (iii), *in the case $n = \aleph_0$, means T_1 has an $\aleph_0$-fold analytic co-kernel.*

We remark that it will later turn out, as one might expect, that for n finite if T does not have an $(n+1)$-fold analytic co-kernel then T_1 does not have an n-fold analytic co-kernel.

Proof. Using Lemma 2.4 both (i) and (ii) hold for any subspace $\mathcal{M}$ in $\operatorname{Lat}(T)$, so we work on the other two. We consider only the case n finite, as the modifications for $n = \aleph_0$ are obvious. Adopting the notation above for the Uchiyama construction of an n-fold analytic co-kernel for T, define the subspace $\mathcal{M}_*$ by

$$\mathcal{M}_* = \bigvee_{1 \leq j < \aleph_0} Y^*(e_1^j).$$

From (1) and (4) we have

$$\bigvee_{\lambda \in \mathbf{D}} f_1(\lambda) \subseteq \mathcal{M}_*.$$

Also, it is easy to compute using (2) and (3) that $\mathcal{M}_*$ is invariant for T^*.

Set $\mathcal{M} = (\mathcal{M}_*)^\perp$; $\mathcal{M}$ is clearly invariant for T, and we will show that it is satisfactory for (7-(iii)) and (7-(iv)). It is straightforward to check that $f_1(\cdot)$ provides a 1-fold analytic co-kernel for $T_2 = T_{\mathcal{M}^\perp} = T_{\mathcal{M}_*}$.

We must therefore show (7-(iii)), and citing the result of [25, Theorem 2.3] it is enough to show $\dim(\ker(T_1^*)) \geq n-1$. To ease the notation slightly, let $S = T^*$, $S_1 = T_2^*$, and $S_2 = T_1^*$, so with respect to the decomposition $\mathcal{M}_* \oplus \mathcal{M}$ we have

$$S = \begin{pmatrix} S_1 & * \\ 0 & S_2 \end{pmatrix}.$$

We must show $\dim(\ker(S_2)) \geq n-1$. We shall use the following elementary fact: if $v \in \mathcal{H}$, $Sv \in \mathcal{M}_*$, and $P_{\mathcal{M}}(v) \neq 0$, then $P_{\mathcal{M}}(v)$ is a non-zero vector in $\ker(S_2)$.

We will produce vectors $d_2, \dots, d_n$, recursively, satisfying the following:

$$\begin{aligned}
&(1)\ SY^*d_i \in \mathcal{M}_*,\ \ 2 \le i \le n, \\
&(2)\ P_{\mathcal{M}}(Y^*d_i) \ne 0,\ \ 2 \le i \le n, \\
&(3)\ P_{\mathcal{M}}(Y^*d_2),\ \dots,\ P_{\mathcal{M}}(Y^*d_n) \text{ are independent, and} \\
&(4)\ \text{each } d_i \text{ is a finite linear combination of } \{e_k^j\}_{\substack{1\le j<\aleph_0,\\ 2\le k\le i}},\ 2 \le i \le n.
\end{aligned} \tag{8}$$

Condition (8-1) ensures that each Y^*d_i is in the kernel of S_2, and it is then easy to finish.

Construction of d_2

Consider the sequence of vectors $e_2^1, e_2^2, e_2^3, \dots$. Choose m_0 to be the least m, $0 \le m < \aleph_0$, such that

$$Y^*(e_2^m) \in \mathcal{M}_* \qquad \text{and} \qquad Y^*(e_2^{m+1}) \notin \mathcal{M}_*.$$

(Under the assumption that no such m exists we have

$$Y^*(e_2^j) \subseteq \mathcal{M}_*,\ \ 1 \le j < \aleph_0, \qquad \text{and so} \qquad \bigvee_{1\le j<\aleph_0} Y^*(e_2^j) \subseteq \bigvee_{1\le j<\aleph_0} Y^*(e_1^j),$$

contradicting Lemma 2.3.) Set

$$d_2 = e_2^{m_0+1}.$$

Using $S = T^*$, (2), and (3), it is easy to compute $SY^*d_2 = Y^*S_{\mathcal{F}}^*e_2^{m_0+1} \in \mathcal{M}_*$, which is (8-1), and the rest of (8) is trivial in this case, so we have constructed d_2.

Construction of d_k, some $2 < k \le n$

Suppose then that $d_2, \dots, d_{k-1}$ are in place and we seek d_k. We begin with the sequence of vectors $e_k^1, e_k^2, e_k^3, \dots$. As before, there must be some least m, $(0 \le m < \aleph_0)$ so that

$$Y^*(e_k^m) \in \mathcal{M}_* \qquad \text{and} \qquad Y^*(e_k^{m+1}) \notin \mathcal{M}_*.$$

If $P_{\mathcal{M}}(Y^*e_k^{m+1})$ (of course non-zero) is independent from $P_{\mathcal{M}}(Y^*d_2)$, $P_{\mathcal{M}}(Y^*d_3)$, $\dots$, $P_{\mathcal{M}}(Y^*d_{k-1})$, we are done as before with the choice $d_k = e_k^{m+1}$, computing the various parts of (8) easily.

If $P_{\mathcal{M}}(Y^*e_k^{m+1})$ is in the span of the vectors $P_{\mathcal{M}}(Y^*d_2)$, $P_{\mathcal{M}}(Y^*d_3)$, $\dots$, $P_{\mathcal{M}}(Y^*d_{k-1})$, we may find scalars α_ℓ, $2 \le \ell \le k-1$ so that

$$P_{\mathcal{M}}(Y^*e_k^{m+1}) = \sum_{\ell=2}^{k-1} \alpha_\ell P_{\mathcal{M}} Y^*(d_\ell).$$

Observe that we know about the various e_k^j (so far) that

$$Y^*(e_k^j) \in \mathcal{M}_*,\ \ 1 \le j \le m. \tag{9}$$

Also, via (8-4) for $d_2, \ldots, d_{k-1}$, we have

$$P_{\mathcal{M}}(Y^* e_k^{m+1}) \in \bigvee_{\substack{1 \le j < \aleph_0, \\ 2 \le \ell \le k-1}} Y^*(e_\ell^j),$$

and

$$P_{\mathcal{M}_*}(Y^* e_k^{m+1}) \in \mathcal{M}_* = \bigvee_{1 \le j < \aleph_0} Y^*(e_1^j).$$

Therefore,

$$Y^*(e_k^{m+1}) \in \bigvee \Big(\bigvee_{\substack{1 \le j < \aleph_0, \\ 2 \le \ell \le k-1}} Y^*(e_\ell^j) \cup \mathcal{M}_* \Big) = \bigvee_{\substack{1 \le j < \aleph_0, \\ 1 \le \ell \le k-1}} Y^*(e_\ell^j). \tag{10}$$

Consider now the vector

$$v_k^{m+1} \triangleq e_k^{m+1} - \sum_{\ell=2}^{k-1} \alpha_\ell d_\ell. \tag{11}$$

Note that $P_{\mathcal{M}}(Y^* v_k^{m+1}) = 0$, and also that v_k^{m+1}, and, in fact, each $S_{\mathcal{F}}^p v_k^{m+1}$ for any positive integer p, is a finite linear combination of vectors from $\{e_\ell^j\}_{\substack{1 \le j < \aleph_0, \\ 1 \le \ell \le k}}$.

It is obvious from (8-4) that

$$Y^*(S_{\mathcal{F}}^p d_\ell) \in \bigvee_{\substack{1 \le j < \aleph_0, \\ 2 \le i \le k-1}} Y^*(e_i^j), \quad 2 \le \ell \le k-1,\ 1 \le p < \aleph_0.$$

From this, the definition (11) of v_k^{m+1}, and noting also that $Y^*(S_{\mathcal{F}}^p e_k^{m+1}) = Y^*(e_k^{m+p+1})$ for any p, we have for each p, $1 \le p < \aleph_0$,

$$Y^*(S_{\mathcal{F}}^p v_k^{m+1}) \in \mathcal{M}_* \Rightarrow Y^*(e_k^{m+p+1}) \in \bigvee_{\substack{1 \le j < \aleph_0, \\ 2 \le \ell \le k-1}} Y^*(e_\ell^j). \tag{12}$$

There must then be some least p so that both $Y^*(S_{\mathcal{F}}^p v_k^{m+1}) \in \mathcal{M}_*$ and $Y^*(S_{\mathcal{F}}^{p+1} v_k^{m+1}) \notin \mathcal{M}_*$ since if not, from (9), (10), and (12) we would get

$$\bigvee_{1 \le j < \aleph_0} Y^*(e_k^j) \subseteq \bigvee_{\substack{1 \le j < \aleph_0, \\ 1 \le \ell \le k-1}} Y^*(e_\ell^j),$$

a contradiction of Lemma 2.3.

Again there are two possibilities: in the case $P_{\mathcal{M}}Y^*(S_{\mathcal{F}}^{p+1}v_k^{m+1})$ is independent from $P_{\mathcal{M}}(Y^*d_2)$, ..., $P_{\mathcal{M}}(Y^*d_{k-1})$, $d_k = S_{\mathcal{F}}^{p+1}v_k^{m+1}$ is a satisfactory choice for d_k and we may compute what is needed for 1 - 4 of (8). If $P_{\mathcal{M}}Y^*(S_{\mathcal{F}}^{p+1}v_k^{m+1})$ is not independent from $P_{\mathcal{M}}(Y^*d_2)$, $P_{\mathcal{M}}(Y^*d_3)$, ..., $P_{\mathcal{M}}(Y^*d_{k-1})$, we may modify $S_{\mathcal{F}}^{p+1}v_k^{m+1}$ by a linear combination of $d_2, \ldots, d_{k-1}$ to obtain a vector w so that $P_{\mathcal{M}}Y^*(w) = 0$, etc., and examine the $S_{\mathcal{F}}^p w$ for $p = 1, 2, \ldots$ as before. If this construction does not eventually terminate in a satisfactory d_k we will have the usual contradiction of Lemma 2.3, so we may produce a d_k as desired.

So we may produce the $d_2, \ldots, d_n$ to satisfy (8), deduce $\dim(\ker(T_2^*)) \geq n-1$, and arrive as discussed above at a decomposition satisfying (7), thus completing the proof of the proposition. □

Before we can show various sorts of multiplicities coincide, we need two lemmas.

Lemma 2.6. *Suppose T is an absolutely continuous contraction acting on $\mathcal{H}$ and in C_{10} with Hilbert-Schmidt defect D_T and with* $\dim(\ker(T^*)) = n$. *Let $\tilde{T}$ be a restriction of T to an invariant subspace $\mathcal{M}$; then* $\dim(\ker(\tilde{T}^*)) \leq n$.

PROOF. It suffices to consider n finite. We know from Lemma 2.4 that any such restriction is in C_{10} and has Hilbert-Schmidt defect operator; it is then obvious that for both T and $\tilde{T}$ the dimension of the co-kernel is the negative of the Fredholm index. Clearly

$$\tilde{T} \prec^i T, \qquad \text{and} \qquad T \prec S^{(n)}$$

from [21, Theorem 1], so

$$\tilde{T} \prec^i S^{(n)}.$$

Let X be an injection such that $S^{(n)}X = X\tilde{T}$, and let $\mathcal{L} = \overline{X\mathcal{M}}$. Clearly $\mathcal{L} \in Lat(S^{(n)})$, and if we define $Y : \mathcal{M} \to \mathcal{L}$ to be the restriction of X to $\mathcal{M}$ then Y is a quasi-affinity. Further, setting $\tilde{S}^{(n)} = S^{(n)}|\mathcal{L}$, one has

$$\tilde{S}^{(n)}Y = Y\tilde{T}.$$

But the restriction $\tilde{S}^{(n)}$ is a shift of some multiplicity $k \leq n$, and then from $\tilde{T} \prec S^{(k)}$ and [22, Theorem 1] we have

$$\dim(\ker(\tilde{T}^*)) = -ind(\tilde{T}) = k \leq n. \qquad \square$$

Lemma 2.7. *Suppose T is an absolutely continuous contraction in C_{10} with Hilbert-Schmidt defect D_T and with* $\dim(\ker(T^*)) = n$, $1 \leq n \leq \aleph_0$. *Then* $R_*(T) = B^{(n)}$.

PROOF. We know from [24, Theorem 2] that

$$T \prec S^{(n)}. \tag{13}$$

We know from [16, Section 1.1] that there is a map X_+ intertwining T and a certain isometry $T_+^{(a)}$, and that X_+ has dense range. Also from that paper (page 176) is the fact that the minimal unitary extension of $T_+^{(a)}$ is $R_*(T)$, and that since $T \in C_{10}$, X_+ is one-to-one. Thus X_+ is a quasi-affinity, and

$$T \prec T_+^{(a)}. \tag{14}$$

Assume that $T_+^{(a)}$ is of the form $S^{(m)} \oplus U$, where U is unitary; we will show U is in fact absent. By [22, Theorem 1] we have from (13) and (14) respectively that

$$T \sim^{ci} S^{(n)} \qquad \text{and} \qquad T \sim^{ci} S^{(m)} \oplus U.$$

Since $\sim^{ci}$ is an equivalence relation, these yield

$$S^{(m)} \oplus U \sim^{ci} S^{(n)}, \qquad \text{and thus} \qquad S^{(m)} \oplus U \prec^{ci} S^{(n)}.$$

Now of course $U \prec^{i} S^{(m)} \oplus U$, so finally

$$U \prec^{i} S^{(n)}.$$

An argument like that in the proof of Lemma 2.6 shows that this leads to the contradiction U has kernel, so U must be absent.

Then $T_+^{(a)} = S^{(m)}$ for some m, and yet another repetition of the argument, using $S^{(m)} \prec^{i} S^{(n)}$ and $S^{(n)} \prec^{i} S^{(m)}$, shows that contradictions involving the dimensions of the co-kernels arise unless $m = n$. Thus $T_+^{(a)} = S^{(n)}$, so of course $R_*(T)$, the minimal unitary extension of $S^{(n)}$, is $B^{(n)}$ as desired. □

Recall next a result which is a combination of [25, Theorem 2.1 and Proposition 2.5].

Theorem 2.8. *Let T be a contraction with Hilbert-Schmidt defect operator D_T. Then $T \in C_{10}$ if and only if T has a full p-fold analytic co-kernel for some p. In this case, $p = -\operatorname{ind}(T)$.*

We have the following, where the equivalence of (i), (ii), and (iii) is due to Uchiyama ([25], especially Proposition 2.5).

Theorem 2.9. *Suppose T is an absolutely continuous contraction acting on $\mathcal{H}$ and in C_{10} with Hilbert-Schmidt defect D_T. Then the following are equivalent for each positive integer n:*

(i)$_n$ $\dim(\ker(T^*)) \geq n$, (15)
(ii)$_n$ *T has a full k-fold analytic co-kernel for some $k \geq n$,*
(iii)$_n$ *T has an n-fold analytic co-kernel,*
(iv)$_n$ $T \in \mathbb{A}_{n,\aleph_0}$,
(v)$_n$ $T \in \mathbb{A}_{n,n}$,
(vi)$_n$ *T dilates 0_n,*
(vii)$_n$ $R_*(T) = B^{(k)}$ *for some $k \geq n$,*
(viii)$_n$ *T has a restriction $\tilde{T}$ satisfying any of the conditions* i)$_n$–vii)$_n$.

PROOF. If all but (vii)$_n$ and the appearance of its condition in (viii)$_n$ are shown to be equivalent, then Lemma 2.7 will show that these are equivalent to the rest. That (i)$_n$, (ii)$_n$ and (iii)$_n$ are equivalent is from [25], (iv)$_n$ implies (v)$_n$ is direct from the definitions, and (v)$_n$ implies (vi)$_n$ is [3, Cor. 4.14]. If T dilates 0_n, then clearly T has a restriction with co-kernel of dimension at least n. If some restriction $\tilde{T}$ of T dilates 0_n, then by Lemmas 2.6 and 2.4 we have $\dim(\ker(\tilde{T}^*)) \geq n$ and therefore $\dim(\ker(T^*)) \geq n$, so we return to (i)$_n$.

To conclude the proof it remains to show that (iii)$_n$ implies (iv)$_n$. Consider first the case in which $\dim(\ker(T^*)) = k \geq n$ with $k < \aleph_0$. The equivalence of (i)$_k$ and iii)$_k$ and a recursive use of Proposition 2.5 shows that T has, with respect to some decomposition $\mathcal{H} = \mathcal{M}_1 \oplus \mathcal{M}_2 \oplus \ldots \oplus \mathcal{M}_k$, an upper triangular matrix form

$$T = \begin{pmatrix} T_1 & & & \\ & T_2 & * & \\ & & \ddots & \\ & 0 & & T_k \end{pmatrix}, \qquad \text{where} \quad T_i \triangleq T_{\mathcal{M}_i}, \quad 1 \leq i \leq k. \tag{16}$$

Further, each T_i has a 1-fold analytic co-kernel, and therefore is in $\mathbb{A}_{1,\aleph_0}$ by the result previously mentioned from [9]. But now from [4], [8] and [7] it follows that $T \in \mathbb{A}_{k,\aleph_0}$. The idea is observe that the multiplicity of the unitary piece of the minimal coisometric extension of each T_i is at least one on $\mathbf{T} \setminus X_{T_i}$ using [8, Theorem 4.6] (see the questions at the end of this paper for a definition of the set X_S for an absolutely continuous contraction S). One may then deduce from the dual of [4, Theorem 1.7] that the multiplicity of the unitary piece of the minimal coisometric extension of T is at least k on $\mathbf{T} \setminus (\bigcup X_{T_i})$. It follows from [8, Proposition 3.5 a)] that multiplicity of the unitary piece of the minimal coisometric extension of T is at least k on $\mathbf{T} \setminus X_T$, and then from [7] that $T \in \mathbb{A}_{k,\aleph_0}$). Since $k \geq n$, $T \in \mathbb{A}_{n,\aleph_0}$, and this completes the proof if the dimension of $\ker(T^*)$ is finite.

To show that (iii)$_n$ implies (iv)$_n$ if $\dim(\ker(T^*)) = \aleph_0$, observe that a recursive use of Proposition 2.5 shows in this case that for any $k < \aleph_0$, T has a decomposition

$$T = \begin{pmatrix} T_0 & & & \\ & T_k & * & \\ & & \ddots & \\ & 0 & & T_1 \end{pmatrix}, \tag{17}$$

where each T_i, $1 \leq i \leq k$, has a 1-fold analytic co-kernel, and therefore is in $\mathbb{A}_{1,\aleph_0}$. By the argument sketched above, we may deduce that $T \in \mathbb{A}_{k,\aleph_0}$. Thus T is in the intersection of all the $\mathbb{A}_{k,\aleph_0}$, and hence in $\mathbb{A}_{\aleph_0}$, so surely in $\mathbb{A}_{n,\aleph_0}$. □

It should be no surprise that there is an $n = \aleph_0$ version of the theorem; its proof may be obtained from the one above using standard facts like $\mathbb{A}_{\aleph_0} = \bigcap_{1 \leq n < \aleph_0} \mathbb{A}_n$, so we omit it.

Theorem 2.10. *Suppose T is an absolutely continuous contraction acting on $\mathcal{H}$ and in C_{10} with Hilbert-Schmidt defect D_T. Then the following are equivalent:*

$$\begin{aligned}
&\text{(i)}\ \dim(\ker(T^*)) = \aleph_0, \\
&\text{(ii)}\ T \text{ has a full } \aleph_0\text{-fold analytic co-kernel,} \\
&\text{(iii)}\ T \text{ has an } \aleph_0\text{-fold analytic co-kernel,} \\
&\text{(iv)}\ T \text{ has, for each } n \in \mathbb{N}, \text{ an } n\text{-fold analytic co-kernel,} \\
&\text{(v)}\ T \in \mathbb{A}_{\aleph_0}, \\
&\text{(vi)}\ T \text{ dilates } 0_{\aleph_0}, \\
&\text{(vii)}\ R_*(T) = B^{(\aleph_0)}, \\
&\text{(viii)}\ T \text{ has a restriction } \tilde{T} \text{ satisfying any of the conditions in (i)–(vii).}
\end{aligned} \tag{18}$$

Remarks

By virtue of the equivalences above, it is easy to see that Theorem 2.9 could be rephrased in forms like '$\dim(\ker(T^*)) = n$ if and only if $T \in \mathbb{A}_{n,\aleph_0} \backslash \mathbb{A}_{n+1,\aleph_0}$'. Thus the results amount to the assertion, for absolutely continuous contractions T in C_{10} with Hilbert-Schmidt defect D_T, that the 'multiplicity' measures $\dim(\ker(T^*))$, size of analytic co-kernel, subscript of maximal dual algebra class(es), size of zero operator dilatable, and multiplicity of $R_*(T)$ all coincide. It includes as well a characterization of membership for these operators in the classes $\mathbb{A}_{n,\aleph_0}$ or $T \in \mathbb{A}_{n,n}$. It seems worth remarking that both such characterizations and results of the form 'class information yields multiplicity information' are fairly rare in the theory of dual operator algebras.

We may as well here also dispose of a trivial consideration. Based on what happens in the case D_T is finite rank, one might suspect that D_{T^*} plays some role in the case D_T is Hilbert-Schmidt (see, for example, [13], where the common value n similar to that in Theorem 2.9 is $d_{T^*} - d_T$, where $d_T = \dim\{(I - T^*T)^{\frac{1}{2}}\mathcal{H}\}^-$). It does, but the role does not yield much new information; after we recall a lemma, this is evident from the subsequent proposition. The following is [25, Lemma 1.1].

Lemma 2.11. *Let Y be a bounded operator and F a Fredholm operator such that FY is trace-class. Then Y is trace-class.*

Proposition 2.12. *Suppose T is a contraction in $C_{1\cdot}$. If D_{T^*} is Hilbert-Schmidt, then*

$$\begin{aligned}
&\text{(i)}\ \dim(\ker(T^*)) < \aleph_0, \\
&\text{(ii)}\ T \text{ is Fredholm, and} \\
&\text{(iii)}\ D_T \text{ is Hilbert-Schmidt.}
\end{aligned} \tag{19}$$

If D_{T^*} is not Hilbert-Schmidt and D_T is Hilbert-Schmidt, then

$$\begin{aligned}
&\text{(i)}\ \dim(\ker(T^*)) = \aleph_0, \\
&\text{(ii)}\ T \text{ is semi-Fredholm, and} \\
&\text{(iii)}\ T \in \mathbb{A}_{\aleph_0}.
\end{aligned} \tag{20}$$

PROOF. Consider the assertions in (19); (i) is a computation and (ii) then follows from the definitions. For iii), recall that

$$D_{T^*}^2 T = T D_T^2.$$

Then $D_{T^*}^2$ Hilbert-Schmidt implies $D_{T^*}^2 T$ Hilbert-Schmidt, so TD_T^2 is Hilbert-Schmidt, and so D_T^2 is Hilbert-Schmidt, where the last implication uses T Fredholm and Lemma 2.11.

For the assertions in (20), suppose first that $\dim(\ker(T^*)) < \aleph_0$. Then T and hence T^* are Fredholm, and a computation using the intertwining of D_T^2 and $D_{T^*}^2$ like that just made produces a contradiction of D_{T^*} not Hilbert-Schmidt. Then (20-(ii)) follows from the definitions and (20-(iii)) from Theorem 2.10. □

With another lemma we may gain a little more information about the matrix form obtained in (16). The following is based on computations from [25, Lemmas 1.1 and 1.4].

Lemma 2.13. *Suppose T is an absolutely continuous contraction acting on $\mathcal{H}$ and in C_{10} with Hilbert-Schmidt defect D_T. Let $\mathcal{M}$ be some subspace invariant for T, and decompose T with respect to $\mathcal{M} \oplus \mathcal{M}^\perp$ as*

$$T = \begin{pmatrix} T_1 & F \\ 0 & T_2 \end{pmatrix}.$$

Then if T_1 is Fredholm, and, in particular, if $\dim(\ker(T^*)) = n < \infty$, D_{T_2} is Hilbert-Schmidt.

PROOF. By considering $I - T^*T$, clearly $I - T_1^*T_1$, T_1^*F and $I - (F^*F + T_2^*T_2)$ are all trace-class. If T_1 is assumed Fredholm we finish directly by Lemma 2.11. And if $\dim(\ker(T^*)) = n < \infty$, citing Lemmas 2.4 and 2.6 it is easy to deduce T_1^* is Fredholm. □

We then have the following decompositions, which seem in some sense best possible. Note that the second removes the apparently separate cases of the decomposition obtained from Proposition 2.5 used in the proof of $(\text{iii})_n$ implies $\text{iv})_n$ of Theorem 2.9.

Corollary 2.14. *Let T be an absolutely continuous contraction in C_{10} with Hilbert-Schmidt defect D_T and with $\dim(\ker(T^*)) = n$, n a positive integer. Then T has an operator matrix of the form*

$$T = \begin{pmatrix} T_1 & & & \\ & T_2 & * & \\ & & \ddots & \\ & 0 & & T_n \end{pmatrix}, \tag{21}$$

where (i) $T_i \in C_{10}$, $1 \leq i \leq n$,
(ii) D_{T_i} *is Hilbert-Schmidt*, $1 \leq i \leq n$,
(iii) $\dim(\ker(T_i^*)) = 1$, $1 \leq i \leq n$, *and*
(iv) T_i *has a full* 1*-fold analytic co-kernel*, $1 \leq i \leq n$.

PROOF. Remark in passing that (iv) is equivalent to the assertion $\mathcal{M}_i$ is a full analytic invariant subspace for T_i, $1 \leq i \leq n$, in the language of [9]. As for the proof, we produced some such upper triangular form in the proof of Theorem 2.9, and it remains to check its features. By citing Lemma 2.13 repeatedly during the construction, we obtain (ii). For $T_2, \ldots, T_n$, observe that the construction shows that each has a full 1-fold analytic co-kernel; by Theorem 2.8, each is necessarily in C_{10}. Then by the theorem, $\dim(\ker(T_i^*)) = 1$, $2 \leq i \leq n$.

From Lemma 2.4 we have $T_1 \in C_{10}$ and D_{T_1} Hilbert-Schmidt, and thus via Theorem 2.8 again that T_1 has a full p-fold analytic co-kernel for some p (necessarily finite by the theorem). But from either $\dim(\ker(T_1^*)) > 1$ or $p > 1$ we may deduce from the theorem that $T \in \mathbb{A}_{n+1,\aleph_0}$, which is a contradiction as observed in the remark following that theorem. □

Corollary 2.15. *Let* T *be an absolutely continuous contraction in* $\mathcal{L}(\mathcal{H})$, *in* C_{10}, *with Hilbert-Schmidt defect* D_T *and with* $\dim(\ker(T^*)) = \aleph_0$. *Then* T *has an operator matrix of the form*

$$T = \begin{pmatrix} T_1 & & & \\ & T_2 & * & \\ & & \ddots & \\ & 0 & T_n & \\ & & & \ddots \end{pmatrix}, \tag{22}$$

where (i) $T_i \in C_{10}$, $1 \leq i < \aleph_0$,
(ii) D_{T_i} *is Hilbert-Schmidt*, $1 \leq i < \aleph_0$,
(iii) $\dim(\ker(T_i^*)) = 1$, $1 \leq i < \aleph_0$, *and*
(iv) T_i *has a full* 1*-fold analytic co-kernel*, $1 \leq i < \aleph_0$.

PROOF. The result is almost contained in [7], in which is produced, for any $T \in \mathbb{A}_{\aleph_0}$, a decomposition satisfying (iv); we sketch the construction and deduce along the way additional information in this special case. We will use repeatedly the fact from [9, Theorem 6.2] that an operator in $\mathbb{A}_{1,\aleph_0}$ has a dense set of vectors each of which generates a cyclic full analytic invariant subspace (in our terminology, a (cyclic) subspace so the restriction to that subspace has a full 1-fold analytic co-kernel). Under our assumptions we have $T \in \mathbb{A}_{\aleph_0}$. Let $\{z_i\}_{i=1}^{\infty}$ be some collection of non-zero vectors dense in $\mathcal{H}$ and $\{\epsilon_i\}_{i=1}^{\infty}$ some sequence of positive numbers tending to zero.

We may choose some subspace $\mathcal{M}_1$ invariant for T and so that the restriction $T_1 \triangleq T|\mathcal{M}_1$ has a full 1-fold analytic co-kernel, and further so that $\|(I-P_{\mathcal{M}_1})z_1\| < \epsilon_1$. Write T in the decomposition

$$T = \begin{pmatrix} T_1 & * \\ 0 & T_1^{\perp} \end{pmatrix}$$

with respect to $\mathcal{M}_1 \oplus \mathcal{N}_1$, where $\mathcal{N}_1 \triangleq \mathcal{M}_1^{\perp}$. Of course T_1 has Hilbert-Schmidt defect operator by Lemma 2.4; by Theorem 2.8 it is then in C_{10}. Also, T_1 has a 1-fold analytic co-kernel and satisfies $\dim(\ker(T_1^*)) = 1$, and so satisfies what is required for $i = 1$ in (i)-(iv). Note for future use that also T_1 is Fredholm.

We claim that $T_1^{\perp} \in \mathbb{A}_{\aleph_0}$. Observe that from Lemma 2.7 we have $R_*(T) = B^{(\aleph_0)}$ and $R_*(T_1) = B^{(1)}$. We may then cite the dual of [4, Theorem 1.7] to deduce that $R_*(T_1^{\perp}) = B^{(\aleph_0)}$, and [18, Corollaire 4.7] to finish.

We turn to the construction of $\mathcal{M}_2$. We may choose some subspace $\mathcal{M}_2$ of $\mathcal{N}_1$ invariant for $T_1^{\perp}$ and so that the restriction $T_2 \triangleq T_1^{\perp}|\mathcal{M}_2$ has a full 1-fold analytic co-kernel, and further so that $\|(I - P_{\mathcal{M}_2})P_{\mathcal{M}_1^{\perp}} z_1\| < \epsilon_2$. Clearly then T has the decomposition

$$T = \begin{pmatrix} T_1 & & * \\ & T_2 & \\ 0 & & T_2^{\perp} \end{pmatrix},$$

with respect to $\mathcal{M}_1 \oplus \mathcal{M}_2 \oplus \mathcal{N}_2$ where $\mathcal{N}_2 \triangleq (\mathcal{M}_1 \oplus \mathcal{M}_2)^{\perp}$. Clearly T_2 satisfies what is required for $i = 2$ of (i), (iii), and (iv); for (ii), use T_1 Fredholm and apply Lemma 2.13 to

$$\begin{pmatrix} T_1 & * \\ 0 & T_2 \end{pmatrix}.$$

Note that (via Theorem 2.8 and a computation) $T|(\mathcal{M}_1 \oplus \mathcal{M}_2)$ is Fredholm.

Repetition of this argument, and the use of the vectors $\{z_i\}_{i=1}^{\infty}$ in the order $z_1, z_1, z_2, z_1, z_2, z_3, \ldots$ to ensure $\bigvee \mathcal{M}_i = \mathcal{H}$ yields the decomposition desired. □

Observe that easy examples with shifts show that not every upper triangular decomposition of a C_{10} operator T with D_T Hilbert-Schmidt has these properties. Note also that a proof along the lines of that for Corollary 2.15 works as well to obtain Corollary 2.14, but uses noticeable extra machinery. A proof for Corollary 2.15 relying essentially on a decomposition of an $\aleph_0$-fold co-kernel would be interesting. Finally, in the case $n = \aleph_0$, it is not possible to deduce, for general $T \in \mathbb{A}_{\aleph_0}$ with a decomposition as in Corollary 2.15 but with only (iv) assumed, that $T \in C_{\cdot 0}$, as is shown by $B^{(\aleph_0)}$. For n finite, from the decomposition as in Corollary 2.14 but with only (iv) assumed it is possible to deduce $T \in C_{\cdot 0}$, as is shown by an easy computation.

3. $C_{1\cdot}$ AND $C_{\cdot 0}$ OPERATORS WITH HILBERT-SCHMIDT DEFECT OPERATOR

In this section we explore some consequences for $C_{1\cdot}$ and $C_{\cdot 0}$ operators with Hilbert-Schmidt defect operator of the results obtained in Section 2. Since C_{10} operators occur in the upper triangular decompositions of each of the more general types (see [20, Equations II.4.6 and II.4.7]) it is not surprising that structure theorems carry over, although there are substantial differences between the two cases.

We have the following for the $C_{1\cdot}$ case; the results again constitute a characterization, and imply that for these operators a number of different multiplicity measures coincide.

Theorem 3.1. *Suppose T is a contraction in $C_{1\cdot}$ with D_T a Hilbert-Schmidt operator. Then the following are equivalent for each n, $1 \leq n < \aleph_0$.*

$$\begin{aligned}
&(\mathrm{i})_n \quad T \textit{ has a restriction with a full } n\textit{-fold analytic co-kernel,} \qquad (23)\\
&(\mathrm{ii})_n \quad T \textit{ has a restriction with an } n\textit{-fold analytic co-kernel,}\\
&(\mathrm{iii})_n \quad T \in \mathbf{A}_{n,\aleph_0},\\
&(\mathrm{iv})_n \quad T \in \mathbf{A}_{n,n},\\
&(\mathrm{v})_n \quad T \textit{ dilates } 0_n,\\
&(\mathrm{vi})_n \quad R_*(T) \textit{ contains } B^{(n)},\\
&(\mathrm{vii})_n \quad T \textit{ has a compression } \tilde{T} \textit{ in } C_{10} \textit{ with } D_{\tilde{T}} \textit{ Hilbert-Schmidt}\\
&\qquad\qquad \textit{satisfying any of the conditions } (\mathrm{i})_n\text{–}(\mathrm{viii})_n \textit{ in Theorem 2.9.}
\end{aligned}$$

Further, for each n, the following condition implies each of $(\mathrm{i})_n$–$(\mathrm{vii})_n$ *in* (23), *but not conversely:*

$$(\mathrm{viii})_n \quad T \textit{ has an } n\textit{-fold analytic co-kernel.} \qquad (24)$$

PROOF. It is obvious that $(\mathrm{viii})_n$ implies $(\mathrm{ii})_n$ and the bilateral shift of multiplicity n shows that the reverse implication need not hold. Supposing the equivalences excluding $(\mathrm{vi})_n$ shown, note that $(\mathrm{vi})_n$ is equivalent to the others since from the dual of [4, Theorem 1.7] it follows that if some compression $\tilde{T}$ of T satisfies $R_*(\tilde{T})$ contains $B^{(n)}$ then $R_*(T)$ does, and from [18, Theoreme 4.5] that if T satisfies $(\mathrm{vi})_n$ then $T \in \mathbf{A}_n$. In the chain of implications $(\mathrm{i})_n \Rightarrow (\mathrm{ii})_n \Rightarrow \ldots \Rightarrow (\mathrm{v})_n \Rightarrow (\mathrm{vii})_n \Rightarrow (\mathrm{i})_n$ only $(\mathrm{ii})_n \Rightarrow (\mathrm{iii})_n$, $(\mathrm{v})_n \Rightarrow (\mathrm{vii})_n$, and $(\mathrm{vii})_n \Rightarrow (\mathrm{i})_n$ are non-trivial, so we consider these.

$(\mathrm{ii})_n \Rightarrow (\mathrm{iii})_n$: Suppose $\mathcal{M}$ is the subspace of the restriction guaranteed in $(\mathrm{ii})_n$, and let $\tilde{T} = T|\mathcal{M}$. Then $\tilde{T}$ is in $C_{1\cdot}$ and $D_{\tilde{T}}$ is Hilbert-Schmidt by Lemma 2.4. Consider $\tilde{T}$ in its canonical triangulation

$$\begin{pmatrix} C_{\cdot 1} & * \\ 0 & C_{\cdot 0} \end{pmatrix}$$

from [20, Theorem II.4.1], say,

$$\tilde{T} = \begin{pmatrix} \tilde{T}_1 & * \\ 0 & \tilde{T}_2 \end{pmatrix}$$

with $\tilde{T}_1$ acting on $\mathcal{H}_1$ and $\tilde{T}_2$ acting on $\mathcal{H}_2$. By [25, Lemma 1.4] we have $\tilde{T}_1 \in C_{11}$, $D_{\tilde{T}_1}$ Hilbert-Schmidt and $\tilde{T}_2 \in C_{10}$, $D_{\tilde{T}_2}$ Hilbert-Schmidt. Since $\tilde{T}$ had an n-fold analytic co-kernel, $\dim(\ker(\tilde{T}^*)) \geq n$. An easy calculation using $\tilde{T}_1$ in C_{11} shows that $\dim(\ker(\tilde{T}_2^*)) \geq n$. Citing Theorems 2.9 or Theorem 2.10 (respectively as $\dim(\ker(\tilde{T}_2^*)) < \aleph_0$ or $\dim(\ker(\tilde{T}_2^*)) = \aleph_0$), we get $\tilde{T}_2$ and hence T in $\mathbb{A}_{n,\aleph_0}$.

$(\mathrm{v})_n \Rightarrow (\mathrm{vii})_n$: the argument to produce the desired compression with some p-fold analytic co-kernel, $p \geq n$, is a small variation of the one just given, so we omit it.

$(\mathrm{vii})_n \Rightarrow (\mathrm{i})_n$: the assumption, and Theorem 2.9 or Theorem 2.10 as appropriate, show that T has a restriction T' in $\mathbb{A}_{n,\aleph_0}$ and with $D_{T'}$ Hilbert-Schmidt via Lemma 2.4. We use again the result from [7] employed in the proof of Corollary 2.15. From that work T' has a decomposition

$$T' = \begin{pmatrix} T_1' & & & \\ & T_2' & * & \\ & & \ddots & \\ & 0 & & T_n' \end{pmatrix} \tag{25}$$

with respect to some subspaces $\mathcal{M}_i$, $i = 1, 2, \ldots, n$, and where $T_i' \triangleq T'_{\mathcal{M}_i}$, $i = 1, \ldots, n$; further, the decomposition has the property

$$T_i' \in \mathbb{A}_{1,\aleph_0}, \quad 1 \leq i \leq n.$$

The argument used in the proof of Corollary 2.15 (repeated, however, but n times) shows that T' has a decomposition

$$T' = \begin{pmatrix} \hat{T}_1 & & & & \\ & \hat{T}_2 & & * & \\ & & \ddots & & \\ & 0 & & \hat{T}_n & \\ & & & & \hat{T}_{n+1} \end{pmatrix} \tag{26}$$

with respect to some subspaces $\mathcal{N}_i$, $i = 1, 2, \ldots, n, n+1$, and where $\hat{T}_i \triangleq \hat{T}_{\mathcal{N}_i}$, $i = 1, \ldots, n, n+1$, and with the additional property that $\mathcal{N}_i$ is a full analytic invariant subspace for $\hat{T}_i$, $i = 1, 2, \ldots, n$. It follows from [10, Prop. 2.8] that

$$\hat{T}_i \in C_{\cdot 0}, \quad i = 1, 2, \ldots, n. \tag{27}$$

Finally, set $\mathcal{N} = \bigvee_{1 \leq i \leq n} \mathcal{N}_i$; it is clear that $\mathcal{N}$ is invariant for T', and hence for T. Let $\hat{T} = T|\mathcal{N}$. The usual application of Lemma 2.4 shows that $\hat{T} \in C_{1\cdot}$ and has Hilbert-Schmidt defect, and a computation using (27) shows that $\hat{T} \in C_{\cdot 0}$. From [7] and (26) we deduce $\hat{T} \in \mathbb{A}_{n,\aleph_0}$, and then from Theorem 2.9 and [25, Prop. 2.5] that $\hat{T}$ has a full p-fold analytic co-kernel for some $p \geq n$.

By repeated use of Lemma 2.13 we may show that each of the $\hat{T}_i$, $1 \leq i \leq n$, has Hilbert-Schmidt defect. It then follows from Theorem 2.8 that each of the $\hat{T}_i$, $1 \leq i \leq n$, is in C_{10}. Since we are now in the C_{10} case with Hilbert-Schmidt defect for both $\hat{T}$ and each of the $\hat{T}_i$, we deduce from Lemma 2.7 and the dual of [4, Theorem 1.7] again that necessarily $p = n$. □

There is a version of the theorem above for $n = \aleph_0$, with one change. We omit the proof, which is a straightforward modification of that just given.

Theorem 3.2. *Suppose T is a contraction in $C_{1\cdot}$ with D_T a Hilbert-Schmidt operator. Then the following are equivalent.*

(i) *T has a restriction with an $\aleph_0$-fold analytic co-kernel,* (28)
(ii) $T \in \mathbb{A}_{\aleph_0}$,
(iii) *T dilates $0_{\aleph_0}$,*
(iv) $R_*(T) = B^{(\aleph_0)}$,
(v) *T has a compression $\tilde{T}$ in C_{10} with $D_{\tilde{T}}$ Hilbert-Schmidt satisfying any of the conditions in Theorem 2.10.*

Further, the following condition implies each of those in (28), but not conversely:

(vi) *T has an $\aleph_0$-fold analytic co-kernel.* (29)

Finally, the following implies each of those in (28):

(vii) *T has a restriction with a full $\aleph_0$-fold analytic co-kernel.* (30)

We do not know whether the condition in (30) follows from those in (28); the difficulty is that one may not, as observed in the remarks following Corollary 2.15, deduce in general from an operator matrix even with each diagonal entry in $C_{\cdot 0}$ that the operator as a whole is in that class. It is evident that the following weaker condition is equivalent to the others:

(viii) *For each n, $1 \leq n < \aleph_0$, T has a restriction with a full n-fold analytic co-kernel.* (31)

Recall from Theorem 2.10 that (30) and (31) are equivalent in the C_{10} case.

We now turn to operators in the class $C_{\cdot 0}$. As for those in the class $C_{1\cdot}$ discussed above, there are C_{10} terms in an upper triangular decomposition, so Theorem 2.9 is relevant. But whereas a C_{11} term can make a contribution toward membership in some class $\mathbb{A}_{n,\aleph_0}$ (the bilateral shift, for example), a C_{00} operator with Hilbert-Schmidt defect cannot. One way to see this is that a C_{00} operator with Hilbert-Schmidt defect is in the class C_0 (see [20, Section III.4]) and such operators do not make such contributions (they can't even be in the class $\mathbb{A}$). Another way is to note that $C_{00} \cap \mathbb{A} \subseteq \mathbb{A}_{\aleph_0}$, and so such a contribution must be "all or nothing". But a little computation with dilations shows that the additional condition Hilbert-Schmidt defect makes "all" impossible. Thus the addition of a C_{00} term (which can contribute to co-kernel or dilation multiplicity) doesn't change the dual operator algebra class. On the other hand, since an operator in C_{00} has no unitary part of its minimal co-isometric extension, via [4, Theorem 1.7] it can make no contribution to $R_*(T)$ of some T for which it is a compression, so the results for multiplicity count of $B^{(n)}$ in $R_*(T)$ are sharpened slightly.

We require some computational lemmas; in the first lemma, the first assertion is from [23, Theorem 1], and the second from [12, Chapter 5].

Lemma 3.3. *Suppose T is a contraction with Hilbert-Schmidt defect. If $T \in C_{00}$, then $T \in C_0$ and $\sigma(T) \cap \mathbb{D} \neq \mathbb{D}$; further, $\dim(\ker(T^*)) < \aleph_0$. If $T \in C_{1\cdot}$, then $\dim(\ker(T-\lambda)^*))$ is constant on $\mathbb{D}$.*

Lemma 3.4. *Suppose T is a contraction and D_T is Hilbert-Schmidt. Then $\sigma_{le}(T) \cap \mathbb{D} = \emptyset$.*

PROOF. Suppose not, and let $\lambda \in \sigma_{le}(T) \cap \mathbb{D}$. Then there exists an orthonormal sequence $\{x_n\}_{n=1}^{\infty}$ such that

$$\|(T-\lambda)x_n\| \to 0.$$

Thus there clearly exist $\delta > 0$ and $N \in \mathbb{N}$ so that

$$\|Tx_n\| \leq 1-\delta, \quad n \geq N.$$

A contradiction of D_T Hilbert-Schmidt follows easily. □

Lemma 3.5. *Suppose T is a contraction and D_T is Hilbert-Schmidt. Then either*

(1) *$\mathbb{D}$ is a hole in $\sigma_e(T)$ and $\operatorname{ind}(T-\lambda) \equiv 0$ on $\mathbb{D}$, or* (32)
(2) *$\mathbb{D}$ is a "pseudo-hole" in $\sigma_e(T)$, $\mathbb{D} \subseteq \sigma_{re}(T)$, and $\operatorname{ind}(T-\lambda) \equiv -\infty$ on $\mathbb{D}$.*

PROOF. If there were $\lambda_1 \in \mathbb{D} \setminus \sigma_e(T)$ and $\lambda_2 \in \sigma_e(T) \cap \mathbb{D}$, clearly $\partial\sigma_e(T) \cap \mathbb{D} \neq \emptyset$. But since $\partial\sigma_e(T) \subseteq \sigma_{le}(T)$ this contradicts Lemma 3.4. The index assertion in (2) follows from [19, Propositions 1.20 and 1.21]. □

The following gives characterizations, including multiplicity ones, for membership of $C_{\cdot 0}$ operators in the various $\mathbb{A}_{n,\aleph_0}$. There is as well a modification of this theorem for $n = \aleph_0$ which we leave to the interested reader.

Theorem 3.6. *Suppose T is a contraction acting on $\mathcal{H}$, in $C_{\cdot 0}$, D_T is Hilbert-Schmidt, and T has the decomposition*

$$T = \begin{pmatrix} T_0 & * \\ 0 & T_1 \end{pmatrix}$$

where T_0 acting on $\mathcal{H}_0$ is in C_{00} and T_1 acting on $\mathcal{H}_1$ is in C_{10}. Consider the following conditions, one for each positive integer n:

$$\begin{aligned}
&(\mathrm{i})_n \;\; T \textit{ has an } n\textit{-fold analytic co-kernel,} \\
&(\mathrm{ii})_n \;\; T \in \mathbb{A}_{n,\aleph_0}, \\
&(\mathrm{iii})_n \;\; T \in \mathbb{A}_{n,n}, \\
&(\mathrm{iv})_n \;\; R_*(T) = B^{(k)} \textit{ for some } k \geq n, \\
&(\mathrm{v})_n \;\; T \textit{ has a restriction } \tilde{T} \textit{ satisfying any of the conditions (33)} \\
&\qquad (\mathrm{i})_n\text{–}(\mathrm{iv})_n,
\end{aligned} \tag{33}$$

and

$$\begin{aligned}
&(\mathrm{vi})_n \;\; T_1 \in \mathbb{A}_{n,\aleph_0} \textit{ (equivalently, } T_1 \textit{ satisfies any of the conditions} \\
&\qquad (\mathrm{i})_n\text{–}(\mathrm{viii})_n \textit{ in Theorem 2.9).}
\end{aligned} \tag{34}$$

Then the conditions (33) $(\mathrm{i})_n$-$(\mathrm{v})_n$ are equivalent. Further, (33) and (34) are equivalent, in the sense that T satisfies any (hence all) of (33) $(\mathrm{i})_n$–$(\mathrm{v})_n$ if and only if $T_1 \in \mathbb{A}_{n,\aleph_0}$. Finally, any of these imply $\dim(\ker(T^*)) \geq n$*, but not conversely.*

Proof. The assertion concerning $\dim(\ker(T^*)) \geq n$ is obvious. In view of the equivalence in Theorem 2.9, and that (34) yields any of the (33) *i)*$_n$–$(\mathrm{v})_n$ is immediate (needed only is the result following from [4, Theorem 1.7] that $R_*(T) = R_*(T_1) \oplus R_*(T_0)$).

That some of (33 $(\mathrm{i})_n$–$(\mathrm{v})_n$ yield things equivalent to (34) is obvious. In particular, $T \in \mathbb{A}_{n,\aleph_0}$ implies $T_1 \in \mathbb{A}_{n,\aleph_0}$ from the techniques of [7] and since $T_0 \in C_0$ citing Lemma 3.3. We will use this implication to prove the rest; observe that each of the conditions in (33) implies that T dilates $\lambda \cdot I_n$ for each $\lambda \in \mathbb{D}$ (to show that $(\mathrm{iv})_n$ will do this one needs [18, Theoreme 4.5]). We shall show that this dilation ability alone implies $T_1 \in \mathbb{A}_{n,\aleph_0}$, which is easily seen to complete the remainder of the proof.

Since $\sigma(T_0) \cap \mathbb{D} \neq \mathbb{D}$ via Lemma 3.3, we may choose $\lambda \in \mathbb{D}$ such that $\ker(T_0 - \lambda) = (0)$. By assumption T dilates $\lambda \cdot I_n$, so suppose with respect to some decomposition T has the form

$$T = \begin{pmatrix} * & * & * \\ 0 & \lambda \cdot I_n & * \\ 0 & 0 & * \end{pmatrix}.$$

Consider the compression $\tilde{T}$ of T corresponding to

$$\tilde{T} = \begin{pmatrix} * & * \\ 0 & \lambda \cdot I_n \end{pmatrix}$$

in its C_{00}-C_{10} upper triangular form, say

$$\tilde{T} = \begin{pmatrix} A_0 & F \\ 0 & A_1 \end{pmatrix},$$

where A_0 acts on $\mathcal{H}_0$ and A_1 on $\mathcal{H}_1$. (Here we have used $\tilde{T} \in C_{\cdot 0}$.) Note that $D_{\tilde{T}}$ and D_{A_0} are Hilbert-Schmidt by a computation, and since easily A_0 is Fredholm via Lemma 2.13 we get D_{A_1} Hilbert-Schmidt as well.

We know that $\dim(\ker(\tilde{T} - \lambda)^*) \geq n$; we will show $\dim(\ker(A_1 - \lambda)^*) \geq n$. Consider some v in $\ker(\tilde{T} - \lambda)^*$ and let $v = v_0 \oplus v_1 \in \mathcal{H}_0 \oplus \mathcal{H}_1$. A computation shows $(A_0 - \lambda)^* v_0 = 0$ and $F v_0 + (A_1 - \lambda)^* v_1 = 0$. Suppose that $v_0 \neq 0$, so

$$\ker(A_0 - \lambda)^* \neq (0).$$

It is then easy to deduce from Lemmas 3.3 and 3.5 that

$$\ker(A_0 - \lambda) \neq (0).$$

Let $w \in \ker(A_0 - \lambda)$, $w \neq 0$. Since A_0 is a restriction of T, of course $w \in \ker(T - \lambda)$, and it is easy to show w is in the C_{00} space of T and thus $w \in \ker(T_0 - \lambda)$, a contradiction of our choice of λ. It then follows that $\dim(\ker(A_1 - \lambda)^*) \geq n$.

Since $A_1 \in C_{10}$ and D_{A_1} is Hilbert-Schmidt, we deduce from this last, Lemma 3.3, and Theorem 2.9 that $A_1 \in \mathbb{A}_{n,\aleph_0}$. Now T has the form

$$T = \begin{pmatrix} A_0 & * & * \\ 0 & A_1 & * \\ 0 & 0 & * \end{pmatrix}.$$

Consider the portion T' of this decomposition given by

$$T' = \begin{pmatrix} A_1 & * \\ 0 & * \end{pmatrix}$$

in its own C_{00}-C_{10} upper triangular form, say

$$T' = \begin{pmatrix} A_0' & * \\ 0 & A_1' \end{pmatrix},$$

where we have used $T' \in C_{\cdot 0}$. Since A_0 is Fredholm, $D_{T'}$ is Hilbert-Schmidt, and thus $A_0' \in C_0$ as usual. Then citing the result from [7] previously mentioned, we get from $T' \in \mathbb{A}_{n,\aleph_0}$ that $A_1' \in \mathbb{A}_{n,\aleph_0}$. Now

$$T = \begin{pmatrix} T_0 & * \\ 0 & T_1 \end{pmatrix} = \begin{pmatrix} A_0 & * & * \\ 0 & A_0' & * \\ 0 & 0 & A_1' \end{pmatrix},$$

with $A_1' \in C_{10}$ and both A_0 and A_0' in C_{00}. By the uniqueness of the C_{00}-C_{10} decomposition of a T in $C_{\cdot 0}$, an easy computation shows $T_1 = A_1'$, so $T_1 \in \mathbb{A}_{n,\aleph_0}$ as desired. □

We give two sample results for contractions with Hilbert-Schmidt defect operator, after recalling a decomposition due to Uchiyama ([25, Theorem 1.5]). Such a T has a matrix form

$$T = \begin{pmatrix} T_{01} & & & * \\ & T_0 & & \\ & & T_{11} & \\ 0 & & & T_{10} \end{pmatrix}, \tag{35}$$

where the diagonal entries are in the indicated classes $C_{\alpha\beta}$ and each diagonal entry has Hilbert-Schmidt defect operator. The following are then apparent from Theorems 3.1, 3.6, and 2.9; one route is to count copies of the bilateral shift in minimal coisometric extensions, use $R_*(T) \cong R(T)$ for $T \in C_{11}$, and the result obtaining from [4, Theorem 1.7] about R_*'s of upper triangular forms.

Corollary 3.7. *Suppose T is a contraction with Hilbert-Schmidt defect operator and adopt the notation indicated in* (35). *For any p, q, and r, if $T_{01} \in \mathbb{A}_{p,p}$, $T_{11} \in \mathbb{A}_{q,q}$, and $T_{10} \in \mathbb{A}_{r,r}$, then*

$$T \in \mathbb{A}_{p+q,\aleph_0} \cap \mathbb{A}_{\aleph_0,q+r}.$$

The next corollary follows upon counting multiplicities of shifts in $R_*(T)$ using [7].

Corollary 3.8. *Suppose T is a contraction with Hilbert-Schmidt defect operator and with the T_{01} term absent in the decomposition* (35), *so*

$$T = \begin{pmatrix} T_0 & * & * \\ 0 & T_{11} & * \\ 0 & 0 & T_{10} \end{pmatrix}.$$

Let T_1 be the compression of T given by

$$T_1 = \begin{pmatrix} T_{11} & * \\ 0 & T_{10} \end{pmatrix}.$$

Then, for any n, $1 \le n \le \aleph_0$, $T \in \mathbb{A}_{n,\aleph_0}$ if and only if $T_1 \in \mathbb{A}_{n,\aleph_0}$.

Remarks

An improvement of the first corollary above awaits resolution of an old problem of increasing importance: if S is the unilateral shift of multiplicity 1, is $S \oplus S^*$ in $\mathbb{A}_{2,2}$? The absence of an answer to this apparently trivial problem is completely indicative of our inability to come to grips with "multipliticy versus dual algebra class" descriptions of operators having both C_{01} and C_{10} parts.

Readers familiar with previous work on $C_{\cdot 0}$ contractions with finite defect indices (see [13] and [14]) will note that conspicuously absent from Theorem 3.6 and even Theorem 2.9 is the condition $T \in \mathbb{A}_{n,1}$, which in the finite defect case is equivalent to the others. The missing ingredient is a Jordan Model theory for $C_{\cdot 0}$ contractions with Hilbert-Schmidt defect; the theorems above and their similarity to those in the finite defect case give evidence that such a theory might be possible.

Observe that many of the results herein could be obtained by use of the decomposition information from [7] which was used in the proof of Corollary 2.15. The important element we cannot obtain by those methods is the relation of possessing an n-fold analytic co-kernel to the other conditions; indeed, we believe that our results indicate that some relationship holds in more general settings.

We close with some questions.

1. In [8, Proposition 3.1] a set $X_T \subseteq \mathbb{T}$ is constructed for any absolutely continuous contraction T and which captures in some sense local $\mathbb{A}_{\aleph_0}$ boundary behavior of T (one result is that $T \in \mathbb{A}_{\aleph_0}$ if and only if $X_T = \mathbb{T}$). More precisely, if T is any absolutely continuous contraction in $\mathcal{L}(\mathcal{H})$, then there exists a (unique) maximal Borel subset X_T of $\mathbf{T}$ such that for each element in

$$\{[f]_{L^1/H_0^1} : f \in L^1(X_T), \|f\|_1 \leq 1\}$$

there exist sequences $\{x_n\}$ and $\{y_n\}$ in the (closed) unit ball of $\mathcal{H}$ satisfying

(a) $\limsup \|[f] - \varphi_T([x_n \otimes y_n]_T)\| \leq \theta$, and
(b) $\|\varphi_T([x_n \otimes w]_T)\| + \|\varphi_T([w \otimes y_n]_T)\| \to 0$, $w \in \mathcal{H}$.

(Recall that φ_T is the mapping from Q_T to L^1/H_0^1 whose dual is the Sz.-Nagy–Foiaş Functional Calculus.)

For T in C_{10} with Hilbert-Schmidt defect operator it is possible to have $X_T = \mathbb{T}$ (this occurs when $T \in \mathbb{A}_{\aleph_0}$ - see Theorem 2.10), and it turns out the unilateral shift S has $X_S = \emptyset$. Are these the only possibilities? Without the Hilbert-Schmidt defect condition they are not, as shown by [8, Proposition 4.5 and Theorem 4.6] and constructions used for the proof of [15, Theorem 2].

Note that there is an argument which goes part way, which is as follows: if there exists some T in C_{10} with Hilbert-Schmidt defect operator with X_T containing an open interval, but not the whole circle, then T is not in $\mathbb{A}_{\aleph_0}$ and $\dim(\ker(T^*))$ would be finite. But clearly one could obtain from a finite direct sum of rotations of T an operator T' in C_{10}, with Hilbert-Schmidt defect, and with $X_{T'} = \mathbb{T}$ and so in $\mathbb{A}_{\aleph_0}$ but with $\dim(\ker(T'^*))$ still finite, a contradiction. Thus if any such T can have any X_T neither empty nor full, X_T must be somewhat pathological.

It is worth making a remark about the argument just given; note that it could have been used to give a weaker version of Lemma 2.7, in which $R_*(T)$ was deduced only of the form $B^{(n)} \oplus U$, where U had spectral measure supported on some set containing no open subset of $\mathbb{T}$. We know of no situation in which this sort of weaker argument succeeds but there can exist nonetheless a non-empty (and somewhat pathological) boundary subset of $\mathbb{T}$ associated with some contraction T and of the given type. Observe that the weaker argument may be improved by the use of Möbius transforms, finite direct sums of operators, and so on.

2. We mention again the question raised after Theorem 3.2: must a $C_1.$ operator with Hilbert-Schmidt defect and in $\mathbb{A}_{\aleph_0}$ have a restriction with a *full* $\aleph_0$-fold analytic co-kernel?

3. Suppose T_1, T_2 are in C_{10} and have Hilbert-Schmidt defects. What can be deduced from $T_1 \oplus T_2^* \in \mathbb{A}_{n,n}$? For example, may one deduce $T_1 \in \mathbb{A}_{k,\aleph_0}$, $T_2 \in \mathbb{A}_{m,\aleph_0}$ with $k+m \geq n$? (This does hold if T_1 and T_2 are each a unilateral shift of some multiplicity, as is shown by considering a dilation to a bilateral shift.) What about $T_1 \oplus T_1^*$?

References

[1] H. Bercovici, *Factorization theorems and the structure of operators on Hilbert space*, Ann. of Math. **128** (1988), 399–413.

[2] ———, *Operator theory and arithmetic in H^∞*, Math. Surveys no. 26, Amer. Math. Soc., Providence, Rhode Island, 1988.

[3] H. Bercovici, C. Foiaş, and C. M. Pearcy, *Dual Algebras with Applications to Invariant Subspaces and Dilation Theory*, CBMS Regional Conf. Ser. in Math. no. 56, Amer. Math. Soc., Providence, Rhode Island, 1985.

[4] H. Bercovici and L. Kérchy, *Quasisimilarity and properties of the commutant of C_{11} contractions*, Acta Sci. Math. (Szeged) **45** (1983), 67–74.

[5] S. Brown, *Full analytic subspaces for contractions with rich spectrum*, Pacific J. of Math. **132** (1988), 1–10.

[6] B. Chevreau, *Sur les contractions à calcul fonctionnel isométrique, II*, J. Operator Theory **20** (1988), 269–293.

[7] B. Chevreau and G. R. Exner, *Multiplicity and dual algebra class*, in preparation.

[8] B. Chevreau, G. R. Exner, and C. M. Pearcy, *Boundary sets for a contraction*, J. Operator Theory **34** (1995), 347–380.

[9] ———, *On the structure of contraction operators, III*, Michigan Math. J. **36** (1989), 29–62.

[10] B. Chevreau and C. M. Pearcy, *On the structure of contraction operators with applications to invariant subspaces*, J. Funct. Anal. **67** (1986), 360–379.

[11] M. J. Cowan and R. G. Douglas, *Complex geometry and operator theory*, Acta Math. **141** (1978), 187–261.

[12] R. G. Douglas, *Banach algebra techniques in operator theory*, Academic Press, New York, 1972.

[13] G. R. Exner, Y. S. Jo, and I. B. Jung, *$C_{\cdot 0}$ contractions: Dual operator algebras, Jordan models, and multiplicity*, J. Operator Theory **33** (1995), 381–394.

[14] G. R. Exner and I. B. Jung, *$C_{\cdot 0}$ and C_{11} contractions with finite defects in the classes $\mathbf{A}_{m,n}$*, Acta Sci. Math. (Szeged) **59** (1994), 555–573.

[15] L. Kérchy, *On the spectra of contractions belonging to special classes*, J. Funct. Anal. **67** (1986), 153–166.

[16] ———, *Isometric asymptotes of power bounded operators*, Indiana University Math. J. **38** (1989), no 1, 173–188.

[17] R. Olin and J. Thompson, *Algebras of subnormal operators*, J. Funct. Anal. **37** (1980), 271–301.

[18] M. Ouannasser, *Sur les contractions dans la classe $\mathbf{A}_n$*, J. Operator Theory **28** (1992), 105–120.

[19] C. M. PEARCY, *Some recent developments in operator theory*, CBMS Regional Conf. Ser. in Math. no. 36, Amer. Math. Soc., Providence, Rhode Island, 1978.
[20] B. SZ.-NAGY AND C. FOIAŞ, *Harmonic analysis of operators on Hilbert space*, North Holland, Amsterdam, 1970.
[21] K. TAKAHASHI, $C_{1\cdot}$ *contractions with Hilbert-Schmidt defect operators*, J. Operator Theory **12** (1984), 331–347.
[22] ———, *On quasiaffine transforms of unilateral shifts*, Proc. Amer. Math. Soc. **100** (1987), no. 4, 683–687.
[23] K. TAKAHASHI AND M. UCHIYAMA, *Every C_{00} operator with Hilbert-Schmidt defect operator is of class C_0*, J. Operator Theory **10** (1983), 331–335.
[24] M. UCHIYAMA, *Contractions and unilateral shifts*, Acta Sci. Math. (Szeged) **46** (1983), 345–356.
[25] ———, *Contractions with (σ, c) defect operators*, J. Operator Theory **12** (1984), 221–233.

G. EXNER
Department of Mathematics
Bucknell University
Lewisburg, PA 17837
E-MAIL: exner@bucknell.edu

I. B. JUNG
Department of Mathematics
Kyungpook National University
Taegu, 702-701, Korea
E-MAIL: ibjung@bh.kyungpook.ac.kr

Received: August 23rd, 1995.

Operator Theory:
Advances and Applications, Vol. 104

Strong Limits of Similarities

DONALD W. HADWIN AND DAVID R. LARSON*

Dedicated to Carl Pearcy on the occasion of his sixtieth birthday

ABSTRACT. We prove that a linear map φ from a linear subspace $\mathcal{R}$ of $B(H)$ into $B(H)$ is a strong operator topology limit of similarities if and only if the restriction $\varphi|_{\mathcal{R}\cap\mathcal{F}(H)}$ to the set of finite-rank operators in $\mathcal{R}$ is a strong limit of similarities. Analogous results for skew-compressions (mappings $X \to AXB$) are obtained, and relations between limits of similarities and skew-compressions are determined.

1. INTRODUCTION

Suppose H is an infinite-dimensional complex Hilbert space, $B(H)$ is the set of bounded linear operators on H, and $\mathcal{F}(H)$ is the set of finite-rank operators in $B(H)$. let $\mathbb{C}$ denote both the complex numbers and the set of scalar multiples of the identity operator, and write $\mathbb{C} + \mathcal{F}(H) = \{\lambda + F\colon \lambda \in \mathbb{C}, F \in \mathcal{F}(H)\}$. In [HNRR] it was shown that if T is an operator not in $\mathbb{C}+\mathcal{F}(H)$, then the similarity orbit of T,

$$\mathcal{S}(T) = \{A^{-1}TA\colon\ A \in B(H),\ A \text{ is invertible}\}$$

is dense in $B(H)$ in the strong operator topology. It was later shown in [L] that if $\{T_1, T_2, \ldots, T_n\}$ is a set of operators that is linearly independent modulo $\mathbb{C}+\mathcal{F}(H)$, then

$$\mathcal{S}(T_1, T_2, \ldots, T_n) = \{(A^{-1}TA, A^{-1}TA, \ldots, A^{-1}TA)\colon\ A \in B(H),\ A \text{ is invertible}\}$$

is dense in $B(H) \times B(H) \times \cdots \times B(H)$ (n factors) with the strong operator product topology.

Suppose K is another Hilbert space with $\dim H = \dim K, \mathcal{R}$ is a linear subspace of $B(H)$, and $\varphi\colon\ \mathcal{R} \to B(K)$ is a (not necessarily bounded) linear map. We say that φ is a *strong* (resp., *weak*) *limit of similarities* if there is a net $\{A_\lambda\}$ of invertible operators from K to H such that, for every S in $\mathcal{R}$,

$$\varphi(S) = \lim_\lambda A_\lambda^{-1} S A_\lambda,$$

where the limit is in the strong (resp., weak) operator topology. It is the purpose of this note to attempt to characterize the maps that are strong limits of similarities.

*Both authors were partially supported by grants from the NSF.

It is clear that a necessary condition for φ to be of this type is that if $1 \in \mathcal{R}$, then φ must be *unital*, i.e., $\varphi(1) = 1$. It is clear that even if 1 is not in $\mathcal{R}$ and φ is a strong limit of similarities on $\mathcal{R}$ and we extend φ to $\mathcal{R} + \mathbb{C}$ by defining $\varphi(1) = 1$, then the extended map is still a strong limit of similarities. We therefore always assume that $1 \in \mathcal{R}$.

Another way to look at strong limits of similarities is to consider the set $\mathcal{V}$ of all functions from $\mathcal{R}$ into $B(K)$ with the topology of pointwise convergence in the strong operator topology. A similarity on $\mathcal{R}$ is a map $\varphi: \mathcal{R} \to B(K)$ defined in terms of an invertible operator A by $\varphi(S) = A^{-1}SA$. Then the set of strong limits of similarities on $\mathcal{R}$ is precisely the closure in $\mathcal{V}$ of the set of similarities. This point of view makes it clear that the set of strong limits of similarities is closed under pointwise strong limits. Another consequence that we often use is that if φ: $\mathcal{R} \to B(K)$ is a strong limit of similarities on $\mathcal{R}$ and ψ: $\varphi(\mathcal{R}) \to B(K)$ is a strong limit of similarities on $\varphi(\mathcal{R})$, then the composition $\psi \circ \varphi$ is a strong limit of similarities on $\mathcal{R}$. (Proof: This is clearly true when ψ is a similarity, and if $\{\psi_\lambda\}$ is a net of similarities on $\varphi(\mathcal{R})$ converging pointwise strongly to ψ, then the net $\{\psi_\lambda \circ \varphi\}$ converges pointwise strongly to $\psi \circ \varphi$.)

Our main result shows that the only obstruction to a unital map being a strong limit of similarities involves the finite-rank operators. More precisely, we show that a unital linear map φ on $\mathcal{R}$ is a strong limit of similarities if and only if the restriction $\varphi|\mathcal{R} \cap \mathcal{F}(H)$ is a strong limit of similarities. This reduces the problem of determining strong limits of similarities to the case of finite-dimensional subspaces of finite-rank operators. Although strong and weak convergence of similarities is not the same, we show that the set of strong limits of similarities always coincides with the set of weak limits of similarities.

We begin with the result that got us started on this problem. It is an immediate consequence of Theorem 5.1 in [L].

Lemma 1. *If $\mathcal{R}$ is a unital linear subspace of $B(H)$ and $\mathcal{R} \cap \mathcal{F}(H) = 0$, then every unital linear mapping on $\mathcal{R}$ is a strong limit of similarities.*

PROOF. Suppose $\mathcal{R} \cap \mathcal{F}(H) = 0$ and φ: $\mathcal{R} \to B(H)$ is a linear map with $\varphi(1) = 1$. Suppose $\{1, S_1, S_2, \ldots, S_n\}$ is a linearly independent subset of $\mathcal{R}, F$ is a finite subset of H and $\varepsilon > 0$. We must find an invertible operator A so that, for $1 \leq k \leq n$, $f \in F$, we have $\|[A^{-1}S_kA - \varphi(S_k)]f\| < \varepsilon$. This is an immediate consequence of Theorem 5.1 in [L], which asserts that $(\varphi(S_1), \ldots, \varphi(S_n))$ is in the strong closure of $\mathcal{S}(S_1, \ldots, S_n)$. □

The proof of the next result is based on an idea given to the first author in a personal communication from P.R. Halmos.

Lemma 2. *Suppose X is a set and $\mathcal{W}$ is a collection of functions from X to $B(H)$ that is closed under similarity. If φ is the pointwise limit in the weak operator topology of a net of functions in $\mathcal{W}$, then φ is a pointwise limit in the strong operator topology of a (possibly different) net of functions in $\mathcal{W}$.*

PROOF. Suppose φ is a pointwise limit in the weak operator topology of a net of functions in $\mathcal{W}$. Suppose E is a finite subset of X, F is a finite subset of the unit ball of H and $\varepsilon > 0$. We must find a ψ in $\mathcal{W}$ such that, for $x \in E$, $f \in F$, we have $\|[\varphi(x) - -\psi(x)]f\| < \varepsilon$.

Let M be the finite-dimensional space spanned by $F \cup \{\varphi(x)f\colon x \in E, f \in F\}$, and let P be the projection onto M. Since φ is a pointwise limit in the weak operator topology of a net of functions in $\mathcal{W}$, it follows that there is a γ in $\mathcal{W}$ such that for every x in E,

$$\|P[\varphi(x) - \gamma(x)]P\| < \varepsilon/2.$$

(To see this, if we let $\{e_1, \ldots, e_m\}$ be an orthonormal basis for M, and choose γ in $\mathcal{W}$ so that, for every x in E and $1 \le i, j \le m$, $|([\varphi(x) - \gamma(x)]e_i, e_j)| < \epsilon/(2m^2)$ holds, then the desired inequality holds.)

Next choose a $\delta > 0$ so that

$$\delta \cdot \max\{\|(1 - P)\gamma(x)f\|\colon x \in E, f \in F\} < \varepsilon/2.$$

Let $A = P + \delta(1 - P)$, and define ψ on X by $\psi(x) = A\gamma(x)A^{-1}$. Since $\mathcal{W}$ is closed under similarity, $\psi \in \mathcal{W}$. Moreover, for $x \in E$ and $f \in F$ we have

$$\begin{aligned}
\|[\varphi(x) - \psi(x)]f\| &= \|[\varphi(x) - \psi(x)]Pf\| \\
&\le \|P[\varphi(x) - \psi(x)]Pf\| + \|(1 - P)[\varphi(x) - \psi(x)]Pf\| \\
&< \varepsilon/2 + \|(1 - P)\varphi(x)f - (1 - P)A\gamma(x)A^{-1}Pf\| \\
&= \varepsilon/2 + \|0 - \delta(1 - P)\gamma(x)f\| \\
&< \varepsilon/2 + \varepsilon/2 = \varepsilon.
\end{aligned}$$

□

Corollary 3. *Every weak limit of similarities on a subspace $\mathcal{R}$ of $B(H)$ is a strong limit of similarities.*

The next lemma shows that one-sided inverses still yield strong limits of similarities.

Lemma 4. *If $A, B \in B(H)$ and $AB = 1$, then the mapping φ on $B(H)$ defined by $\varphi(T) = ATB$ is a strong limit of similarities on $B(H)$.*

PROOF. It follows from [H2, Lemma 1] that there is a positive invertible operator P and an isometry V such that $A = V^*P^{-1}$ and $B = PV$. It follows that φ is the composition of the similarity ($T \mapsto P^{-1}TP$) and the map ($T \mapsto V^*TV$). Hence we can assume that $A = V^*$ and $B = V$. It follows from [H1] that φ is a weak limit of similarities (using unitaries). (Another way to see this is to use the result of P.R. Halmos [Hal] that the isometry V is a strong limit of a sequence $\{U_n\}$ of unitary operators. It clearly follows that, for each T in $B(H)$, $\varphi(T)$ is the limit in the weak operator topology of $\{V_n^* T V_n\}$.) It then follows from Lemma 2 that φ is a strong limit of similarities on $B(H)$. □

The next lemma is reminiscent of a key ingredient Theorem 1.3 in the proof of D. Voiculescu's theorem on approximate unitary equivalence [V]. We let $id_{\mathcal{R}}\colon \mathcal{R} \to B(H)$ denote the identity representation on $\mathcal{R}$, i.e., $id_{\mathcal{R}}(T) = T$ for every T in $\mathcal{R}$.

Lemma 5. *Suppose $\mathcal{R}$ is a unital linear subspace of $B(H)$ and φ is a unital linear map on $\mathcal{R}$ such that $\varphi|\mathcal{R}\cap\mathcal{F}(H) = 0$. Then $id_{\mathcal{R}} \oplus \varphi$ is a strong limit of similarities on $\mathcal{R}$.*

PROOF. Since the proof that a map is a strong limit of similarities involves only finitely many operators in $\mathcal{R}$ at a time, we can assume that $\mathcal{R}$ is finite-dimensional.

We can write $\mathcal{R}$ as a linear direct sum

$$\mathcal{R} = \mathbb{C} + \mathcal{R}_\infty + \mathcal{R}_\mathcal{F}, \text{ where } \mathcal{R}_\mathcal{F} = \mathcal{R} \cap \mathcal{F}.$$

Suppose P is any finite-rank projection such that $P\mathcal{R}_\mathcal{F}P = \mathcal{R}_\mathcal{F}$, and, using the finite-dimensionality of $\mathcal{R}$, choose a finite-rank projection $Q \geq P$ so that $(1 - Q)SP = 0$ for every S in $\mathcal{R}$.

Let V be any isometry on H such that $V|\text{ran } P$ is the identity, and V maps $\ker P$ onto $\ker Q$. Let $W = V(1 - P)$. Thus $VP = P$ and $(1 - P)V^* = (1 - P)W^*(1 - Q)$. It follows from Lemma 4 that the map $\alpha(S) = V^*SV$ is a strong limit of similarities. However, for each S in $\mathcal{R}$, we have

$$(1 - P)\alpha(S)P = (1 - P)V^*SVP = (1 - P)V^*(1 - Q)SP = 0.$$

For each positive integer n, let $A_n = nP + (1 - P)$. If we define maps β_1 and β_2 on $\mathcal{R}$ by $\beta_1(S) = PSP|\text{ran } P$ and $\beta_2(S) = W^*SW|\text{ran}(1 - P)$, then, for every S in $\mathcal{R}$, $A_n^{-1}SA_n \to \beta_1(S) \oplus \beta_2(S)$ in norm. Thus $\beta_1 \oplus \beta_2$ is a strong limit of similarities.

It follows from the choice of P and Q that $\ker \beta_2 = \mathcal{R} \cap \mathcal{F}(H)$ and $\beta_2(\mathcal{R})$ contains no nonzero finite-rank operators. It then follows from Lemma 1 that the mapping on $\beta_2(\mathcal{R})$ that sends $\beta_2(S)$ to $[(1 - P)S(1 - P)|\text{ran}(1 - P)] \oplus \varphi(S)$ is a strong limit of similarities on $\beta_2(\mathcal{R})$. Hence the map $\psi_P\colon \mathcal{R} \to B(H \oplus H)$ defined by

$$\psi_P(S) = [PSP + (1 - P)S(1 - P)] \oplus \varphi(S)$$

is a strong limit of similarities on $\mathcal{R}$. If we let $P \to 1$ in the strong operator topology, then $\psi_P(S)$ converges to $(id_{\mathcal{R}} \oplus \varphi)(S)$ in the strong operator topology for every S in $\mathcal{R}$. Hence $id_{\mathcal{R}} \oplus \varphi$ is a strong limit of similarities on $\mathcal{R}$. □

Corollary 6. *$\mathcal{R}$ is a unital linear subspace of $B(H)$, φ is a strong limit of similarities on $\mathcal{R}$, ψ is a unital linear map on $\mathcal{R}$ such that $\psi|\mathcal{R}\cap\mathcal{F}(H) = 0$, and $0 < t < 1$. Then $(1 - t)\varphi + t\psi$ is a strong limit of similarities.*

PROOF. It follows from Lemma 5 that $id_{\mathcal{R}} \oplus \psi$ is a strong limit of similarities. It is now clear that $\varphi \oplus \psi$ is a strong limit of similarities. Suppose $0 < t < 1$, and define an isometry $V\colon H \to H \oplus H$ by $Vh = \sqrt{1 - t}\, h \oplus \sqrt{t}\, h$. It follows from Lemma 4 that the map γ on $\mathcal{R}$ defined by $\gamma(S) = V^*[\varphi(S) \oplus \psi(S)]V$ is a strong limit of similarities. However, for every f, g in H,

$$\begin{aligned}(\gamma(S)f, g) &= ((\varphi \oplus \psi)(S)Vf, Vg) \\ &= (\varphi(S)\sqrt{1-t}\,f, \sqrt{1-t}\,g) + (\psi(S)\sqrt{t}\,f, \sqrt{t}\,g) \\ &= ([(1-t)\varphi + t\psi](S)f, g).\end{aligned}$$

Thus $\gamma = (1-t)\varphi + t\psi$. □

We are now ready to prove our main result.

Theorem 7. *Suppose $\mathcal{R}$ is a unital linear subspace of $B(H)$ and φ is a unital linear map from $\mathcal{R}$ to $B(H)$. The following are equivalent:*

1. *φ is a strong limit of similarities on $\mathcal{R}$.*
2. *$\varphi|\mathcal{R} \cap \mathcal{F}(H)$ is a strong limit of similarities on $\mathcal{R} \cap \mathcal{F}(H)$.*

PROOF. The implication(1) $\Rightarrow$ (2) is obvious. To prove the reverse implication, suppose (2) is true. As in the proof of Lemma 5, write $\mathcal{R}$ as a linear direct sum

$$\mathcal{R} = \mathbb{C} + \mathcal{R}_\infty + \mathcal{R}_\mathcal{F}, \text{ where } \mathcal{R}_\mathcal{F} = \mathcal{R} \cap \mathcal{F}.$$

For each invertible operator A, define maps φ_A, ρ_A on $\mathcal{R}$ by $\rho_A(S) = A^{-1}SA$, and

$$\varphi_A(S) = \begin{cases} A^{-1}SA & \text{if } S \in \mathcal{R}_\mathcal{F} \\ \varphi(S) & \text{if } S \in \mathbb{C} + \mathcal{R}_\infty. \end{cases}$$

Also define a map γ on $\mathcal{R}$ by $\gamma(1) = 1$ and $\gamma = 0$ on $\mathcal{R}_\infty + \mathcal{R}_\mathcal{F}$. Since $\varphi|\mathcal{R}_\mathcal{F}$ is a strong limit of similarities, φ is a pointwise strong limit of maps of the form φ_A. Hence we just need to show that each φ_A is a strong limit of similarities.

However, for each $t, 0 < t < 1$, the map

$$\frac{1-t}{t}\left[\varphi_A - \rho_A + \frac{t}{1-t}\gamma\right]$$

is unital and is 0 on $\mathcal{R}_\mathcal{F}$, and it follows from Corollary 6 that

$$(1-t)\varphi_A + t\gamma = (1-t)\rho_A + t\frac{1-t}{t}\left[\varphi_A - \rho_A + \frac{t}{1-t}\gamma\right]$$

is a strong limit of similarities. Letting $t \to 0^+$, we see that φ_A is a strong limit of similarities. □

We now turn our attention to characterizing limits of similarities on linear spaces $\mathcal{R}$ of finite-rank operators. Although this problem seems difficult, we will see that on such spaces the strong limits of similarities coincide with the strong limits of skew-compressions. If $A, B \in B(H)$, we call the mapping $C_{A,B}$ defined by $C_{A,B}(T) = ATB$, a *skew-compression*. Note that the norm of $C_{A,B}$ on $B(H)$ is $\|A\|\,\|B\|$; we call this the compression norm of $C_{A,B}$. The only unital skew compressions are the mappings like the ones in Lemma 4. We give an analogue of Theorem 7 for skew compressions.

Theorem 8. *Every skew-compression on $\mathcal{F}(H)$ is a strong limit of similarities. If $\mathcal{R}$ is a linear subspace of $\mathcal{F}(H)$, then a mapping on $\mathcal{R}$ is a strong limit of similarities if and only if it is a strong limit of skew-compressions.*

PROOF. It is sufficient to prove that $C_{A,B}$ is a strong limit of similarities on a finite-dimensional linear subspace $\mathcal{R}$ of $\mathcal{F}(H)$. First note that the identity representation on $\mathcal{R}$, $id_{\mathcal{R}}$, is unitarily equivalent to $id_{\mathcal{R}} \oplus 0$ on $H \oplus H$. Using Lemma 2 in [H2], we can find an invertible operator D in $B(H \oplus H) = \mathcal{M}_2(B(H))$ such that the (1,1) entries of the matrices for D^{-1} and D are, respectively, A and B. If $W\colon H \to H \oplus H$ is the isometry defined by $Wf = f \oplus 0$, then, for every S in $\mathcal{R}$,

$$C_{A,B}(S) = W^* D^{-1}(id_{\mathcal{R}} \oplus 0)(S) DW.$$

It follows from Lemma 4 that $C_{A,B}$ is a strong limit of similarities on $\mathcal{R}$. □

Remark 9. *In [H3] it was shown that a pointwise strong limit of a net of skew-compressions on a C^*-subalgebra of $B(H)$ for which the compression norms are bounded is a pointwise norm limit of another bounded net of skew-compressions. It was also shown in [H3, Proposition 7] that such maps are precisely the completely bounded maps whose restriction to the set of compact operators in the algebra is a skew-compression. A similar result was proven for completely positive maps on C^*-subalgebras of $B(H)$ in [H2].*

We prove an analogue of Theorem 7 for strong limits of skew-compressions.

Theorem 10. *Suppose $\mathcal{R}$ is a linear subspace of $B(H)$ and φ is a linear map from $\mathcal{R}$ to $B(H)$. The following are equivalent:*

1. *φ is a strong limit of skew-compressions on $\mathcal{R}$.*
2. *$\varphi|\mathcal{R} \cap \mathcal{F}(H)$ is a strong limit of skew-compressions on $\mathcal{R} \cap \mathcal{F}(H)$.*

PROOF. The implication (1) ⇒ (2) is obvious. To prove the reverse implication suppose (2) holds. Since showing that (1) holds only involves finitely many operators at a time, we can assume that $\mathcal{R}$ is finite-dimensional. In this case $\mathcal{R} + \mathcal{F}(H) + \mathbb{C}$ is not all of $B(H)$, and we can choose an operator B not in this set. Replacing B with an appropriate $B - \lambda$, if necessary, we can assume that B is invertible. It follows that 1 is not in $\mathcal{R}B^{-1} + \mathcal{F}(H)$. Let $\mathcal{R}_1 = \mathcal{R}B^{-1} + \mathbb{C}$. It follows that $\mathcal{F}(H) \cap \mathcal{R}_1 = \mathcal{F}(H) \cap \mathcal{F}B^{-1}$. Define the map ψ on $\mathcal{R}_1$ by $\psi(1) = 1$, and $\psi(T) = \varphi(TB)$ for T in $\mathcal{R}B^{-1}$. It is clear that φ is a strong limit of skew-compressions on $\mathcal{R}$ if and only if $\psi|\mathcal{R}B^{-1}$ is a strong limit of skew-compressions on $\mathcal{R}B^{-1}$.

However, ψ is unital, and $\psi|\mathcal{F}(H) \cap \mathcal{R}_1 = \psi|\mathcal{F}(H) \cap \mathcal{R}B^{-1}$ is a strong limit of skew-compressions. By Theorem 8, $\psi|\mathcal{F}(H) \cap \mathcal{R}_1$ is a strong limit of similarities. It follows from Theorem 7 that ψ is a strong limit of similarities on $\mathcal{R}_1$. Thus φ is a strong limit of skew-compressions on $\mathcal{R}$. □

Corollary 11. *Suppose $\mathcal{R}$ is a subspace of $B(H)$ such that $\mathcal{R} \cap [\mathcal{F}(H) + \mathbb{C}] = \mathcal{R} \cap \mathcal{F}(H)$. The set of strong limits of similarities on $\mathcal{R}$ equals the set of strong limits of skew-compressions on $\mathcal{R}$.*

PROOF. Suppose φ is a strong limit of skew-compressions on $\mathcal{R}$. Let $\mathcal{R}_1 = \mathcal{R} + \mathbb{C}$, and extend φ to $\mathcal{R}_1$ by defining $\varphi(1) = 1$. Since $\mathcal{R}_1 \cap \mathcal{F}(H) = \mathcal{R} \cap \mathcal{F}(H)$, it follows from Theorem 9 that $\varphi|\mathcal{R}_1 \cap \mathcal{F}(H)$ is a strong limit of similarities. Hence, by Theorem 7, φ is a strong limit of similarities on $\mathcal{R}_1$, and thus on $\mathcal{R}$. □

We conclude by showing that the problem of characterizing strong limits of similarities (equivalently, skew-compressions) on subspaces of $\mathcal{F}(H)$ can be reduced to characterizing norm limits of similarities on subspaces of $\mathcal{M}_n$ for finite values of n. The problem reduces to characterizing norm limits of skew-compressions on subspaces of $\mathcal{M}_n$, and we show, for each positive integer n, how norm limits of skew-compressions on subspaces of $\mathcal{M}_n$ can be viewed as norm limits of similarities on subspaces of $\mathcal{M}_{2n}$.

Theorem 12. *Suppose $\mathcal{R}$ is a finite-dimensional linear subspace of $\mathcal{F}(H)$ and $\varphi\colon \mathcal{R} \to \mathcal{F}(H)$ is a linear mapping. Suppose P is a finite-rank projection such that $PTP = T$ and $P\varphi(T)P = \varphi(T)$ for all $T \in \mathcal{R}$. Let $n =$ rank P, and let $V\colon \mathbb{C}^n \to H$ be a linear isometry whose range is PH. Define a mapping α: $PB(H)P \to \mathcal{M}_n$ by $\alpha(T) = V^*TV$, and define β: $PB(H)P \to \mathcal{M}_{2n} = \mathcal{M}_2(\mathcal{M}_n)$ by*

$$\beta(T) = \begin{pmatrix} 0 & \alpha(T) \\ 0 & 0 \end{pmatrix}.$$

Then the mappings α and β are one-to-one. Furthermore, the following statements are equivalent.

1. *φ is a strong limit of similarities on $\mathcal{R}$.*
2. *φ is a norm limit of skew-compressions on $\mathcal{R}$.*
3. *$\alpha \circ \varphi \circ \alpha^{-1}$ is a norm limit of skew-compressions on $\alpha(\mathcal{R})$.*
4. *$\beta \circ \varphi \circ \beta^{-1}$ is a norm limit of similarities on $\beta(\mathcal{R})$.*

PROOF. It is clear that the mappings α and β are one-to-one.

(1) ⇒ (2). Suppose $\{S_\lambda\}$ is a net of invertible operators such that, for every T in $\mathcal{R}$, $S_\lambda^{-1}TS_\lambda \to \varphi(T)$ in the strong operator topology. Since rank $P < \infty$, we conclude that $PS_\lambda^{-1}TS_\lambda P \to P\varphi(T)P = \varphi(T)$ in norm for every T in $\mathcal{R}$. Hence (2) is true.

(2) ⇒ (3). Suppose C_{A_λ, B_λ} converges pointwise in the norm topology to φ on $\mathcal{R}$. Suppose $T \in \mathcal{R}$. Then, since $VV^* = P$, we have

$$T = \alpha^{-1}(\alpha(T)) = \alpha^{-1}(V^*TV) = C_{V,V^*}(\alpha(T)).$$

Hence $C_{V^*A_\lambda V, VB_\lambda V^*}$ converges pointwise in the norm topology to $\alpha \circ \varphi \circ \alpha^{-1}$ on $\alpha(\mathcal{R})$. Hence (3) holds.

(3) $\Rightarrow$ (4). Suppose $\{A_\lambda\}, \{B_\lambda\}$ are nets in $\mathcal{M}_n$ and C_{A_λ,B_λ} converges pointwise to $\alpha \circ \varphi \circ \alpha^{-1}$ on $\alpha(\mathcal{R})$. Since the set of invertible matrices is dense in $\mathcal{M}_n$, we can choose the A_λ's and B_λ's to be invertible. For each λ define the invertible matrix S_λ in $\mathcal{M}_{2n}$ by

$$S_\lambda = \begin{pmatrix} A_\lambda^{-1} & 0 \\ 0 & B_\lambda \end{pmatrix}.$$

A direct matrix computation shows that, for every T in $\mathcal{R}$, $S_\lambda^{-1}\beta(T)S_\lambda \to \beta(\varphi(T))$. This proves (4).

(4) $\Rightarrow$ (1). Suppose $\{S_\lambda\}$ is a net of invertible operators in $\mathcal{M}_{2n}$ such that, for every T in $\mathcal{R}$, $S_\lambda^{-1}\beta(T)S_\lambda \to \beta(\varphi(T))$. Identifying $\mathcal{M}_{2n}$ with $\mathcal{M}_2(\mathcal{M}_n)$, suppose the (1,1)-entry of S_λ^{-1} is A_λ and the (1,1)-entry of S_λ is B_λ. Another computation shows that, for every T in $\mathcal{R}$, $C_{A_\lambda,B_\lambda}(\alpha(T))$ converges to $\alpha(\varphi(T)) = V^*\varphi(T)V$. Hence $C_{VA_\lambda V^*,VB_\lambda V^*}$ converges pointwise in norm to φ. It follows from Theorem 8 that φ is a strong limit of similarities. Thus (1) holds. □

In a sequel to this paper we will address in more detail the problem of characterizing limits of similarities on subspaces of $\mathcal{F}(H)$.

REFERENCES

[1] D.W. HADWIN, *Completely positive maps and approximate equivalence*, Indiana Univ. Math. J., 36 (1987), 211–228.

[2] D.W. HADWIN, *Dilations and Hahn decompositions for linear maps*, Can. J. Math., 33 (1981), 826–839.

[3] D.W. HADWIN, *Approximately hyperreflexive algebras*, J. Oper. Th., 28 (1992), 51–64.

[4] D.W. HADWIN, E.A. NORDGREN, H. RADJAVI AND P. ROSENTHAL, *Most similarity orbits are strongly dense*, Proc. Amer. Math. Soc., 76 (1979), 250–252.

[5] P.R. HALMOS, *Ten problems in Hilbert space*, Bull. Amer. Math. Soc., 76 (1970), 887–933.

[6] D. LARSON, *Reflexivity, algebraic reflexivity, and linear interpolation*, Amer. J. Math., 110 (1988), 283–299.

[7] D. VOICULESCU, *A non-commutative Weyl-von Neumann theorem*, Rev. Roum. Math. Pures et Appl., 21 (1976), 97–113.

DONALD W. HADWIN
Department of Mathematics
University of New Hampshire
Durham, NH 03824

DAVID R. LARSON
Department of Mathematics
Texas A&M University
College Station, TX 77843

Received: August 23rd, 1995.

Operator Theory:
Advances and Applications, Vol. 104

L^p Multipliers and Nested Sigma-Algebras

ALAN LAMBERT

With best wishes to Carl Pearcy on the occasion of his sixtieth birthday

Throughout this note $(X, \mathcal{F}, \mu)$ denotes a complete probability space. All sub sigma algebras of $\mathcal{F}$ considered are assumed to be complete with respect to μ. We shall explore the relationship between a sigma algebra $\mathcal{B} \subset \mathcal{F}$ and the set of multiplication operators which map $L^p(X, \mathcal{B}, \mu)$ into $L^p(X, \mathcal{F}, \mu)$. (All vector spaces encountered are with respect to the scalar field $\mathbb{C}$.) These operators are closely related to averaging operators on order ideals in Banach lattices and to operators called conditional expectation-type operators in [1]. However, our primary interest in studying these operators lies in their use in investigating chains of sigma algebras. The next section contains some basic notation and several predominantly measure-theoretic facts frequently referred to in the sequel. Since conditional expectations play a central role in this investigation, a subsection of Section I is devoted to a discussion of these operators.

1. PRELIMINARIES

- The notation $\mathcal{A} \leq \mathcal{B}$ is meant to convey the information that both $\mathcal{A}$ and $\mathcal{B}$ are sub sigma algebras of $\mathcal{F}$, and that $\mathcal{A}$ is contained in $\mathcal{B}$. A^+ denotes the collection of subsets of $\mathcal{A}$ of positive measure.
- All set and function statements are to be interpreted as being valid modulo $\mu-$ null sets. In particular, the statement "$S = \emptyset$" simply means that $S \in \mathcal{A} - \mathcal{A}^+$. All sets and functions encountered are either by construction or assumption measurable with respect to $\mathcal{F}$. Sub sigma algebra measurability will, of course, be explicitly noted when applicable.
- For a given function f, the symbol $\{f > \alpha\}$ denotes the set $\{x \in X : f(x) > \alpha\}$, etc.
- For a given function f, S_f is defined to be $\{f \neq 0\}$, and is referred to as the support of f (supp f). It is important to note that although this is only defined up to $\mu-$ null sets, at no single time in this article is this concept applied to more than a countable collection of functions.
- For $\mathcal{B} \leq F$, $L^p(\mathcal{B}) = L^p(X, \mathcal{B}, \mu|_{\mathcal{B}})$. Hereafter we will abbreviate $\mu|_{\mathcal{B}}$ to μ.

We view $L^p(\mathcal{B})$ as a subspace of $L^p(\mathcal{F})$ and as a Banach space in its own right.

1.1. Conditional Expectation. Let $\mathcal{B} \leq \mathcal{F}$. For each $f \in L^p(\mathcal{F})$ $(1 \leq p \leq \infty)$ or $f \geq 0$, there is a unique $\mathcal{B}$–measurable function $E(f|\mathcal{B})$ such that, $\forall B \in \mathcal{B}$

$$\int_B E(f|\mathcal{B})\, d\mu = \int_B f\, d\mu.$$

We shall use the notation $E^{\mathcal{B}} f$ for $E(f|\mathcal{B})$.

- $E^{\mathcal{B}}$ maps $L^p(\mathcal{F})$ onto $L^p(\mathcal{B})$ and is a contractive projection.
- The averaging property: $E^{\mathcal{B}}\ (\phi f) = \phi \cdot E^{\mathcal{B}}(f)$ whenever ϕ is $\mathcal{B}$-measurable and both sides of this equation are meaningful; in particular, if ϕ is in $L^q(\mathcal{B})$ and f is in $L^p(\mathcal{F})$, 1/p + 1/q=1.
- $f \geq g \Rightarrow E^{\mathcal{B}} f \geq E^{\mathcal{B}} g$. If $f \geq 0$ and $f \neq 0$, then $E^{\mathcal{B}} f \neq 0$.
- For 1/p + 1/q = 1, $|E^B(fg)| \leq (E^B|f|^p)^{1/p} \cdot (E^B|g|^q)^{1/q}$.
- $S_{E^{\mathcal{B}}|f|}$ is the smallest $\mathcal{B}$-set containing S_f ([2]).
- If $\mathcal{B} \leq \mathcal{C} \leq \mathcal{F}$, then $E^{\mathcal{B}}\left(E^{\mathcal{C}} f\right) = E^{\mathcal{C}}\left(E^{\mathcal{B}} f\right) = E^{\mathcal{B}} f$

2. $L^p(\mathcal{B})$ TO $L^p(\mathcal{F})$ MULTIPLIERS

Let $\mathcal{B} \leq \mathcal{F}$ and define $\mathcal{K}_p = \{f : f \cdot L^p(\mathcal{B}) \subset L^p(\mathcal{F})\}$. When several sigma algebras are being considered we will refer to $\mathcal{K}_p$ as $\mathcal{K}_p(\mathcal{B})$ or $\mathcal{K}_p(\mathcal{B}, \mathcal{F})$. Note that for $1 \leq p < \infty$,

$$L^\infty(\mathcal{F}) \subset \mathcal{K}_p \subset L^p(\mathcal{F}) \subset L^1(\mathcal{F})$$

and that $\mathcal{K}_p$ is a vector space. For $\phi \in \mathcal{K}_p$, let K_ϕ be the corresponding multiplication operator. Since sequential L^p convergence always leads to an a.e. convergent subsequence, a routine application of the closed graph theorem shows that for $\phi \in \mathcal{K}_p$, K_ϕ is bounded. One of the more commonly used properties of multiplication operators is that $\mathcal{K}_p(\mathcal{F}, \mathcal{F}) = L^\infty(\mathcal{F})$. Our first task is to establish an analytic description of the members of $\mathcal{K}_p$.

Proposition 2.1. *$\phi \in \mathcal{K}_p$ if and only if $E^{\mathcal{B}}|\phi|^p \in L^\infty(\mathcal{B})$. (The emphasis in this statement should be placed on the essential boundedness of $E^{\mathcal{B}}|\phi|^p$; such an expression is automatically $\mathcal{B}$-measurable).*

PROOF. Suppose that for some constant C, $E^{\mathcal{B}}|\phi| \leq C$. Then for each $g \in L^p(\mathcal{B})$,

$$\begin{aligned} \|\phi g\|^p &= \int_x |\phi g|^p\, d\mu = \int_x E^{\mathcal{B}}\left(|\phi|^p \cdot |g|^p\, d\mu\right) \\ &= \int_x E^B\left(|\phi|^p\right) \cdot |g|^p\, d\mu \leq C^p \|g\|^p. \end{aligned}$$

Thus $\phi \in \mathcal{K}_p$.

Now suppose only that $\phi \in \mathcal{K}_p$. Then for each $B \in \mathcal{B}$,

$$\int_B E^B(|\phi|^p)\, d\mu = \int_B |\phi|^p\, d\mu = \|\phi\chi_B\|^p \le \|K_\phi\|^p \mu(B).$$

But $E^{\mathcal{B}}(|\phi|^p)$ is $\mathcal{B}$-measurable, so

$$\|E^{\mathcal{B}}(|\phi|^p)\|_\infty = \sup\left\{\left(\int_B E^{\mathcal{B}}(|\phi|^p)\, d\mu\right)/\mu(B) : B \in \mathcal{B}^+\right\} \le \|K_\phi\|^p. \qquad \square$$

Assuming $\phi \in \mathcal{K}_p$, let $\|\phi\|_{\mathcal{K}_p} = \left(\|E^B(|\phi|^p)\|_\infty\right)^{1/p}$. It follows that $\|\phi\|_{\mathcal{K}_p} \le \|K_\phi\|$. On the other hand, for $g \in L^p(\mathcal{B})$ the same string of inequalities used in the preceding argument shows that

$$\|K_\phi g\|^p = \int_x \left(E^{\mathcal{B}}|\phi|^p\right)|g|^p\, d\mu \le \left(\|\phi\|_{\mathcal{K}_p}\right)^p \cdot \|g\|^p,$$

so that $\|\phi\|_{\mathcal{K}_p} = \|K_\phi\|$. Consequently $\|\ \|_{\mathcal{K}_p}$is a norm and $\mathcal{K}_p$ is a Banach space with respect to this norm.

Recall that $L^\infty(\mathcal{F}) \subset \mathcal{K}_p \subset L^p(\mathcal{F}) \subset L^1(\mathcal{F})$. Also, if $\phi \in \mathcal{K}_p$ and $\rho \le \phi$ then $\rho \in \mathcal{K}_p$. Thus $\mathcal{K}_p$ is an order ideal ([1]). A straightforward calculation shows that for $1 \le p < \infty$, the adjoint operator $K_\phi^* : L^q(\mathcal{F}) \to L^q(\mathcal{B})$ is given by

$$K_\phi^* f = E^{\mathcal{B}}(\phi f).$$

Such operators played a central role in the classification project undertaken in [1].

Our next task is to give a complete classification of those sigma algebras $\mathcal{B} \le \mathcal{F}$ for which $\mathcal{K}_p(\mathcal{B})$ is extreme; that is, either $L^\infty(\mathcal{F})$ or $L^p(\mathcal{F})$.

Proposition 2.2. (a) *$\mathcal{K}_p(\mathcal{B}) = L^p(\mathcal{F})$ if and only if $\mathcal{B}$ is generated by a finite partition of x.*

(b) *$\mathcal{K}_p(\mathcal{B}) = L^\infty(\mathcal{F})$ if and only if there is a constant C so that for every $f \in L^1(\mathcal{F})$, $|f| \le C \cdot E^{\mathcal{B}}|f|$.*

PROOF. We present the case for $p = 1$; the proof for the general case differing only in the tedious use of superscripts p and $1/p$.

(a) Suppose $\mathcal{B}$ is generated by $\{B_1, ..., B_n\}$, where the B_i's are pairwise disjoint and of positive measure. Then $E^{\mathcal{B}}$ is given by the formula

$$E^{\mathcal{B}} f = \sum_{i=1}^n \left(\frac{1}{\mu(B_i)} \int_{B_i} f\, d\mu\right)\chi_{B_i}$$

and consequently for $f \in L^1(\mathcal{F})$,

$$E^{\mathcal{B}}|f| \le \left(\sum_{i=1}^n \frac{1}{\mu(B_i)}\right)\|f\| \quad \text{a.e.}$$

Now suppose that B is not finitely generated. We may choose an infinite sequence $\{B_i\} \subset \mathcal{B}^+$ so that for each i, $\mu(B_i) < 2^{-i}$. Let $f = \sum_{i=1}^{\infty} (2^i/i^2) \cdot \chi_{B_i}$. Then $f \in L^1(\mathcal{B})$ and $f \notin L^\infty(\mathcal{B})$. But since f is $\mathcal{B}$-measurable, $f \cdot L^1(\mathcal{B})$ consists of $\mathcal{B}$-measurable functions. If this set were back in $L^1(\mathcal{F})$ it would actually be in $L^1(\mathcal{B})$. But this could only be true if f were in $L^\infty(\mathcal{B})$, which it is not. We have exhibited an L^1 function which is not in $\mathcal{K}_1(\mathcal{B})$.

(b) Suppose that there is a constant C so that for each $f \in L^1(\mathcal{F})$, $|f| \leq C \cdot E^{\mathcal{B}}|f|$. It then follows from Proposition 1 that $\mathcal{K}_1(\mathcal{B}) \subset L^\infty(\mathcal{F})$ while the reverse inclusion is valid in any case.
Now suppose that $\mathcal{K}_1(\mathcal{B}) = L^\infty(\mathcal{F})$. Let $0 \neq f \in L^1(\mathcal{F})$ and set $S = \{E^{\mathcal{B}}|f| \neq 0\} \in \mathcal{B}$. Define $g = (f/(E^{\mathcal{B}}|f|))\chi_s$. Then $E^{\mathcal{B}}|g| = \chi_s \in L^\infty(\mathcal{B})$ and so $g \in \mathcal{K}_1(\mathcal{B})$. By hypothesis, then, g is in $L^\infty(\mathcal{F})$.Now $S \supset S_f$ so $f = g \cdot E^{\mathcal{B}}|f|$, and consequently $|f| \leq \|g\|_\infty \cdot E^{\mathcal{B}}|f|$ a.e. Define

$$C_n = \{f \in L^1_+(\mathcal{F}) : f \leq n \cdot E^{\mathcal{B}} f \text{ a.e.}\}, \quad n = 1, 2, \dots .$$

Since $E^{\mathcal{B}}$ is continuous on L^1 (and L^1 convergence yields subsequential a.e. convergence), each C_n is closed in $L^1_+(\mathcal{F})$. Moreover the discussion immediately preceding the introduction of the C_n's shows that $\bigcup_{n=1}^{\infty} C_n = L^1_+(\mathcal{F})$. By the Baire Category Theorem, there is an integer N so that C_N has non empty interior. Thus there is an $f_0 \in L^1_+(\mathcal{F})$ and an $r > 0$ so that $\{f \in L^1_+(\mathcal{F}) : \|f - f_0\| \leq r\} \subset C_N$. Moreover, since $L^\infty_+(\mathcal{F})$ is dense in $L^1_+(\mathcal{F})$, we may and do assume that $f_0 \in L^\infty_+(\mathcal{F})$. Let g be a non zero member of $L^1_+(\mathcal{F})$. Then $r \cdot g/\|g\| + f_0 \in C_N$, so

$$\begin{aligned} \frac{r}{\|g\|} g + f_0 &\leq N \cdot E^{\mathcal{B}} \left(\frac{r}{\|g\|} g + f_0 \right) \\ &= \frac{N \cdot r}{\|g\|} E^{\mathcal{B}} g + N E^{\mathcal{B}} f_0. \end{aligned}$$

It follows that

$$g \leq \left(\frac{N \cdot E^{\mathcal{B}} f_0 - f_0}{r} \right) \|g\| + N \cdot E^{\mathcal{B}} g.$$

Let $\rho = (N \cdot E^{\mathcal{B}} f_0 - f_0)/r$. Then since f_0 is in L^∞, so are $E^{\mathcal{B}} f_0$ and ρ. It then follows that $g \leq D \cdot \|g\| + N \cdot E^{\mathcal{B}} g$, where $D = \|\rho\|_\infty$ (which is independent of g). Now let G be any strictly positive function in $L^1(\mathcal{F})$. By replacing g in the last inequality by $G/E^{\mathcal{B}}G$ and noting that $\|G/E^{\mathcal{B}}G\| = \|E^{\mathcal{B}}(G/E^{\mathcal{B}}G)\| = 1$, we see that $G \leq (D + N) \cdot E^{\mathcal{B}} G$. This inequality extends easily to all $G \in L^1_+(\mathcal{F})$, which completes the proof. □

Remark 2.3. Certainly the boundedness criterion in part (b) of Proposition 2.2 is more mysterious than its counterpart in part (a). One would not expect this condition to hold in most examples. Indeed a possible message in the preceding result is that the extreme cases are rare. In fact, if $\mathcal{B}$ is not finitely generated and there exists a non finitely generated sigma algebra independent from $\mathcal{B}$ (in the statistical sense) then $K_1(\mathcal{B})$ differs from both extremes. We now present an example where B is completely non-atomic, $K_1(\mathcal{B}) = L^\infty(\mathcal{F})$, and $\mathcal{B}$ is in some sense far removed from $\mathcal{F}$.

Example 2.4. Let $X = [-1,1]$, $d\mu = \frac{1}{2}\,dx, \mathcal{F}$ the Lebesgue sets, and $\mathcal{B}$ the sigma subalgebra of $\mathcal{F}$ consisting of sets symmetric about the origin. One easily verifies that $E^{\mathcal{B}}f\ (x) = (f(x) + f(-x))/2$, so that $|f| \le 2E^{\mathcal{B}}|f|$ a.e.

Proposition 2.2 shows that in general many subalgebras of $\mathcal{F}$ have the same $\mathcal{K}_p$ spaces, so that analysis of these spaces is of limited value in discriminating among sub sigma algebras. However, when attention is paid to certain somewhat natural subsets, the relationship is brought to focus. As previously, we concentrate on the case $p = 1$.

For $\mathcal{B} \le \mathcal{F}$, define $\mathcal{G}(\mathcal{B}) = \{g \ge 0 : E^{\mathcal{B}}g = 1 \text{ a.e.}\}$ This is precisely the set of members g of $K_1(\mathcal{B})$ for which K_g is a positive isometry. Note that $\mathcal{G}(\mathcal{B})$ is a convex, norm and weakly closed subset of the unit ball of $L^1(\mathcal{F})$.

Proposition 2.5. *The mapping $\Lambda : \mathcal{B} \to \mathcal{G}(\mathcal{B})$ is injective and order reversing. Moreover, Λ is order continuous in the sense that if $\{\mathcal{B}_n\}$ and $\{\mathcal{C}_n\}$ are sequences of sigma algebras such that $\mathcal{B}_n \downarrow \mathcal{B}$ and $\mathcal{C}_n \uparrow \mathcal{B}$ then $\bigcap \mathcal{G}(\mathcal{C}_n) = \mathcal{G}(\mathcal{B}) = L^1$ closure of $\bigcup \mathcal{G}(\mathcal{B}_n)$.*

PROOF. If $\mathcal{B} \le \mathcal{C} \le \mathcal{F}$ and $\phi \in \mathcal{G}(\mathcal{C})$ then $E^{\mathcal{B}}\phi = E^{\mathcal{C}}E^{\mathcal{B}}\phi = E^{\mathcal{B}}E^{\mathcal{C}}\phi = E^{\mathcal{B}}1 = 1$, so $\phi \in \mathcal{G}(\mathcal{B})$. This shows that Λ is order reversing.

Suppose now that $\mathcal{B}$ and $\mathcal{C}$ are sub sigma algebras for which $\mathcal{G}(\mathcal{C}) \subset \mathcal{G}(\mathcal{B})$. We shall show that $\mathcal{B} \subset \mathcal{C}$, which will complete the proof of the first part of the stated proposition. Let $0 < s \le t < \infty$ and let ϕ be a $\mathcal{B}$-measurable function for which $s \le \phi \le t$ a.e. Then $\phi/(E^{\mathcal{C}}\phi) \in \mathcal{G}(\mathcal{C})$ and consequently $\phi/(E^{\mathcal{C}}\phi) \in \mathcal{G}(\mathcal{B})$, i.e., $E^{\mathcal{B}}(\phi/(E^{\mathcal{C}}\phi)) = 1$. But $1/(E^{\mathcal{C}}\phi)$ is essentially bounded and $\phi \in L^\infty(\mathcal{B})$, so

$$1 = E^B\left(\frac{\phi}{E^{\mathcal{C}}\phi}\right) = \phi \cdot E^B\left(\frac{1}{E^{\mathcal{C}}\phi}\right).$$

Noting that for any positive h and any sigma algebra $\mathcal{D}$,

$$1 = \left(E^{\mathcal{D}}(\sqrt{h}\cdot\frac{1}{\sqrt{h}})\right)^2 \le (E^{\mathcal{D}}h)(E^{\mathcal{D}}\frac{1}{h}),$$

we have

$$1 = \phi \cdot E^{\mathcal{B}}\left(\frac{1}{E^{\mathcal{C}}\phi}\right) \ge \phi \cdot \frac{1}{E^{\mathcal{B}}(E^{\mathcal{C}}\phi)};$$

that is, $E^{\mathcal{B}}(E^{\mathcal{C}}\phi)) \geq \phi$. But $\int_X E^{\mathcal{B}}(E^{\mathcal{C}}\phi)\, d\mu = \int_X E^{\mathcal{C}}\ \phi\ d\mu = \int_X \phi f\, d\mu$, so that $E^{\mathcal{B}}(E^{\mathcal{C}}\phi)) = \phi$. A standard approximation argument shows that this equality remains valid for all ϕ in $L^1(\mathcal{B})$. Equivalently, $E^{\mathcal{B}}E^{\mathcal{C}}E^{\mathcal{B}} = E^{\mathcal{B}}$. Now $L^2(\mathcal{F}) \subset L^1(\mathcal{F})$ and the L^2 restrictions of the conditional expectations are the corresponding orthogonal projections. But for Hilbert space orthogonal projections P and Q, $PQP = P$ if and only if $QP = P$. Thus we have $E^{\mathcal{C}}E^{\mathcal{B}} = E^{\mathcal{B}}$ on $L^2(\mathcal{F})$. This means that $L^2(\mathcal{B}) \subset L^2(\mathcal{C})$ and consequently $\mathcal{B} \subset \mathcal{C}$.

Now suppose that $\mathcal{C}_n \uparrow \mathcal{B}$, i.e., the $\mathcal{C}_n$'s form an increasing sequence of sigma algebras and $\mathcal{B}$ is the smallest (complete) sigma algebra containing all of them. Since each $\mathcal{C}_n \subset \mathcal{B}$ and Λ is order reversing, $\bigcap \mathcal{G}(\mathcal{C}_n) \supset \mathcal{G}(\mathcal{B})$.

Let $f \in \cap\mathcal{G}(\mathcal{C}_n)$. Then for each n, $E^{\mathcal{C}_n}(f) = 1$. By the (increasing) matringale convergence theorem [3; p. 29] $E^{\mathcal{B}}f = 1$. Thus $f \in \mathcal{G}(\mathcal{B})$. This shows that $\cap\mathcal{G}(\mathcal{C}_n) = \mathcal{G}(\mathcal{B})$.

Finally, suppose that $B_n \downarrow B$. Then (again by the order reversing nature of Λ) we see that $\mathcal{G}(\mathcal{B}) \supset \bigcup \mathcal{G}(\mathcal{B}_n)$; and since $\mathcal{G}(\mathcal{B})$ is closed in L^1, we have $\mathcal{G}(\mathcal{B}) \supset L^1$ closure of $\bigcup \mathcal{G}(\mathcal{B}_n)$. Let $f \in \mathcal{G}(\mathcal{B})$ and let $t \in (0,1)$. Define $f_t = (1-t)1 + tf$. Since $\mathcal{G}(\mathcal{B})$ is convex, $f_t \in \mathcal{G}(\mathcal{B})$. For each n define f_{tn} to be $f_t/E^{\mathcal{B}_n}(f_t)$. By the (decreasing; a.e. convergence) martingale convergence theorem [3; p. 29] (a.e.) $\text{limit}_{n\to\infty}E^{\mathcal{B}_n}(f_t) = E^{\mathcal{B}}(f_t) = 1$. This shows that (a.e.) $\text{limit}_{n\to\infty} f_{tn} = f_t$. Now

$$
\begin{aligned}
f_{tn} &= \frac{1}{E^{\mathcal{B}_n}(f_t)} \cdot f_t = \frac{1}{(1-t) + tE^{\mathcal{B}_n}(f)} \cdot f_t \\
&\leq \frac{1}{1-t} \cdot f_t = 1 + \frac{t}{1-t} \cdot f \in L^1.
\end{aligned}
$$

The dominated convergence theorem guarantees that $(L^1)\text{limit}_{n\to\infty} f_{tn} = f_t$. But $E^{\mathcal{B}_n}(f_t) = 1$ a.e. so $f_{tn} \in \mathcal{G}(\mathcal{B}_n)$, and consequently $f_t \in L^1$ closure of $\bigcup \mathcal{G}(\mathcal{B}_n)$. Thus $f = (L^1)\ \text{limit}_{t\to 1} f_t \in L^1$ closure of $\bigcup \mathcal{G}(\mathcal{B}_n)$, and consequently $\mathcal{G}(\mathcal{B}) \subset L^1$ closure of $\bigcup \mathcal{G}(\mathcal{B}_n)$. □

We conclude this note with results relating chains of sigma algebras and functional factorization.

Proposition 2.6. *Suppose $\mathcal{B} \leq \mathcal{C} \leq \mathcal{F}$. Then $\mathcal{G}(\mathcal{B},\mathcal{F}) = \mathcal{G}(\mathcal{B},\mathcal{C}) \cdot \mathcal{G}(\mathcal{C},\mathcal{F})$.*

PROOF. The product in the statement of the proposition is with respect to pointwise products of functions. The right-most sigma algebra in each term indicates measurability; so that, for example, $\mathcal{G}(\mathcal{B},\mathcal{C})$ consists of those non negative $\mathcal{C}$-measurable functions f for which $E^{\mathcal{B}}f = 1$ a.e. Let $\gamma \in \mathcal{G}(\mathcal{B},\mathcal{C})$ and $\alpha \in \mathcal{G}(\mathcal{C},\mathcal{F})$. Then

$$
\begin{aligned}
E^{\mathcal{B}}(\alpha\gamma) &= E^{\mathcal{B}}E^{\mathcal{C}}(\alpha\gamma) = E^{\mathcal{B}}(\gamma E^{\mathcal{C}}(\alpha)) \\
&= E^{\mathcal{B}}(\gamma \cdot 1) = E^{\mathcal{B}}(\alpha) = 1.
\end{aligned}
$$

Thus $\mathcal{G}(\mathcal{B},\mathcal{F}) \supset \mathcal{G}(\mathcal{B},\mathcal{C}) \cdot \mathcal{G}(\mathcal{C},\mathcal{F})$. Now let $\phi \in \mathcal{G}(\mathcal{B},\mathcal{F})$. Define

$$\gamma = E^{\mathcal{C}}\phi, \qquad \chi = \chi_{\text{supp } E^{\mathcal{C}}\phi}, \qquad \alpha = \frac{\phi}{E^{\mathcal{C}}\phi}\chi + (1-\chi).$$

Then (since supp $\phi \subset$ supp $E^{\mathcal{C}}\phi$) $\alpha \cdot \gamma = \phi \cdot \chi + (1-\chi) \cdot E^{\mathcal{C}}\phi = \phi$. Now $\gamma \in L^1(\mathcal{C})$ and $E^{\mathcal{B}}\gamma = E^{\mathcal{B}}(E^{\mathcal{C}}\phi) = E^{\mathcal{B}}\phi = 1$, so that $\gamma \in \mathcal{G}(\mathcal{B},\mathcal{C})$. We must show that $\alpha \in \mathcal{G}(\mathcal{C},\mathcal{F})$. Noting that both χ and 1-χ are $\mathcal{C}$-measurable, we have

$$E^{\mathcal{C}}\alpha = E^{\mathcal{C}}\left(\frac{\phi}{E^{\mathcal{C}}\phi}\chi + (1-\chi)\right) = \chi + (1-\chi) = 1;$$

hence $\alpha \in \mathcal{G}(\mathcal{C},\mathcal{F})$. □

Remark 2.7. 1. The γ part of the above factorization is unique, but the α constructed in the above proof may be replaced by $\alpha + \delta$, where δ is any non negative function with integral 1 supported off supp $E^{\mathcal{C}}\phi$.

2. Proposition 2.5 leads directly to the following:

Suppose $\mathcal{B} \leq \mathcal{C} \leq \mathcal{F}$. Then $\mathcal{K}_1(\mathcal{B},\mathcal{F}) = \mathcal{K}_1(\mathcal{B},\mathcal{C}) \cdot \mathcal{K}_1(\mathcal{C},\mathcal{F})$.

References

[1] P. Dodds, C. Huijsmans, and P. DePagter, *Characterizations of conditional expectation-type operators*, Pacific J. Math. **141** (1990), 55–77.

[2] A. Lambert, *Localising sets for sigma-algebras and related point transformations*, Proc. Royal Soc. Edinburgh **118A** (1991), 111–118.

[3] W. Parry, *Topics in Ergodic Theory*, Cambridge Tracks in Mathematics No. 75, Cambridge University Press, London, 1981.

Department of Mathematics
University of North Carolina at Charlotte
Charlotte, North Carolina 28223
E-mail: fma00all@unccvm.uncc.edu

Received: August 23rd, 1995.

Operator Theory:
Advances and Applications, Vol. 104

On Isometric Intertwining Liftings

W. S. LI AND D. TIMOTIN *

Dedicated to Professor Carl M. Pearcy on the occasion of his sixtieth anniversary

0. INTRODUCTION

The commutant lifting theorem of Sz.-Nagy and Foias ([19]), a cornerstone result in the theory of contractive operators on Hilbert space, has found a large variety of applications in most distinct areas of pure and applied analysis (see [13] and the references within). The existence theorem is complemented by different descriptions of the class of all commutant liftings of a given triple. Subsequently, a major direction of studies have concentrated on describing the properties of distinguished liftings; as interesting cases, let us note the central lifting([14], [8]), or the liftings which correspond in engineering applications to rational realizations([18]).

The purpose of the present paper is the investigation of certain isometric liftings. For Hankel operators, an investigation has been pursued in [11], yielding a rather complete characterization of the "non-degenerate" case. In the general situation, Foias, Frazho and Tannenbaum have obtained in [16] existence theorems by a careful analysis of the first step in the standard construction of an intertwining lifting. We have been able to extend their results, mainly by removing the finite dimensionality assumptions, and thence settling some conjectures left open in [16]. A main tool used has been the alternate approach to commutant lifting provided by the theory of couplings (as described in [4]; see also [13] and the references within).

The paper is organized as follows. Section 1 contains the basic notations. Sections 2 and 3 are devoted to the coupling approach to intertwining. Section 4 presents some examples which help to understand the main obstructions towards the existence of an isometric lifting. After a slight reduction of the problem (section 5), the main result (theorem 6.1) is proved in section 6. A few consequences and remarks appear in section 7. Section 8 contains an application to the case of generalized Hankel operators.

1. BASIC NOTATION

Suppose T, T' are two contractions on the Hilbert spaces $\mathcal{H}, \mathcal{H}'$ respectively. The defect operators of T and T' are defined, as usual, by $D_T = (I - T^*T)^{1/2}$ and

*The first named author is supported by grant DMS-9303702 from the National Science Foundation. The second named author wishes to express his gratitude to the School of Mathematics of Georgia Institute of Technology for its kind hospitality while this paper was written.

$D_{T'} = (I - T'^*T')^{1/2}$, and the defect spaces by $\mathcal{D}_T = \overline{D_T\mathcal{H}}$, $\mathcal{D}_{T'} = \overline{D_{T'}\mathcal{H}'}$. The minimal unitary dilations of T and T' will be denoted by U and U', acting on $\mathcal{K}$ and $\mathcal{K}'$ respectively. We will also use the notations

$$\mathcal{K}_+ = \bigvee_{k=0}^{\infty} U^k\mathcal{H}, \quad \mathcal{K}_- = \bigvee_{k=0}^{\infty} U^{*k}\mathcal{H}$$

and similarly for $\mathcal{K}'_+, \mathcal{K}'_-$. Note that $U \mid \mathcal{K}_+$ and $U'^* \mid \mathcal{K}'_-$ are the minimal isometric dilations of T and T'^* respectively. The projection onto any closed subspace $\mathcal{E}_1$ of a Hilbert space $\mathcal{E}$ is denoted by the customary notation $P_{\mathcal{E}_1}$ (or, incidentally, $P^{\mathcal{E}}_{\mathcal{E}_1}$).

Suppose $A : \mathcal{H} \to \mathcal{H}'$ is a contraction which satisfies the commutation relation $AT = T'A$; we will call then (T, T', A) an *intertwining triple*. The Sz.Nagy-Foias commutant lifting theorem ([19]) states that there exists $B : \mathcal{K} \to \mathcal{K}'$, $\|B\| \leq 1$, such that $BU = U'B$ and $P_{\mathcal{H}'}B \mid \mathcal{K}_+ = AP_{\mathcal{H}} \mid \mathcal{K}_+$. The collection of all such operators B is called $CID(A)$. Subsequent work of Arsene, Ceausescu and Foias has led to a detailed description of $CID(A)$. To state their results, we need some more terminology. Thus, define $\mathcal{F} \subset \mathcal{D}_A \oplus \mathcal{D}_T$ by $\mathcal{F} = \overline{\{D_ATh \oplus D_Th \mid h \in \mathcal{H}\}}$, $\mathcal{F}' \subset \mathcal{D}_A \oplus \mathcal{D}_{T'}$ by $\mathcal{F}' = \overline{\{D_Ah \oplus D_{T'}Ah \mid h \in \mathcal{H}\}}$, and $\mathcal{G} = (\mathcal{D}_A \oplus \mathcal{D}_T) \ominus \mathcal{F}$, $\mathcal{G}' = (\mathcal{D}_A \oplus \mathcal{D}_{T'}) \ominus \mathcal{F}'$. A $\{\mathcal{G}, \mathcal{G}'\}$-choice sequence is a sequence of contractions $(\Gamma_n)_{N\geq 1}$ such that $\Gamma_1 : \mathcal{G} \to \mathcal{G}'$, while, for $n \geq 1$, $\Gamma_{n+1} : \mathcal{D}_{\Gamma_n} \to \mathcal{D}_{\Gamma_n^*}$.

Theorem 1.1. ([6]; *see also* [13]) *There exists a one-to-one correspondence between all intertwining liftings B of A and all $\{\mathcal{G}, \mathcal{G}'\}$-choice sequences.*

The actual form of the correspondence is rather intricate; sections 2 and 3 will present an alternate approach to this parametrization, more convenient for our purposes.

The main problem we are interested in is the existence, for a given A, of an isometric intertwining lifting B. The relation between the choice sequence and the lifting is quite intricate, and thus it is not generally possible to deduce properties of one of them from the other; however, in our situation there are good reasons to obtain sufficient conditions by considering a particular type of choice sequence, namely the case in which Γ_1 is an isometry. Such liftings already appear in [2], where they are called *canonical*; they have been subsequently analyzed in [16]. Note that the next choice operators have to be 0, and thus the lifting B is determined by Γ_1 alone. Thus, we will try to find sufficient conditions on T, T' and A, such that an isometric Γ_1 would give rise to an isometric lifting B.

2. Unitary extensions of partial isometries

In this section, we will recall some basic facts about unitary extensions, Schur parameters and choice sequences that will be later needed. For a more detailed account, we refer the interested readers to [3], [7] or [13].

Let $W_0 : \mathcal{M}_0 \to \mathcal{M}_0$ be a partial isometry with initial space $\mathcal{L}_0'$ and range space $\mathcal{L}_0$; denote also, for further use, $\mathcal{G}_0' = \mathcal{M}_0 \ominus \mathcal{L}_0'$, $\mathcal{G}_0 = \mathcal{M}_0 \ominus \mathcal{L}_0$ (the defect spaces of W_0). $(W, \mathcal{M})$ is called a *unitary extension* of $(W_0, \mathcal{M}_0)$ if the following are satisfied:

a) $\mathcal{M}_0 \subset \mathcal{M}$, and $W : \mathcal{M} \to \mathcal{M}$ is unitary,
b) $\mathcal{M} = \bigvee_{n\in\mathbb{Z}} W^n \mathcal{M}_0$,
c) $W|\mathcal{L}_0' = W_0|\mathcal{L}_0'$.

Unitary extensions $(W_1, \mathcal{M}_1)$ and $(W_2, \mathcal{M}_2)$ of $(W_0, \mathcal{M}_0)$ are said to be unitarily equivalent if W_1 and W_2 are unitarily equivalent via an isomorphism which is the identity on $\mathcal{M}_0$.

Suppose $(W, \mathcal{M})$ is a unitary extension of $(W_0, \mathcal{M}_0)$. Denote $\mathcal{M}_j = \mathcal{M}_0 \vee W\mathcal{M}_0 \vee W^2\mathcal{M}_0 \ldots \vee W^j\mathcal{M}_0$, for $j \geq 1$, and

$$\mathcal{G}_j = \mathcal{M}_j \ominus W\mathcal{M}_{j-1}, \qquad \mathcal{G}_j' = \mathcal{M}_j \ominus \mathcal{M}_{j-1},$$

$$\gamma_1 = P_{\mathcal{G}_0}^{\mathcal{M}} W|\mathcal{G}_0', \qquad \gamma_{j+1} = P_{\mathcal{G}_j}^{\mathcal{M}} W|\mathcal{G}_j'.$$

The contractions (γ_j) are called the *Schur parameters* associated to the unitary extension $(W, \mathcal{M})$ of $(W_0, \mathcal{M}_0)$.

If $(W, \mathcal{M})$ is a unitary extension of $(W_0, \mathcal{M}_0)$, then $(W^*, \mathcal{M})$ is a unitary extension of $(W_0^*, \mathcal{M}_0)$. Let $(\hat{\gamma}_j)$, with $\hat{\gamma}_j : \hat{\mathcal{G}}_j' \to \hat{\mathcal{G}}_j$, be the sequence of Schur parameters associated with the unitary extension $(W^*, \mathcal{M})$ of $(W_0^*, \mathcal{M}_0)$. Then γ_j^* and $\hat{\gamma}_j$ are unitarily equivalent for each $j \geq 1$:

$$W^j \hat{\mathcal{G}}_j' = \mathcal{G}_j, \qquad W^{*j}\mathcal{G}_j' = \hat{\mathcal{G}}_j, \quad \text{and} \quad W^j \hat{\gamma}_{j+1} W^{*j}|\mathcal{G}_j = \gamma_{j+1}^*.$$

Recall that a $\{\mathcal{G}_0', \mathcal{G}_0\}$-choice sequence is a sequence of contractions $(C_j)_{j\geq 1}$ such that $C_1 : \mathcal{G}_0' \to \mathcal{G}_0$, and for each $j \geq 1$, $C_{j+1} : \mathcal{D}_{C_j} \to \mathcal{D}_{C_j^*}$.

The following proposition (essentially [3], proposition IV.1) says that a sequence of Schur parameters can be identified as a choice sequence.

Proposition 2.1. *Let $(W, \mathcal{M})$ be a unitary extension of $(W_0, \mathcal{M}_0)$ with Schur parameters (γ_j). For each $j \geq 1$, there exist unitary operators $\phi_j \in \mathcal{L}(\mathcal{G}_j', \mathcal{D}_{\gamma_j})$ and $\psi_j \in \mathcal{L}(\mathcal{G}_j, \mathcal{D}_{\gamma_j^*})$, such that*

$$C_1 = \gamma_1, \qquad C_{j+1} = \psi_j \gamma_{j+1} \phi_j^*, \qquad j \geq 1, \tag{1}$$

is a $\{\mathcal{G}_0', \mathcal{G}_0\}$-choice sequence. Moreover, if (C_j) and $(\hat{C}_j)$ are the choice sequences determined by (γ_j) and $(\hat{\gamma}_j)$, then $\hat{C}_j = C_j^*$ for every $j \geq 1$.

Actually, as has been shown in [9] (see also [7]), a unitary extension $(W, \mathcal{M})$ of $(W_0, \mathcal{M}_0)$ is determined (up to unitary equivalence) by its Schur parameters, and consequently by a choice sequence.

Theorem 2.2. *There is a one-one correspondence between the set of (classes of unitary equivalent) unitary extensions $(W, \mathcal{M})$ of $(W_0, \mathcal{M}_0)$ and the set of $\{\mathcal{G}_0', \mathcal{G}_0\}$-choice sequences.*

We will recall the correspondence as presented in theorem IV.2 of [3], whose form is more convenient for our purposes. Suppose (C_j) is a $\{\mathcal{G}_0', \mathcal{G}_0\}$-choice sequence. Let $\mathcal{M}_+ = \mathcal{M}_0 \oplus (\oplus_{j\geq 1}\mathcal{D}_{C_j})$. We will define inductively an isometry $W_+ : \mathcal{M}_+ \to \mathcal{M}_+$. The first summand of $\mathcal{M}_+$, $\mathcal{M}_0 = \mathcal{L}' \oplus \mathcal{G}_0'$, is mapped into $\mathcal{M}_0 \oplus \mathcal{D}_{C_1}$ according to the formulas $W_+h_0' = W_0h_0' \oplus 0$ for $h_0' \in \mathcal{L}_0'$, while $W_+g_0' = C_1g_0' \oplus D_{C_1}g_0'$ for $g_0' \in \mathcal{G}_0'$. If $j \geq 1$, the induction step will also include the existence of a unitary operator η_j^* from

$$(\mathcal{M}_0 \oplus (\bigoplus_{k=1}^{j} \mathcal{D}_{C_k}) \oplus \{0\}) \ominus W_+(\mathcal{M}_0 \oplus (\bigoplus_{k=1}^{j} \mathcal{D}_{C_k}) \oplus \{0\})$$

onto $\mathcal{D}_{C_j^*}$. Then W_+ is defined inductively on $\{0\} \oplus \ldots \oplus \mathcal{D}_{C_j} \oplus \{0\}$ as

$$W_+(0 \oplus \ldots \oplus d_j \oplus 0) = 0 \oplus \ldots \oplus \eta_j^* C_{j+1} d_j \oplus D_{C_{j+1}} d_j \oplus 0 \tag{2}$$

Finally, take $W : \mathcal{M} \to \mathcal{M}$ to be the minimal unitary extension of W_+. Then $(W, \mathcal{M})$ is a unitary extension of $(W_0, \mathcal{M}_0)$ and $(W, \mathcal{M})$ is determined uniquely (up to unitary equivalence) by (C_j).

Corollary 2.3. *Suppose (C_j) is the corresponding choice sequence of $(W, \mathcal{M})$, a particular unitary extension of $(W_0, \mathcal{M}_0)$. Then $\mathcal{M} = \mathcal{M}_0$ if and only if C_1 is unitary (which implies that $C_j = 0$ for all $j > 1$).*

PROOF. This follows immediately from formula (2). □

3. THE COUPLING APPROACH TO INTERTWINING

The constructions in the preceeding section lead to an alternate, more geometrical approach to the description of all intertwining liftings. Let us first recall (e.g. [13], VII.7.1) that, if $C : \mathcal{H} \to \mathcal{H}'$ is any contraction, then $\mathcal{H}$ and $\mathcal{H}'$ can be embedded in a larger space, denoted $\mathcal{H} \vee_C \mathcal{H}'$, such that $C = P_{\mathcal{H}'} \mid \mathcal{H}$. A convenient description is $\mathcal{H} \vee_C \mathcal{H}' = \mathcal{H}' \oplus \mathcal{D}_C$, via the embeddings

$$h \mapsto Ch \oplus D_C h \quad (h \in \mathcal{H}), \qquad h' \mapsto h' \oplus 0 \quad (h' \in \mathcal{H}') \tag{3}$$

Formulas (3) have some useful consequences, which we will state as a separate corollary.

Corollary 3.1. (a) *In $\mathcal{H} \vee_C \mathcal{H}'$ we have $\ker D_C \subset \mathcal{H}'$; C is isometric iff $\mathcal{H} \subset \mathcal{H}'$.*
(b) *If $\|C\| < 1$, then $\mathcal{D}_C = \mathcal{H}$, and the operator $S : \mathcal{H} \oplus \mathcal{H}' \to \mathcal{H} \vee_C \mathcal{H}'$ defined by $S(h \oplus h') = h + h'$ is invertible.*

Consider now an intertwining triple (T, T', A), $\|A\| \leq 1$; $U \in \mathcal{L}(\mathcal{K})$, $U' \in \mathcal{L}(\mathcal{K}')$ are, as usually, the minimal unitary dilations of T, T' respectively.

With the same notation as in Section 1, let $U_+ \in \mathcal{L}(\mathcal{K}_+)$ and $U'^*_- \in \mathcal{L}(\mathcal{K}'_-)$ be the minimal isometric dilations of T and T'^* respectively. Define $\tilde{A} : \mathcal{K}_+ \to \mathcal{K}'_-$ as

$$\tilde{A} = i_{\mathcal{H}' \to \mathcal{K}'_-} A P_{\mathcal{H}}^{\mathcal{K}_+}$$

where $P_{\mathcal{H}}^{\mathcal{K}_+}$ is the orthogonal projection of $\mathcal{K}_+$ onto $\mathcal{H}$ and $i_{\mathcal{H}' \to \mathcal{K}'_-}$ is the inclusion map of $\mathcal{H}'$ into $\mathcal{K}'_-$. It is easy to check that $\tilde{A}U_+ = U'_-\tilde{A}$. We will denote, for further use, $\mathcal{M}_0(A) = \mathcal{K}'_- \vee_{\tilde{A}} \mathcal{K}_+$.

Now, the partial isometry W_0 of interest to us acts on the space $\mathcal{M}_0 = \mathcal{M}_0(A)$; it has as initial space $\mathcal{L}'_0 = U'^*_-\mathcal{K}'_- \vee_{\tilde{A}} \mathcal{K}_+$ and as range space $\mathcal{L}_0 = \mathcal{K}'_- \vee_{\tilde{A}} U_+\mathcal{K}_+$, and is defined on a dense subset by

$$W_0(U'^*_- k'_- + k_+) = k'_- + U_+k_+$$

for $k'_- \in \mathcal{K}'_-$ and $k_+ \in \mathcal{K}_+$.

Consider now a unitary extension $(W, \mathcal{M})$ of $(W_0, \mathcal{M}_0)$. If $\mathcal{K} = \vee\{W^n\mathcal{K}_+ : n \in \mathbb{Z}\}$ and $\mathcal{K}' = \vee\{W^n\mathcal{K}'_- : n \in \mathbb{Z}\}$, then $U = W|\mathcal{K}$ and $U' = W|\mathcal{K}'$ are minimal unitary dilations of T and T' respectively. Denote $B_W = P_{\mathcal{K}'}^{\mathcal{M}}|\mathcal{K}$. The connection with the theory of intertwining liftings is made by the following theorem ([4], see also [8]).

Theorem 3.2. *The map $W \mapsto B_W$ gives a one-one correspondence between the set of unitary extensions $(W, \mathcal{M})$ of $(W_0, \mathcal{M}_0)$ and the set of contractive intertwining liftings W_B of A. Furthermore, suppose $B \in CID(A)$, and (Γ_j) is the $\{\mathcal{G}, \mathcal{G}'\}$-choice sequence associated with B (in the sense of Theorem 1.1), while (C_j) is the $\{\mathcal{G}'_0, \mathcal{G}_0\}$-choice sequence associated with the extension W_B. Then there exist unitary operators $\Phi : \mathcal{G} \to \mathcal{G}_0$, $\Phi' : \mathcal{G}' \to \mathcal{G}'_0$, such that, for any $j \geq 1$, Φ maps $\mathcal{D}_{\Gamma_j}$ onto $\mathcal{D}_{C^*_j}$, Φ' maps $\mathcal{D}_{\Gamma^*_j}$ onto $\mathcal{D}_{C_j}$, and*

$$\Gamma_j = \Phi'^* C^*_j \Phi \tag{4}$$

The following corollary of proposition 2.1 and formula (4) notices a relation between adjoint intertwining liftings which is implicit in [5].

Corollary 3.3. *Let (T, T', A) be an intertwining triple. Suppose (Γ_j) is the $\{\mathcal{G}, \mathcal{G}'\}$-choice sequence associated with $B \in CID(A)$ and $(\hat{\Gamma}_j)$ is the $\{\hat{\mathcal{G}}, \hat{\mathcal{G}}'\}$-choice sequence associated with $B^* \in CID(A^*)$. Then there exist unitary operators $\Psi : \mathcal{G} \to \hat{\mathcal{G}}'$ and $\Psi' : \mathcal{G}' \to \hat{\mathcal{G}}$ such that for any $j \geq 1$, Ψ maps $\mathcal{D}_{\Gamma_j}$ onto $\mathcal{D}_{\hat{\Gamma}^*_j}$, Ψ' maps $\mathcal{D}_{\Gamma^*_j}$ onto $\mathcal{D}_{\hat{\Gamma}_j}$, and*

$$\hat{\Gamma}_j = \Psi \Gamma^*_j \Psi'^* \tag{5}$$

Recalling that ([13]) the central lifting B corresponds to all choice operators equal to 0, we obtain as an immediate consequence theorem 3.1 of [8].

Corollary 3.4. *The adjoint of the central lifting of a contraction A coincides with the central lifting of A^*.*

4. Some counterexamples

It is not always true that an isometric Γ_1 corresponds to an isometric B. The following counterexamples point out the main obstructions.

Proposition 4.1. *Let T and T' be contractions on $\mathcal{H}$ and $\mathcal{H}'$ respectively, and suppose $T^{*n} \not\to 0$ strongly (in the terminology of [19], $T \notin C_{.0}$). If $A = 0$, then the intertwining triple (T, T', A) has no isometric liftings.*

Proof. If we denote by $\mathcal{R}$ the subspace of $\mathcal{K}_+$ on which U_+ is unitary, it is known ([19]) that $T \notin C_{.0}$ is equivalent to $\mathcal{R} \neq \{0\}$. If B is an isometry, then, according to corollary 3.1 a), there exists an embedding of $\mathcal{K}$ into $\mathcal{K}'$, such that $U' \mid \mathcal{K} = U$. If $\xi \in \mathcal{R}$, then, for any n we may decompose $U'^{*n}\xi = \zeta_n + \eta_n$, $\zeta_n \in \mathcal{K}'_-$, $\eta_n \in \mathcal{K}'^{\perp}_-$. Since $U' \mid \mathcal{K}'^{\perp}_-$ is a pure isometry, $\eta_n \to 0$. We have

$$\|\xi\|^2 = \|U'^{*n}\xi\|^2 = \langle U'^{*n}\xi, U'^{*n}\xi\rangle = \langle U'^{*n}\xi, \zeta_n\rangle + \langle U'^{*n}\xi, \eta_n\rangle$$

The first term is zero, since $A = 0$ implies $\mathcal{K}_+ \perp \mathcal{K}'_-$, and $U'^{*n}\xi = U^{*n}\xi \in \mathcal{R} \subset \mathcal{K}_+$. The second tends to 0 since $\eta_n \to 0$. Thus, $\xi = 0$, contradicting the assumption $\mathcal{R} \neq \{0\}$. □

We are thus led to suppose $T \in C_{.0}$. We will now present an example that appears in [1]; the details are provided for completeness.

Example 4.2. Let $\mathcal{H} = \mathcal{K}_+ = H^2(\mathbb{T})$, $\mathcal{H}' = \mathcal{K}'_- = H^2_-(\mathbb{T}) = L^2(\mathbb{T}) \ominus H^2(\mathbb{T})$, $\mathcal{K} = \mathcal{K}' = L^2(\mathbb{T})$. If $U = U'$ = multiplication with e^{it} on $L^2(\mathbb{T})$, define $T = U|\mathcal{H}$ and $T' = P_{\mathcal{H}'}U'|\mathcal{H}'$ (the notation is consistent with our usual one, since U and U' are indeed minimal unitary dilations of T and T' respectively). We will then reverse the usual order of definition, by taking $B : \mathcal{K} \to \mathcal{K}'$ to be multiplication with the function $f = \chi b$, where χ is the characteristic function of some interval $(\pi - \alpha, \pi + \alpha)$, while $b \in H^\infty$ is defined by $b(z) = e^{(z+1)/(z-1)}$. Consequently, A will be defined by $Ag = P_{H^2_-} fg$. We claim that in this case $\mathcal{G} = \{0\}$.

Indeed, consider $\mathcal{M} = \mathcal{K} \vee_B \mathcal{K}'$. According to formula (3), we can identify $\mathcal{M}$ with $L^2 \oplus \overline{(1 - |f|^2)^{1/2} L^2}$, with the embeddings of $\mathcal{K}$ and $\mathcal{K}'$ being defined by $k \mapsto fk \oplus (1 - |f|^2)^{1/2} k$ and $k' \mapsto k' \oplus 0$ ($k \in \mathcal{K}$, $k' \in \mathcal{K}'$). Then $\mathcal{G} = \{0\}$ is equivalent to $\mathcal{K}'_- \vee U\mathcal{K}_+ = \mathcal{K}'_- \vee \mathcal{K}_+$.

We will prove the stronger equality $\mathcal{K}'_- \vee U\mathcal{K}_+ = \mathcal{M}$. We have $\mathcal{K}'_- = H^2_- \oplus \{0\}$ and $U\mathcal{K}_+ = \{e^{it} f\xi \oplus e^{it}(1 - |f|^2)^{1/2}\xi \mid \xi \in H^2\}$. Take $\psi \oplus \phi \in \mathcal{M}$, orthogonal to $\mathcal{K}'_- \vee U\mathcal{K}_+$ ($\psi \in L^2(\mathbb{T})$, $\phi \in \overline{(1 - |f|^2)^{1/2} L^2(\mathbb{T})}$). We should have then $\psi \in H^2$ and

$$\langle \psi, e^{it} f\xi\rangle + \langle \phi, e^{it}(1 - |f|^2)^{1/2}\xi\rangle = 0$$

for any $\xi \in H^2$, or, equivalently, $\overline{f}\psi + (1 - |f|^2)^{1/2}\phi \in e^{it} H^2_-$. Using the definition of f, this becomes

$$\overline{b}\psi + (1 - \chi)((1 - |b|^2)^{1/2}\phi - \overline{b}\psi) \in e^{it} H^2_-$$

Denoting $g = (1-|b|^2)^{1/2}\phi$, we should thus have

$$b\psi + (1-\chi)(g - b\psi) = a + \zeta \tag{6}$$

where $\psi \in H^2$, $\zeta \in H^2_-$, $g \in L^2(\mathbb{T})$, $a \in \mathbb{C}$.

Consider the conformal map $w = i(1+z)/(1-z)$, which maps the unit disc onto the upper half-plane, and the unit circle onto the extended real line; its inverse is given by $z = (w-i)/(w+i)$. The corresponding transformation on function spaces, namely

$$g(z) \mapsto \frac{1}{w+i} g(\frac{w-i}{w+i}),$$

maps $L^2(\mathbb{T})$ onto $L^2(\mathbb{R})$, H^2 onto the $\mathbb{H}^2$, the Hardy space of the upper half-plane and H^2_- onto $\mathbb{H}^2_-$, the Hardy space of the lower half-plane. Applying this transformation to (6), we obtain

$$e^{iw}\Psi + (1-X)(G - e^{iw}\Psi) = \frac{a}{w+i} + Z \tag{7}$$

where $\Psi \in \mathbb{H}^2$, $Z \in \mathbb{H}^2_-$, $G \in L^2(\mathbb{R})$, while $1-X$ is the characteristic function of the interval $(-\tan\alpha/2, \tan\alpha/2)$. Denote $G_1 = (1-X)(G - e^{iw}\Psi)$; G_1 is an L^2 function with compact support in $\mathbb{R}$, and we can rewrite (7) as

$$e^{iw}\Psi + G_1 = \frac{a}{w+i} + Z$$

Applying the Fourier transform to both sides, we obtain

$$\hat{\Psi}(t-1) + \hat{G_1} = a' e^{-t}\chi_{[0,\infty)}(t) + \hat{Z}(t)$$

where, by the Paley-Wiener theorem, $\hat{\Psi}$ is supported on $[0,\infty)$, $\hat{Z}$ is supported on $(-\infty, 0]$, and $\hat{G_1}$ is entire. Since the two entire functions $\hat{G_1}$ and $a'e^{-t}$ agree on the real interval $[0,1]$, they should coincide. But e^{-t} is not in $L^2(\mathbb{R})$; therefore this is possible only for $a' = 0$ and $G_1 \equiv 0$. It follows then that $\hat{\Psi} \equiv \hat{Z} \equiv 0$, whence, turning back to the original assumption, $\psi = \zeta = 0$, $a = 0$. From (6) and the definition of g we must have $(1-|f|^2)^{1/2}\phi = (1-\chi)(1-|b|^2)^{1/2}\phi = 0$, and thence $\phi = 0$, since $\phi \in \overline{(1-|f|^2)^{1/2}L^2}$. We have thus proved that $\psi = \phi = 0$, and thus indeed $\mathcal{K}'_- \vee U\mathcal{K}_+ = \mathcal{M}$, $\mathcal{G} = \{0\}$.

Consequently, there is a unique choice sequence corresponding to the triple (T, T', A), and thus a unique intertwining dilation, which has as first choice operator an isometry, namely the zero operator. (If one feels uneasy about calling it an isometry, a derived "proper" counterexample can always be obtained by adding a direct summand.) This intertwining dilation must coincide with B. But, obviously, multiplication with f is not an isometric operator on $L^2(\mathbb{T})$, since $f = \chi b$ is identically 0 on an arc of $\mathbb{T}$.

It turns out that the behaviour of this example is connected to the fact that the range of D_A is not closed. Thus, we will suppose in the sequel, as in [16], that the range of D_A is closed; this is verified, for example, if A is compact, or strictly contractive.

5. A reduction

Let (T, T', A) be an intertwining triple. We are investigating the liftings of A that correspond to Γ_1 being an isometry; obviously a necessary condition for the existence of such liftings is $\dim \mathcal{G} \leq \dim \mathcal{G}'$. Suppose that $\Gamma_1 : \mathcal{G} \to \mathcal{G}'$ is a *nonunitary* isometry, and B is the corresponding lifting. Let r be the dimension of $\mathcal{G}' \ominus \Gamma_1 \mathcal{G}$; consider the contraction $\breve{T} = T \oplus 0$ acting on the Hilbert space $\breve{\mathcal{H}} = \mathcal{H} \oplus \mathcal{H}_r$, where $\mathcal{H}_r$ is a Hilbert space of dimension r. The unitary dilation of $\breve{T}$ acts on

$$\breve{\mathcal{K}} = \mathcal{K} \oplus M(\mathcal{H}_r) \qquad (8)$$

where we have denoted by $M(\mathcal{H}_r)$ a countable direct sum of copies of $\mathcal{H}_r$, indexed according to $\mathbb{Z}$, and is equal to $U \oplus S_r$, S_r being the bilateral shift of multiplicity r. If $\breve{A} : \breve{\mathcal{H}} \to \mathcal{H}'$ is defined by the operator matrix $(A \quad 0)$, then $\breve{A}\breve{T} = T'\breve{A}$. Moreover, if $\breve{\mathcal{G}}_0, \breve{\mathcal{G}}_0'$ are the corresponding defect spaces for the triple $(\breve{T}, T', A)$, then it is easily checked that $\breve{\mathcal{G}}_0 = \mathcal{G}_0 \oplus \mathcal{H}_r$, while $\breve{\mathcal{G}}_0' = \mathcal{G}_0'$. We may thus define a unitary operator $\breve{\Gamma}_1 : \breve{\mathcal{G}}_0 \to \mathcal{G}_0'$ which coincides with Γ_1 on $\mathcal{G}_0$, and there exists a unique lifting $\breve{B}$ of $\breve{A}$ that has $\breve{\Gamma}_1$ as first choice operator.

Proposition 5.1. *With the above notations, B is an isometry if and only if $\breve{B}$ is an isometry.*

Proof. Consider the space $\mathcal{M}_0(\breve{B})$. The definition of $\breve{A}$ implies that $\mathcal{M}_0(B) \subset \mathcal{M}_0(\breve{B})$. Applying corollary 3.1 a), if $\breve{B}$ is an isometry, then $\mathcal{M}_0(\breve{B}) = \mathcal{K}'$, and thus also B is an isometry.

On the other hand, if B is an isometry, it follows that inside $\mathcal{M}_0(\breve{B})$ we have $\mathcal{K} \subset \mathcal{K}'$. On the other side, $\breve{\Gamma}_1$ unitary implies, according to (1) and (4), that $\breve{\gamma}_1$, the corresponding Schur parameter, is unitary. Therefore $\mathcal{H}_r \subset \breve{\mathcal{G}}_0 \subset W\breve{\mathcal{G}}_0' \subset \mathcal{K} \vee \mathcal{K}' \subset \mathcal{K}'$. Since $\mathcal{K}'$ reduces W, it follows that $W^k\mathcal{H}_r \subset \mathcal{K}'$ for any $k \in \mathbb{Z}$. By formula (8), this implies $\breve{\mathcal{K}} \subset \mathcal{K}'$, and thus $\breve{B}$ is an isometry. □

Proposition 5.1 shows that we may restrict ourselves to the case when Γ_1 is unitary.

6. The main result

In this section we will keep the same notation as in sections 1 and 2. Fix an intertwining triple (T, T', A) as before, and let $U \in \mathcal{L}(\mathcal{K})$ and $U' \in \mathcal{L}(\mathcal{K}')$ be minimal unitary dilations of T and T' respectively. Throughout the section, we will assume that $T \in C_{\cdot 0}$ and D_A has closed range. Recall that $\tilde{A} = P^{\mathcal{M}}_{\mathcal{K}'_-}|\mathcal{K}_+$, $\mathcal{M}_0 = \mathcal{K}'_- \vee_{\tilde{A}} U_+\mathcal{K}_+$, while $W_0 \in \mathcal{L}(\mathcal{M}_0)$ is a partial isometry with initial space $\mathcal{L}_0'$ and range space $\mathcal{L}_0$.

Theorem 6.1. *Let (T, T', A) be an intertwining triple with $\|A\| \leq 1$, $T \in C_{\cdot 0}$, and D_A has closed range. Suppose $B \in CID(A)$ with $\{\mathcal{G}, \mathcal{G}'\}$-choice sequence $(\Gamma_1, 0, 0, \ldots)$. If Γ_1 is an isometry, then B is an isometry .*

PROOF. According to proposition 5.1, we can assume that Γ_1 is unitary. Therefore, by proposition 2.1 and theorem 3.2, $\dim \mathcal{G}_0 = \dim \mathcal{G}_0'$ and C_1 is unitary, while $C_j = 0$ for $j > 1$. Let $(W, \mathcal{M})$ be the unitary extension of $(W_0, \mathcal{M}_0)$ with (C_j) as the corresponding $\{\mathcal{G}_0', \mathcal{G}_0\}$-choice sequence. Since C_1 is unitary, corollary 2.3 implies that $\mathcal{M} = \mathcal{M}_0 = \mathcal{K}_-' \vee_{\tilde{A}} U_+\mathcal{K}_+$.

Recall now that $B = P_{\mathcal{K}'}|\mathcal{K}$. To show that it is unitary, it suffices to check that $\mathcal{K}' = \mathcal{M}$, or $\mathcal{K} \subset \mathcal{K}'$.

Define $\mathcal{K}_{++} = \mathcal{K}_+ \ominus \ker D_{\tilde{A}}$. Since D_A has closed range, therefore $D_{\tilde{A}}$ also has closed range and it follows that 1 is an isolated point in the spectrum of $|\tilde{A}|$. Thus we have

$$\|\tilde{A}|\mathcal{K}_{++}\| = \|P_{\mathcal{K}_-'}|\mathcal{K}_{++}\| < 1.$$

Denote $\mathcal{K}_0 = \mathcal{K}_-' \vee_{(\tilde{A}|\mathcal{K}_{++})} U_+\mathcal{K}_{++}$. From Corollary 3.1 a), we have $\ker D_{\tilde{A}} \subset \mathcal{K}_-'$. Therefore, to complete the proof, we need to show that $\mathcal{K}_0 = \mathcal{K}'$. First we need the following elementary lemma.

Lemma 6.2. *Let $\mathcal{E}$ and $\mathcal{E}'$ be closed linear subspaces of some Hilbert space $\mathcal{N}$ and $\mathcal{E} \vee \mathcal{E}' = \mathcal{N}$. Suppose $C = P_{\mathcal{E}'}|\mathcal{E}$ with $\|C\| < 1$. Then there exists a constant $\alpha > 0$ such that*

$$\|y\| \leq \alpha(\|P_{\mathcal{E}}y\| + \|P_{\mathcal{E}'}y\|) \tag{9}$$

for every $y \in \mathcal{N}$.

PROOF. Consider, in corollary 3.1 b), $\mathcal{H} = \mathcal{E}, \mathcal{H}' = \mathcal{E}'$. The adjoint of the operator S is given by the formula

$$S^*(y) = P_{\mathcal{E}}y \oplus P_{\mathcal{E}'}y$$

Since S is invertible, S^* is also invertible, and formula (9) follows if we take $\alpha = \|S^{*-1}\|$. □

To complete the proof of Theorem 6.1, fix $\xi \in \mathcal{K}_0 \ominus \mathcal{K}'$. For every $n \geq 1$, $P_{\mathcal{K}_-'} W^{*n}\xi = 0$ and

$$\begin{aligned}
\|\xi\| = \|W^{*n}\xi\| &\leq \alpha(\|P_{\mathcal{K}_{++}} W^{*n}\xi\| + \|P_{\mathcal{K}_-'} W^{*n}\xi\|) \\
&= \alpha\|P_{\mathcal{K}_{++}} P_{\mathcal{K}} W^{*n}\xi\| = \alpha\|P_{\mathcal{K}_{++}} W^{*n} P_{\mathcal{K}}\xi\| \\
&\leq \alpha\|P_{\mathcal{K}_+} W^{*n} P_{\mathcal{K}}\xi\| = \alpha\|U_+^{*n} P_{\mathcal{K}}\xi\|.
\end{aligned}$$

Since $T \in C_{\cdot 0}$, the minimal isometric dilation U_+^* of T is a pure isometry, therefore $\|U_+^{*n} P_{\mathcal{K}}\xi\| \to 0$ as $n \to \infty$. Thus $\xi = 0$ and the proof is completed. □

7. Further consequences

It is not hard to obtain by duality some consequences of theorem 6.1. We need first the following simple lemma.

Lemma 7.1. *For any contraction A, the range of D_A is closed if and only if the range of D_{A^*} is closed.*

Proof. Since D_A is selfadjoint, its range is closed if and only if 0 is not an accumulation point of the spectrum of D_A, or, equivalently, of the spectrum of $I - A^*A$. This is the same as saying that 1 is not an accumulation point of the spectrum of A^*A; since $\sigma(A^*A)\cup\{0\} = \sigma(AA^*)\cup\{0\}$, we can reverse the argument to obtain that the range of D_{A^*} is closed. □

Then, combining Theorem 6.1 with Corollary 3.3, we easily obtain the following results.

Corollary 7.2. *Let (T, T', A) be an intertwining triple, with D_A closed and $T' \in C_{0\cdot}$. If Γ_1 is a coisometry, then so is B.*

Corollary 7.3. *Suppose that, moreover, $T \in C_{\cdot 0}$. If Γ_1 is unitary, then so is B.*

Theorem 6.1, together with the above corollaries and with proposition 4.1, answers some questions left open in [16], where similar results are obtained under the restrictive hypothesis of the finite dimensionality of some defect spaces. According to the terminology therein, the intertwining liftings corresponding to Γ_1 isometry or coisometry are said to have *minimal entropy*; the reason is that in the finite dimensional case their entropy integral (as defined in [12]) is $-\infty$.

In the general case, it does not seem easy to decide when a given intertwining triple has isometric lifting. An appealing conjecture would be that whenever an isometric lifting exists at all, we can find also one of minimal entropy. The next simple result also hints towards this direction. Recall that the central intertwining ([14]) is defined by taking the corresponding choice sequence to be identically 0.

Proposition 7.4. *If the central lifting is isometric, then $\mathcal{G} = \{0\}$.*

Proof. We may assume that T is isometric; note that in this case $\mathcal{F} \subset \mathcal{D}_A$ and $\mathcal{G} = \mathcal{D}_A \ominus \mathcal{F}$. Formula (1.10) from [15] says that, if B is the central lifting of A, we have, for any $h \in \mathcal{H}$,

$$\|Bh\|^2 \leq \|Ah\|^2 + \|P_{\mathcal{F}} D_A h\|^2 = \|h\|^2 + \|P_{\mathcal{G}} D_A h\|^2$$

If B is an isometry, we must have $P_{\mathcal{G}} D_A h = 0$ for any $h \in \mathcal{H}$, which implies $\mathcal{G} = \{0\}$. □

Thus, if the central lifting is isometric, it is actually the unique lifting, and obviously has minimal entropy as well.

On the other hand, even the following simple question is not yet settled to our knowledge: if an intertwining triple has an isometric lifting, does it follow that $\dim \mathcal{G} \leq \dim \mathcal{G}'$?

8. An application: generalized Hankel operators

Theorem 6.1 and its corollaries can be applied to several classical problems. In this section we will take a closer look at the generalized Nehari problem ([2]). This can be stated as follows: given Hilbert spaces $\mathcal{E}$ and $\mathcal{E}'$ and a sequence of operators $A_n : \mathcal{E} \to \mathcal{E}'$, find conditions for the existence of a measurable function $f \in L^\infty(\mathcal{E}, \mathcal{E}')$, $\|f\|_\infty \leq 1$, such that

$$\frac{1}{2\pi} \int_0^{2\pi} e^{int} f(e^{it})\, dt = A_n \tag{10}$$

for all $n \geq 1$. It is well known ([2], [13]) that a necessary and sufficient condition is the requirement that $\|A\| \leq 1$, where A is the generalized Hankel operator defined by the matrix

$$A = \begin{pmatrix} A_1 & A_2 & A_3 & \dots \\ A_2 & A_3 & \dots & \\ A_3 & \dots & & \\ \vdots & & & \end{pmatrix} \tag{11}$$

This result can be obtained from the commutant lifting theorem as follows. Define by U and U' the operators of multiplication with e^{it} acting on the spaces $L^2(\mathcal{E})$ and $L^2(\mathcal{E}')$ respectively. Consider A as an operator from $H^2(\mathcal{E})$ to $H^2_-(\mathcal{E}')$ having the matrix (11) with respect to the natural decompositions of these spaces, while $T = U|H^2(\mathcal{E})$, $T' = P_{H^2_-(\mathcal{E}')}U'|H^2_-(\mathcal{E}')$. Then (T, T', A) is an intertwining triple, and any intertwining lifting for A has to be multiplication with an function $f \in L^\infty(\mathcal{E}, \mathcal{E}')$; it is easily seen that (10) is then verified.

Since we have $T \in C_{.0}$, $T' \in C_{0.}$, applying theorem 6.1 and corollary 7.2 yields the next result.

Theorem 8.1. *With the above notations, if D_A has closed range, then there exists a function $f \in L^\infty(\mathcal{E}, \mathcal{E}')$, satisfying (10), which is either isometric almost everywhere or coisometric almost everywhere (on $\mathbb{T}$).*

Obviously the two cases correspond to $\dim \mathcal{G} \leq \dim \mathcal{G}'$ and $\dim \mathcal{G} \geq \dim \mathcal{G}'$ respectively. We may strengthen this result in case one of the dimensions of $\mathcal{E}$ or $\mathcal{E}'$ is finite.

Theorem 8.2. *Suppose $\dim \mathcal{E} < \infty$, and also $\dim \mathcal{E} \leq \dim \mathcal{E}'$. If D_A has closed range, then there exists $f \in L^\infty(\mathcal{E}, \mathcal{E}')$, satisfying (10), almost everywhere isometric.*

Proof. According to the preceeding result, it is enough to show that $\dim \mathcal{G} \leq \dim \mathcal{G}'$. Suppose that, on the contrary, $\dim \mathcal{G} > \dim \mathcal{G}'$. Consider the space $\breve{\mathcal{E}}' = \mathcal{E}' \oplus \mathbb{C}$, and the Hankel operator $\breve{A}$, whose action is the same as that of A, but with range space $H^2_-(\breve{\mathcal{E}}')$. For the new situation we have $\breve{\mathcal{G}} = \mathcal{G}$, while $\dim \breve{\mathcal{G}}' = \dim \mathcal{G}' +$

1, and thus $\dim \breve{\mathcal{G}}' \leq \dim \mathcal{G}$. We may then apply corollary 7.2 to get a coisometric intertwining lifting. But then the corresponding function $f \in L^\infty(\mathcal{E}, \breve{\mathcal{E}}')$ has to be coisometric almost everywhere, which contradicts $\dim \breve{\mathcal{E}}' = \dim \mathcal{E}' + 1 > \dim \mathcal{E}$. □

The condition $\dim \mathcal{E} < \infty$ cannot be deleted from the statement of theorem 8.2. In case $\dim \mathcal{E} = \dim \mathcal{E}' = \infty$, it may happen that only one of the possibilities in the conclusion of theorem 8.1 occurs. The simplest example is obtained by taking $A_n = 0$ for $n \geq 2$ and A_1 is a nonunitary isometry; then A has a unique lifting B which is also a nonunitary isometry.

In case $\|A\| < 1$, the condition $\dim \mathcal{E} < \infty$ is no longer necessary; the corresponding result has been proved by Gadidov ([17]).

REFERENCES

[1] V. M. ADAMJAN, D. Z. AROV AND M. G. KREIN, *Infinite Hankel matrices and generalized Caratheodory-Fejer and I. Schur problems* , Functional Anal. Appl **2** (1968), 269–281.

[2] V. M. ADAMJAN, D. Z. AROV AND M. G. KREIN, *Infinite Hankel block matrices and related extension problems*, Amer. Math. Soc. Trans. **111** (1978), 133–156.

[3] R. AROCENA, *Schur analysis of a class of translation invariant forms* in: "Analysis and partial differential equations. A collection of papers dedicated to Mischa Cotlar" (C. Sadosky, Ed.) (New York: Marcel Dekker, 1989, pages 355–369).

[4] R. AROCENA, *Unitary extensions of isometries and contractive intertwining dilations*, Operator theory: Advances and Applications **41** (1989), 13–23.

[5] GR. ARSENE, Z. CEAUSESCU AND C. FOIAS, *On intertwining dilations VII*. In: "Complex Analysis, Joensuu 1978," LNM 747 (New York: Springer-Verlag, 1979, pages 24–45).

[6] GR. ARSENE, Z. CEAUSESCU AND C. FOIAS, *On intertwining dilations VIII*, J. Operator Theory **4** (1980), 55–91.

[7] M. BAKONYI AND T. CONSTANTINESCU, *Schur's algorithm and several applications*. Pitman Research Notes in Mathematics Series No. 261 (1992).

[8] H. BERCOVICI, C. FOIAS AND A. FRAZHO, *Central commutant liftings in the coupling approach*, (Preprint).

[9] T. CONSTANTINESCU, *A general extrapolation problem*, Rev. Roumaine Math. Pures Appl. **32** (1987), 509–521.

[10] M. COTLAR AND C. SADOSKY, *Transference of matrices induced by unitary couplings, a Sarason theorem for the bidimensional torus, and a Sz.-Nagy – Foias theorem for two pairs of dilations*, J. Functional Anal. **111** (1993), 473–488.

[11] H. DYM AND I. GOHBERG, *Unitary interpolants, factorization indices and infinite block Hankel matrices*, J. Functional Anal.**54** (1983), 229–289.

[12] H. DYM AND I. GOHBERG, *A maximum entropy principle for contractive interpolants*, J. Functional Anal. **65** (1986), 83–125 .

[13] C. FOIAS AND A. FRAZHO, *The commutant lifting approach to interpolation problems*. Boston: Birkhäuser Verlag, 1990.

[14] C. FOIAS, A. FRAZHO AND I. GOHBERG, *Central intertwining lifting, maximum entropy and their permanence*, Integral Equations Operator Theory **18** (1994), 166–201.

[15] C. Foias, A. Frazho and W. S. Li, *The exact H^2 estimate for the central H^∞ interpolant*, Operator Theory: Advances and Applications **64** (1993), 119–156.

[16] C. Foias, A. Frazho and A. Tannenbaum, *On certain minimal entropy extensions appearing in dilation theory*, Linear Algebra Appl. **137/138** (1990), 213–238.

[17] R. Gadidov, *On the commutant lifting theorem and Hankel operators.* In: "Algebraic Methods in Operator Theory" (Boston: Birkhäuser Verlag, 1994, 3–9).

[18] T. T. Georgiou, *Realization of power spectra from partial covariance sequences*, IEEE Transactions Acoustics, Speech and Signal Processing **ASSP-35** (1987), 438–449.

[19] B. Sz.-Nagy, C. Foias, *Harmonic Analysis of Operators on Hilbert space.* Amsterdam: North-Holland, 1970.

W. S. Li
School of Mathematics
Georgia Institute of Technology
Atlanta, GA 30332
E-mail: li@math.gatech.edu

D. Timotin
Inst. of Mathematics of the Romanian Academy
PO-Box 1-764
Bucharest 70700, Romania
E-mail: dtimotin@imar.ro

Received: August 23rd, 1995.

Operator Theory:
Advances and Applications, Vol. 104

The Predual of a Type I Von Neumann Algebra

MICHAEL MARSALLI

For Professor Carl Pearcy on his 60th birthday

ABSTRACT. A von Neumann algebra A on a separable, complex Hilbert space H has property $\mathbf{A}_n$ if for every $n \times n$ array $\{f_{i,j}\}$ of elements in the predual there exist sequences $\{x_i\}, \{y_j\}$ in H such that $f_{i,j}(a) = (ax_i, y_j)$ for all a in A and $0 \leq i, j < n$. We characterize the type I von Neumann algebras with property $\mathbf{A}_n$.

Let $B(H)$ be the algebra of bounded operators on a separable, complex Hilbert space H. Let S be an ultraweakly closed subspace of $B(H)$. The predual of S, denoted by S_*, is the set of all ultraweakly continuous linear functions on S. There has been considerable interest in the structure of S_*, particularly when S is a singly generated algebra (cf. [1]). But the study of the predual of a von Neumann algebra goes back to [4]. In this note we use the results of [3] to obtain more precise information about the structure of the predual of a type I von Neumann algebra.

For x, y in H, let $x \otimes y$ denote the element of S_* defined by $(x \otimes y)(s) = (sx, y)$ for $s \in S$. The following definition plays a central role.

Definition 1. *Let S be an ultraweakly closed subspace of $B(H)$, and let n be a cardinal number with $1 \leq n \leq \aleph_0$. The space S has property $\mathbf{A}_n$ if for every $n \times n$ array $\{f_{i,j}\}, 0 \leq i, j \leq n$, of elements of S_*, there exist sequences $\{x_i\}$ and $\{y_j\}, 0 \leq i, j \leq n$ in H such that $f_{i,j} = x_i \otimes y_j$ for $0 \leq i, j \leq n$.*

We will also make use of the following refinement of property $\mathbf{A}_1$.

Definition 2. *Let S be an ultraweakly closed subspace of $B(H)$, and let $r \geq 1$. The space S has property $\mathbf{A}_1(r)$ if for every f in S_* and for every $s > r$, there exist vectors x and y in H such that $f = x \otimes y$ and $\|x\| \, \|y\| \leq s\|f\|$.*

In [3] the abelian von Neumann algebras with property $\mathbf{A}_n$ are characterized, and it is shown that type III von Neumann algebras have property $\mathbf{A}_{\aleph_0}$. In this note we will characterize the type I von Neumann algebras with property $\mathbf{A}_n$. First we will establish our notation. Henceforth, A will denote a von Neumann algebra. If B is also a von Neumann algebra, we write $A \cong B$ to denote that A is unitarily equivalent to B. It is easy to see that the above properties are preserved by unitary equivalence. Also, the above properties are inherited by ultraweakly closed subspaces of S by [1, Proposition 2.04]. For $1 \leq n \leq \aleph_0$, we use $M_n(A)$ to denote the von Neumann algebra of all $n \times n$ matrices with entries from A which act as bounded operators on $H^{(n)}$,the direct sum of n copies of H. And $\mathbf{I}_n$ will denote the algebra of scalars on an n dimensional Hilbert space.

Let A be a type I von Neumann algebra. It is well known that

$$A \cong \sum_{p,q} \oplus M_p(A_q \otimes \mathbf{I}_q)$$

where each A_q is a maximal abelian von Neumann algebra (cf. [2, Section 9.3]). We are now ready to state the main theorem.

Theorem 1. *Suppose A is a type I von Neumann algebra. So $A \cong \sum_{p,q} \oplus M_p(A_q \otimes \mathbf{I}_q)$, where each A_q is a maximal abelian algebra acting on H_q. Then A has property $\mathbf{A}_n$ if and only if $H_q = \{0\}$ for $q < pn$.*

The proof will require three lemmas. The first lemma is a combination of [1, Proposition 2.3] and [3, Theorem 4]. We include it for convenience.

Lemma 1. *Let A be a von Neumann algebra. For $1 \leq n \leq \aleph_0, A$ has property $\mathbf{A}_n$ if and only if $M_n(A)$ has property $\mathbf{A}_1$.*

The next lemma reduces our problem to characterizing the summands $M_p(A_q \otimes \mathbf{I}_q)$ with property $\mathbf{A}_n$.

Lemma 2. *Let (A_i) be a sequence of von Neumann algebras. Then the von Neumann algebra $A = \sum_i \oplus A_i$ has property $\mathbf{A}_n$ if and only if each A_i has property $\mathbf{A}_n$.*

PROOF. Because each A_i is unitarily equivalent to an ultraweakly closed subspace of A, each A_i has property $\mathbf{A}_n$ if A does. Now assume each A_i has property $\mathbf{A}_n$. Then each $M_n(A_i)$ has property $\mathbf{A}_1$ by Lemma 1. Thus by [3, Theorem 1] each $M_n(A_i)$ has property $\mathbf{A}_1(1)$. Now $M_n(A) \cong \sum_i \oplus M_n(A_i)$. Since each summand has property $\mathbf{A}_1(1)$, the sum has property $\mathbf{A}_1$ by [1, Proposition 2.055]. Thus A has property $\mathbf{A}_n$. □

The last lemma characterizes the summands $M_p(A_q \otimes \mathbf{I}_q)$ with property $\mathbf{A}_n$.

Lemma 3. *Let A be a maximal abelian von Neumann algebra, and let $1 \leq p, q \leq \aleph_0$. Then $M_p(A \otimes \mathbf{I}_q)$ has property $\mathbf{A}_n$ whenever $pn \leq q$. If $q < pn$, and $M_p(A \otimes \mathbf{I}_q)$ does not act on the space $\{0\}$, then $M_p(A \otimes \mathbf{I}_q)$ does not have property $\mathbf{A}_n$.*

PROOF. Because $M_n(M_p(A \otimes \mathbf{I}_q)) \cong M_{pn}(A \otimes \mathbf{I}_q)$, we have that $M_p(A \otimes \mathbf{I}_q)$ has property $\mathbf{A}_n$ if and only if $A \otimes \mathbf{I}_q$ has property A_{pn} by Lemma 1. By [3, Theorem 8] $(A \otimes \mathbf{I}_q)$ has property $\mathbf{A}_q$. Suppose $pn \leq q$. Then $(A \otimes \mathbf{I}_q)$ has property $\mathbf{A}_{pn}$. Thus $M_p(A \otimes \mathbf{I}_q)$ has property $\mathbf{A}_n$.

Now suppose that $q < pn$ and $M_p(A \otimes \mathbf{I}_q)$ does not act on the space $\{0\}$. Then $A \otimes \mathbf{I}_q$ does not act on the space $\{0\}$, so by [3, Theorem 8] $A \otimes \mathbf{I}_q$ does not have property $\mathbf{A}_{pn}$. Thus $M_p(A \otimes \mathbf{I}_q)$ does not have property $\mathbf{A}_n$. □

Theorem 1 is now an easy consequence of the previous lemmas.

1.1. Proof of Theorem 1. First assume that A has property $\mathbf{A}_n$. By Lemma 2, each summand $M_p(A_q \otimes \mathbf{I}_q)$ has property $\mathbf{A}_n$. By Lemma 3, $H_q = \{0\}$ for $q < pn$.

Now assume that $H_q = \{0\}$ for $q < pn$. By Lemma 3, the remaining summands have property $\mathbf{A}_n$. By Lemma 2, A has property $\mathbf{A}_n$.

References

[1] H. Bercovici, C. Foiaş, and C. Pearcy, *Dual algebras with applications to invariant subspaces and dilation theory*, CBMS Regional Conf. Ser. in Math., No. 56, Amer. Math. Soc., Providence, RI, 1985.

[2] R. Kadison and J. Ringrose, *Fundamentals of the Theory of Operator Algebras*, Academic Press, Orlando, FL, 1986.

[3] M. Marsalli, *Systems of equations in the predual of a von Neumann algebra*, Proc. Amer. Math. Soc., **111** (1991), 517–522.

[4] F. Murray and J. von Neumann, *On rings of operators*, Ann. of Math. **37** (1936), 116–229.

Department of Mathematics
Illinois State University
Normal, Illinois 61790–4520
E-mail: marsalli@math.ilstu.edu

Received: August 23rd, 1995.

Operator Theory:
Advances and Applications, Vol. 104

The Canonical Complex Structure of Flag Manifolds in a C^*-algebra

MIRCEA MARTIN AND NORBERTO SALINAS*

Dedicated to Professor Carl Pearcy
on his sixtieth birthday

ABSTRACT. The final objective of this article is to study the space of increasing n-tuples of self-adjoint idempotents in a C^*-algebra—which is called a flag manifold—from a differential geometric point of view. It is proved that a flag manifold has a natural intrinsic complex structure. Some properties of this structure are examined and a generalization of the well-known Gram-Schmidt construction is considered.

0. INTRODUCTION

The objective of this article is to call attention to the study of a natural intrinsic complex structure on flag manifolds of C^*-algebras. More specifically, we shall consider the space of all n-tuples of mutually orthogonal hermitian idempotents in a unital C^*-algebra that decomposes the identity. Such spaces will be called flag manifolds of the given algebra, for the obvious reason that the classical flag manifolds can be alternatively defined in this way. The flag manifolds corresponding to the simplest case $n = 2$ are referred to as Grassmann manifolds.

The generalized flag manifolds have many interesting geometric features resembling those of their classical relatives. A systematic study of the differential geometry of flag manifolds has been initiated in [3]. An alternative point of view was suggested in [10]. Some specific properties of the Grassmann manifolds were described in [4], [9], [13], [14], [17]. For the moment let us just remark that the existence of a canonical complex structure on Grassmann manifolds was proved in [17]. In this respect the present paper provides a generalization.

The existence of invariant complex structures on homogeneous spaces satisfying some restrictive assumptions is a classical and basic problem. An excellent account on the subject is presented in [2] and [16] (for other details see also [15] and [18]). As an example, the flag manifold $\mathrm{U}(n)/\mathrm{U}(1)^n$ possesses $2^{n(n-1)/2}$ invariant almost complex structures and precisely $n!$ of them are integrable. Actually, we shall prove below that a similar conclusion is still true for flag manifolds of C^*-algebras.

*The authors were supported in part by NSF Grant DMS-9301187.

Moreover, we shall exhibit a canonical complex structure as the only one that makes the Gram-Schmidt mapping holomorphic (see Theorems 4.4 and 4.6). This mapping is a generalization of the standard Gram-Schmidt process in finite dimensional spaces (see Section 4 for more details).

The organization of the paper is as follows. In Section 1 we begin with notations and a few preliminaries. The canonical almost complex structure is introduced in Section 2, and its integrability is proved in Section 5. The intermediate Sections 3 and 4 are concerned with holomorphic maps into flag manifolds and the Gram-Schmidt map, respectively.

Our interest in the study of flag manifolds is essentially motivated by some geometric aspects of the Cowen-Douglas theory (see [5], [6], and [7]). It is the purpose of a subsequent article [12] to discuss about this interesting circle of ideas.

We are grateful to D. R. Wilkins for very helpful comments on a preliminary version of our paper.

1. Flag Manifolds

The goal of this section is to fix notations and, in view of our later purposes, to recall briefly the definition and a few differential geometric properties of flag manifolds, as well as some related constructions.

1.1. Throughout the paper A will be a fixed unital C^*-algebra. By A_{h}, A_{sh}, $\mathrm{GL}(A)$, and $\mathrm{U}(A)$, we denote the set of all hermitian, skew-hermitian, invertible, and unitary elements of A, respectively.

Definition. Let $n \geq 2$ be an integer. By *an extended n-flag* (resp. *n-flag*) of A we shall mean any n-tuple of mutually orthogonal not necessarily hermitian idempotents (resp. hermitian idempotents) in A that decomposes the identity of A.

The space of all extended n-flags (resp. n-flags) of A will be denoted by $\mathcal{E}_n(A)$, or just $\mathcal{E}$ (resp. $\mathcal{P}_n(A)$, or $\mathcal{P}$).

Explicitly, $E = (e_1, \ldots, e_n)$ is an element of $\mathcal{E} = \mathcal{E}_n(A)$ if and only if

$$e_j^2 = e_j, \quad e_j e_k = 0 \quad (1 \leq j, k \leq n, j \neq k), \tag{1.1}$$

$$e_1 + e_2 + \cdots + e_n = 1, \tag{1.2}$$

and E belongs to $\mathcal{P} = \mathcal{P}_n(A)$ just when it satisfies the additional condition

$$e_j^* = e_j \quad (1 \leq j \leq n). \tag{1.3}$$

The elements of $\mathcal{P}_n(A)$ will be usually denoted by $P = (p_1, \ldots, p_n)$.

Obviously any flag $P = (p_1, \ldots, p_n)$ determines an increasing n-tuple $\hat{P} = (\hat{p}_1, \ldots, \hat{p}_n)$ of hermitian idempotents of A by the rule

$$\hat{p}_j = p_1 + \cdots + p_j, \quad (1 \leq j \leq n), \tag{1.4}$$

and all increasing n-tuples of hermitian idempotents of A subject to the condition that the last component is 1 occur in this way. This remark motivates our terminology.

1.2. Both $\mathcal{E}$ and $\mathcal{P}$ are subsets of A^n, the direct product of n copies of A. Actually, as we already know, $\mathcal{E}$ is a complex analytic submanifold of A^n, whereas $\mathcal{P}$ is only a real analytic submanifold of A^n.

For more specific details concerning the differential geometry of $\mathcal{E}$ and $\mathcal{P}$ the reader is advised to consult [3], where a straightforward description of the smooth structures of $\mathcal{E}$ and $\mathcal{P}$ is given.

An alternative approach was outlined in [10]. More precisely, let $\mathcal{R}(\mathbb{Z}/n, A)$ (resp. $\mathcal{U}(\mathbb{Z}/n, A)$) be the set of all group homomorphisms from the cyclic group $\mathbb{Z}/n$ into $\mathrm{GL}(A)$ (resp. $\mathrm{U}(A)$). By means of a properly defined Fourier transform we identify $\mathcal{R}(\mathbb{Z}/n, A)$ (resp. $\mathcal{U}(\mathbb{Z}/n, A)$) with $\mathcal{E}_n(A)$ (resp. $\mathcal{P}_n(A)$). Since the former space has a natural smooth structure, the later one becomes a smooth manifold also. An extension of these results in a general framework is discussed in [11].

1.3. We summarize next, in a suitable form, some facts implicitly presented in [10].

The tangent space $T_E\mathcal{E}$ to $\mathcal{E}$ at $E = (e_1, \ldots, e_n) \in \mathcal{E}$ consists of all vectors $T = (t_1, \ldots, t_n) \in A^n$ such that

$$t_j e_j + e_j t_j = t_j, \quad t_j e_k + e_j t_k = 0 \quad (1 \leq j, k \leq n, j \neq k), \tag{1.5}$$

$$t_1 + \cdots + t_n = 0. \tag{1.6}$$

If $E \in \mathcal{P}$ then the tangent space $T_E\mathcal{P}$ to $\mathcal{P}$ at E is the subspace of $T_E\mathcal{E}$ defined by the additional condition

$$t_j^* = t_j \quad (1 \leq j \leq n). \tag{1.7}$$

For each $E = (e_1, \ldots, e_n)$ we define

$$\partial_E : A \to A^n, \quad \partial_E(x) = ([x, e_1], \ldots, [x, e_n]) \quad (x \in A), \tag{1.8}$$

$$\varepsilon_E : A^n \to A, \quad \varepsilon_E(X) = \frac{1}{2}\sum_{j=1}^{n}[x_j, e_j] \quad (X = (x_1, \ldots, x_n) \in A^n), \tag{1.9}$$

where $[a, b] = ab - ba$ denotes the commutator of $a, b \in A$.

The linear maps ∂_E and ε_E play an important role in the description of the tangent space $T_E\mathcal{E}$. More precisely, let A^E and A_E be the complementary subspaces of A defined by

$$A^E = \{x \in A : e_j x e_k = 0 \text{ for any } 1 \leq j, k \leq n, \ j \neq k\}, \tag{1.10}$$

$$A_E = \{x \in A : e_j x e_j = 0 \text{ for any } 1 \leq j \leq n\}. \tag{1.11}$$

Then for any tangent vector $T \in T_E\mathcal{E}$ the element $x = \varepsilon_E(T)$ is the unique element of A_E such that $T = \partial_E(x)$.

At the same time, the map

$$\pi_E : A^n \to A^n, \quad \pi_E = \partial_E \varepsilon_E, \tag{1.12}$$

is a projection of A^n onto $T_E\mathcal{E}$, and the map

$$\chi_E : A \to A, \quad \chi_E = \varepsilon_E \partial_E, \tag{1.13}$$

is a projection of A onto A_E.

Similar conclusions follow for the tangent spaces to $\mathcal{P}$, which are subspaces of A^n_{h}. Given $P = (p_1, \ldots, p_n) \in \mathcal{P}$, we have to consider the maps $\partial_P \mid A_{\mathrm{sh}} : A_{\mathrm{sh}} \to A^n_{\mathrm{h}}$ and $\varepsilon_P \mid A^n_{\mathrm{h}} : A^n_{\mathrm{h}} \to A_{\mathrm{sh}}$, and to use the complementary subspaces $A^P_{\mathrm{sh}} = A^P \cap A_{\mathrm{sh}}$ and $(A_P)_{\mathrm{sh}} = A_P \cap A_{\mathrm{sh}}$ of A_{sh}. The map ∂_P induces an isomorphism from $(A_P)_{\mathrm{sh}}$ onto $T_P\mathcal{P}$.

2. The Canonical Almost Complex Structure

Our major aim at this moment is to introduce a canonical almost complex structure on the flag manifold $\mathcal{P}$. Recall that $\mathcal{P}$ is a real analytic submanifold of the complex manifold $\mathcal{E}$, but not a complex submanifold. As we shall see below, in spite of this unpleasant feature, $\mathcal{P}$ carries a lot of intrinsic almost complex structures, and some of them are in fact integrable (i.e., complex structures). A canonical almost complex structure will be described in this section. Later on we shall prove that this structure is indeed integrable and, moreover, it is the only one that makes the Gram-Schmidt map holomorphic. In the particular case of Grassmann manifolds the existence of this canonical structure was proved in [17].

2.1. We construct now a real analytic tensor field J on $\mathcal{P}$ which associates to each $P \in \mathcal{P}$ a linear operator $J_P : T_P\mathcal{P} \to T_P\mathcal{P}$ such that $J_P(J_P(T)) = -T$ for all $T \in T_P\mathcal{P}$. This tensor field will provide the space $\mathcal{P}$ with an almost complex structure.

Given an element $P = (p_1, \ldots, p_n) \in \mathcal{P}$ we consider the space

$$(A_P)_{\mathrm{sh}} = \{x \in A_{\mathrm{sh}} : p_j x p_j = 0 \text{ for any } 1 \leq j \leq n\}$$

and let N^-_P and N^+_P be the subspaces of A defined by

$$N^-_P = \{z \in A : p_j z p_k = 0 \text{ for any } 1 \leq j, k \leq n \text{ with } j \leq k\}$$

$$N^+_P = \{z \in A : p_j z p_k = 0 \text{ for any } 1 \leq j, k \leq n \text{ with } j \geq k\}.$$

Each $x \in (A_P)_{\mathrm{sh}}$ can be represented as $x = \sum_{j \neq k} p_j x p_k$, therefore it has a unique decomposition $x = x^- + x^+$ with $x^- \in N^-_P$ and $x^+ \in N^+_P$, namely

$$x^+ = \sum_{1 \leq j < k \leq n} p_j x p_k, \quad x^- = x - x^+.$$

It follows easily that $(x^+)^* \in N_P^-$ and $(x^-)^* \in N_P^+$. Actually $(x^+)^* = (x^*)^- = -x^-$, hence $x^+ = -(x^-)^*$. Consequently, any element $x \in (A_P)_{\text{sh}}$ has a unique representation of the form $x = z - z^*$, where $z \in N_P^-$, and for each $z \in N_P^-$ the element $x = z - z^*$ belongs to $(A_P)_{\text{sh}}$. Now the fact that N_P^- is a complex linear subspace of A enables us to define a linear map $I_P : (A_P)_{\text{sh}} \to (A_P)_{\text{sh}}$ as follows. If $x \in (A_P)_{\text{sh}}$ with $x = z - z^*$, where $z \in N_P^-$, then $I_P(x) = w - w^*$, where $w = \mathrm{i}z$, hence $I_P(x) = \mathrm{i}(z + z^*)$. One can readily verify that I_P does map $(A_P)_{\text{sh}}$ into itself, and $(I_P)^2(x) = -x$. Thus I_P induces a complex structure on the vector space $(A_P)_{\text{sh}}$. But $(A_P)_{\text{sh}}$ is isomorphic to the tangent space $T_P\mathcal{P}$. The definition of an almost complex structure J on the manifold $\mathcal{P}$ is now obvious. For any $P \in \mathcal{P}$ one defines

$$J_P : T_P\mathcal{P} \to T_P\mathcal{P}, \quad J_P = \partial_P \circ I_P \circ \varepsilon_P \mid T_P\mathcal{P},$$

where ∂_P and ε_P are given by (1.8) and (1.9), respectively. For each tangent vector $T \in T_P\mathcal{P}$ we have

$$\begin{aligned} J_P(J_P(T)) &= \partial_P I_P \varepsilon_P \partial_P I_P \varepsilon_P(T) = \partial_P I_P \chi_P I_P \varepsilon_P(T) \\ &= \partial_P (I_P)^2 \varepsilon_P(T) = -\partial_P \varepsilon_P(T) \\ &= -\pi_P(T) = -T, \end{aligned}$$

where χ_P and π_P are the projections defined by (1.13) and (1.12), respectively. Moreover, since the correspondence $P \mapsto \partial_P \circ I_P \circ \varepsilon_P$ is a real analytic function from $\mathcal{P}$ into the (real) Banach space of all bounded linear operators on A_{h}^n, we conclude that J defines an almost complex structure on $\mathcal{P}$. For a later use we denote by ν_P^- and ν_P^+ the obviously defined projections from A onto N_P^- and N_P^+, respectively, that is

$$\nu_P^-(x) = \sum_{j>k} p_j x p_k, \quad \nu_P^+(x) = \sum_{j<k} p_j x p_k \quad (x \in A).$$

The map $I_P : A_{\text{sh}} \to A_{\text{sh}}$ defined above is given by $I_P(x) = \mathrm{i}\nu_P^-(x) - \mathrm{i}\nu_P^+(x)$ $(x \in A_{\text{sh}})$.

Consequently, the map $J_P : T_P\mathcal{P} \to T_P\mathcal{P}$ can be written as follows:

$$J_P = \partial_P \circ \left(\mathrm{i}\nu_P^- - \mathrm{i}\nu_P^+\right) \circ \varepsilon_P \mid T_P\mathcal{P}. \tag{2.1}$$

2.2. We refer to the almost complex structure J defined above as the canonical almost complex structure on $\mathcal{P}$. There are a lot of other almost complex structures on $\mathcal{P}$. Some of them arise by using the real analytic action of the group $\mathfrak{S}_n$ of all permutations σ of $\{1, 2, \ldots, n\}$ on $\mathcal{P}$, defined by

$$\sigma \cdot (p_1, \ldots, p_n) = (p_{\sigma(1)}, \ldots, p_{\sigma(n)}).$$

The new almost complex structure J^σ corresponding to a permutation σ is obtained from J by the natural rule

$$J_P^\sigma(T) = (\tilde{\sigma}_P)^{-1} J_{\sigma \cdot P}(\tilde{\sigma}_P(T)) \quad (T \in T_P\mathcal{P})$$

where $\tilde{\sigma}_P : T_P\mathcal{P} \to T_{\sigma \cdot P}\mathcal{P}$ is the derivative of the diffeomorphism induced by σ.

2.3. As a matter of fact J, as well as all the structures J^σ, are complex structures on $\mathcal{P}$. For J this result will be proved in Section 5. The similar assertion for J^σ will follow easily.

Some other almost complex structures on $\mathcal{P}$ can be defined as follows. Fix a subset S of the set of all pairs (j,k) of integers, with $1 \le j < k \le n$. Define next for any $P \in \mathcal{P}$ a linear map $I_P^S : (A_P)_{\mathrm{sh}} \to (A_P)_{\mathrm{sh}}$ in the following way. Given $x \in (A_P)_{\mathrm{sh}}$ with $x = z - z^*$, where $z \in N_P^-$, set $I_P^S(x) = w - w^*$, where $w \in N_P^-$ and

$$p_k w p_j = \mathrm{i}\, p_k z p_j \quad \text{if } (j,k) \in S, \tag{2.2}$$

$$p_k w p_j = -\mathrm{i}\, p_k z p_j \quad \text{if } (j,k) \notin S,\ 1 \le j < k \le n. \tag{2.3}$$

It is easy to check that $\left(I_P^S\right)^2(x) = -x$ for any $x \in (A_P)_{\mathrm{sh}}$. Substituing now I_P in the definition of J_P by I_P^S, we find a new almost complex structure J^S on $\mathcal{P}$.

Clearly the previous construction includes the already mentioned structures J^σ, $\sigma \in \mathfrak{S}_n$. On the other hand it is well-known that in the case of the classical flag manifolds this construction provides all the possible invariant almost complex structures on $\mathcal{P}$, and only those corresponding to a permutation σ in $\mathfrak{S}_n$ are integrable (for details see, for instance, [18, Section 11]).

3. Holomorphic Maps into Flag Manifolds

In this section we give a criterion for determining whether a C^∞ map $P : \Omega \to \mathcal{P}$ from a complex manifold Ω to $\mathcal{P} = \mathcal{P}_n(A)$ is holomorphic. To be more specific, Ω is an arbitrary finite or infinite dimensional complex analytic manifold and we assume that $\mathcal{P}$ is endowed with the canonical almost complex structure J.

3.1. In general, given a complex manifold Ω, a complex Banach space B, and a C^∞ map $f : \Omega \to B$, we denote by ∂f and $\bar\partial f$ the B-valued 1-forms on Ω defined by

$$\partial f = \frac{1}{2}\left(\mathrm{d}f - \mathrm{i}(\mathrm{d}f)\circ J^\Omega\right), \tag{3.1}$$

$$\bar\partial f = \frac{1}{2}\left(\mathrm{d}f + \mathrm{i}(\mathrm{d}f)\circ J^\Omega\right), \tag{3.2}$$

where $\mathrm{d}f$ is the differential of f, and J^Ω denotes the complex structure of Ω.

Assume that $P : \Omega \to \mathcal{P}$ is a C^∞ map. Since $\mathcal{P}$ is a real analytic submanifold of the direct product A^n, we represent P as an n-tuple of A-valued maps on Ω, namely, $P = (p_1, \ldots, p_n)$, and we can consider the ∂ derivative and the $\bar\partial$ derivative of P, as in (3.1) and (3.2) above. Explicitly, $\bar\partial P$ is given by

$$\bar\partial P = \frac{1}{2}\left(\mathrm{d}P + \mathrm{i}(\mathrm{d}P)\circ J^\Omega\right). \tag{3.3}$$

Because $\mathcal{P}$ is not a complex submanifold of A^n, it is out of question to hope that condition $\bar\partial P \equiv 0$ characterizes holomorphic maps. Actually, if Ω is connected, then $\bar{\partial}P \equiv 0$ holds if and only if P is a constant map, due to the fact that the values of P are hermitian elements of the algebra A^n. However, as we shall prove below, there is a characterization of holomorphic maps from Ω into $\mathcal{P}$ in terms of their $\bar{\partial}$ derivative.

We begin by recalling the concept of holomorphic maps in our special setting.

Definition. A C^∞ map $P : \Omega \to \mathcal{P}$ is said to be *holomorphic with respect to the almost complex structure J on $\mathcal{P}$* if and only if

$$\mathrm{d}P(\omega) \circ J_\omega^\Omega = J_{P(\omega)} \circ \mathrm{d}P(\omega) \quad (\omega \in \Omega). \tag{3.4}$$

With a little abuse of notation, already used in (3.1) and (3.2) above, we write (3.4) in the form

$$\mathrm{d}P \circ J^\Omega = J \circ \mathrm{d}P \tag{3.5}$$

The following is an intermediate step towards the characterization of holomorphic maps.

3.2 Lemma. *Let $P = (p_1, \dots, p_n) : \Omega \to \mathcal{P}$ be a C^∞ map. Then P is holomorphic if and only if*

$$\bar{\partial}p_l = \Big(\sum_{j=1}^{l-1} p_j\Big)(\mathrm{d}p_l) + (\mathrm{d}p_l)\Big(\sum_{k=l+1}^{n} p_k\Big) \quad (1 \le l \le n), \tag{3.6}$$

where the product of A-valued 1-forms and A-valued functions is defined in the standard fashion.

Proof. Comparing (3.5) and (3.3) one concludes that P is holomorphic if and only if

$$\bar{\partial}P = \frac{1}{2}(\mathrm{d}P + \mathrm{i}\, J \circ \mathrm{d}P). \tag{3.7}$$

We need to use now equation (2.1) that describes the complex structure J. In the following discussion, we shall use the notation introduced before. For each $\omega \in \Omega$, by (2.1) one finds

$$\mathrm{i}\, J_{P(\omega)} \circ \mathrm{d}P(\omega) = \mathrm{i}\, \partial_{P(\omega)} \circ \Big(\mathrm{i}\, \nu_{P(\omega)}^- - \mathrm{i}\, \nu_{P(\omega)}^+\Big) \circ \varepsilon_{P(\omega)} \circ \mathrm{d}P(\omega) =$$

$$= -\partial_{P(\omega)} \circ \Big(\nu_{P(\omega)}^- - \nu_{P(\omega)}^+\Big) \circ \varepsilon_{P(\omega)} \circ \mathrm{d}P(\omega).$$

But $\varepsilon_{P(\omega)} \circ \mathrm{d}P(\omega)$ takes values in A_{sh}, and

$$\Big(\nu_{P(\omega)}^- - \nu_{P(\omega)}^+\Big) \mid A_{\mathrm{sh}} = \Big(\mathrm{id} - 2\nu_{P(\omega)}^+\Big) \mid A_{\mathrm{sh}}.$$

On the other hand $\partial_{P(\omega)} \circ \varepsilon_{P(\omega)}$ is the projection of A^n onto $T_{P(\omega)}\mathcal{P}$. Using these remarks we have

$$\mathrm{i}\, J_{P(\omega)} \circ \mathrm{d}P(\omega) = -\mathrm{d}P(\omega) + 2\partial_{P(\omega)} \circ \nu^+_{P(\omega)} \circ \varepsilon_{P(\omega)} \circ \mathrm{d}P(\omega),$$

therefore, (3.7) is equivalent with

$$\bar{\partial}P(\omega) = \partial_{P(\omega)} \circ \nu^+_{P(\omega)} \circ \varepsilon_{P(\omega)} \circ \mathrm{d}P(\omega) \quad (\omega \in \Omega). \tag{3.8}$$

The explicit formulas for $\partial_{P(\omega)}$, $\nu^+_{P(\omega)}$, and $\varepsilon_{P(\omega)}$ given in Sections 1 and 2 lead successively to the next equalities, written in terms of the components of the map P:

$$\varepsilon_{P(\omega)} \circ \mathrm{d}P(\omega) = \frac{1}{2}\sum_{k=1}^{n}(\,dp_k(\omega)) \cdot p_k(\omega) - \frac{1}{2}\sum_{j=1}^{n} p_j(\omega) \cdot (\mathrm{d}p_j(\omega)),$$

$$\nu^+_{P(\omega)} \circ \varepsilon_{P(\omega)} \circ \mathrm{d}P(\omega) = \frac{1}{2} \sum_{1\le j<k\le n} \{p_j(\omega)\cdot(\mathrm{d}p_k(\omega))\cdot p_k(\omega) - p_j(\omega)\cdot(\mathrm{d}p_j(\omega))\cdot p_k(\omega)\},$$

and, consequently, for a fixed $1 \le l \le n$,

$$\begin{aligned} \bar{\partial}p_l(\omega) \quad &= \quad \frac{1}{2}\sum_{j=1}^{l-1} p_j(\omega)(\mathrm{d}p_l(\omega) - \mathrm{d}p_j(\omega))p_l(\omega) \\ &\qquad - \frac{1}{2}\sum_{k=l+1}^{n} p_l(\omega)(\mathrm{d}p_k(\omega) - \mathrm{d}p_l(\omega))p_k(\omega). \end{aligned} \tag{3.9}$$

But

$$(\mathrm{d}p_l(\omega))p_l(\omega) + p_l(\omega)(\mathrm{d}p_l(\omega)) = \mathrm{d}(p_l p_l)(\omega) = \mathrm{d}p_l(\omega),$$

and, on the other hand, if $j,\, k \ne l$, then

$$(\mathrm{d}p_j(\omega))p_l(\omega) + p_j(\omega)(\mathrm{d}p_l(\omega)) = \mathrm{d}(p_j p_l)(\omega) = 0$$

and, similarly,

$$p_l(\omega)(\mathrm{d}p_k(\omega)) + (\mathrm{d}p_l(\omega))p_k(\omega) = 0.$$

Using these equalities in (3.9), one obtains

$$\bar{\partial}p_l(\omega) = \Big(\sum_{j=1}^{l-1} p_j(\omega)\Big)(\mathrm{d}p_l(\omega)) + (\mathrm{d}p_l(\omega))\Big(\sum_{k=l+1}^{n} p_k(\omega)\Big) \quad (1 \le l \le n) \tag{3.10}$$

a relation equivalent to (3.8) and, at the same time, an explicit form of (3.6). □

3.3. A further simplification in (3.6) above is possible. The next theorem generalizes results from [9] and [14].

Theorem. *Let $P = (p_1, \ldots, p_n) : \Omega \to \mathcal{P}$ be a C^∞ map. Then P is holomorphic if and only if*

$$p_m \bar{\partial} p_l = 0 \quad (1 \le l < m \le n). \tag{3.11}$$

PROOF. We have to show that (3.6) and (3.11) are equivalent.

Assume that $1 \le l < m \le n$ are fixed. Clearly $p_m\big(\sum_{j=1}^{l-1} p_j\big) = 0$. Since $p_m(\mathrm{d}p_l) = -(\mathrm{d}p_m)p_l$, we also have $p_m(\mathrm{d}p_l)\big(\sum_{k=l+1}^{n} p_k\big) = 0$. Therefore, indeed (3.6) implies (3.11).

Conversely, suppose that (3.11) is true. The involution $x \mapsto x^*$ in A induces an involution in the space of all A-valued 1-forms on Ω. Since $(\bar{\partial} p_l)^* = \partial(p_l^*) = \partial p_l$, from (3.11) one finds

$$(\partial p_l) p_m = 0 \quad (1 \le l < m \le n). \tag{3.12}$$

On the other hand $(\bar{\partial} p_m) p_l = -p_m(\bar{\partial} p_l) = 0$; hence, after a change of notation,

$$(\bar{\partial} p_l) p_m = 0 \quad (1 \le m < l \le n) \tag{3.13}$$

and, by taking adjoints, we have

$$p_m(\partial p_l) = 0 \quad (1 \le m < l \le n). \tag{3.14}$$

In addition, note that $\mathrm{d}p_l = \partial p_l + \bar{\partial} p_l$ for all $1 \le l \le n$. The proof of (3.6) follows now easily. Indeed, if $1 \le l \le n$ is fixed, and we denote

$$E = \Big(\sum_{j=1}^{l-1} p_j\Big)(\mathrm{d}p_l) + (\mathrm{d}p_l)\Big(\sum_{k=l+1}^{n} p_k\Big),$$

then $E = E_1 + E_2 + E_3$ where, respectively,

$$E_1 = \Big(\sum_{j=1}^{l-1} p_j\Big)(\partial p_l) = 0 \quad (\text{cf. } (3.14)),$$

$$E_2 = (\partial p_l)\Big(\sum_{k=l+1}^{n} p_k\Big) = 0 \quad (\text{cf. } (3.12)),$$

and

$$E_3 = \Big(\sum_{j=1}^{l-1} p_j\Big)(\bar{\partial} p_l) + (\bar{\partial} p_l)\Big(\sum_{k=l+1}^{n} p_k\Big).$$

Thus

$$E = E_3. \tag{3.15}$$

On the other hand, equations (3.11) and (3.13) imply

$$\Big(\sum_{j=l+1}^{n} p_j\Big)(\bar{\partial} p_l) + (\bar{\partial} p_l)\Big(\sum_{k=1}^{l-1} p_k\Big) = 0. \tag{3.16}$$

Combining (3.15) and (3.16), one obtains that

$$E = \Big(\sum_{j\neq l} p_l\Big)(\bar\partial p_j) + (\bar\partial p_l)\Big(\sum_{k\neq l} p_k\Big).$$

But $\sum_{j\neq l} p_l = 1 - p_l$, and consequently,

$$E = 2\bar\partial p_l - p_l(\bar\partial p_l) - (\bar\partial p_l)p_l = 2\bar\partial p_l - \bar\partial(p_l p_l) = \bar\partial p_l.$$

The proof is complete. □

3.4. With a little effort we can write condition (3.11) in a more compact form.

Corollary. *Let $P = (p_1, \dots, p_n) : \Omega \to \mathcal{P}$ be a C^∞ map. Then P is holomorphic if and only if*

$$(p_{j+1} + \cdots + p_n)\bar\partial(p_1 + p_2 + \cdots + p_j) = 0 \quad (1 \leq j \leq n-1). \tag{3.17}$$

PROOF. Clearly (3.11) implies (3.17). Conversely, fix $1 \leq l < m \leq n$ and use (3.17) for $j = l$. One obtains

$$p_m(p_{l+1} + \cdots + p_n)(\bar\partial(p_1 + \cdots + p_l))p_l = 0. \tag{3.18}$$

But $p_m(p_{l+1} + \cdots + p_n) = p_m$, and

$$\begin{aligned}(\bar\partial(p_1 + \cdots + p_l))p_l &= \sum_{j=1}^{l-1}(\bar\partial p_j)p_l + (\bar\partial p_l)p_l \\ &= -\Big(\sum_{j=1}^{l-1} p_j\Big)(\bar\partial p_l) + \bar\partial p_l - p_l(\bar\partial p_l),\end{aligned}$$

thus (3.18) reduces to (3.11). □

4. The Gram-Schmidt Map

4.1. The classical Gram-Schmidt construction associates to any linear basis in a finite dimensional Hilbert space $\mathcal{H}$ an orthonormal basis. If $\mathcal{H}$ has dimension n, then each linear basis defines, in an obvious fashion, an extended n-flag in the C^*-algebra $\mathcal{L}(\mathcal{H})$ of all linear operators on $\mathcal{H}$. Under this correspondence any orthonormal basis determines an n-flag. Consequently the Gram-Schmidt procedure yields to a map from the manifold of extended n-flags onto the manifold of n-flags of $\mathcal{L}(\mathcal{H})$.

It is the purpose of this section to discuss a generalization of the Gram-Schmidt procedure. A similar construction was described in [1] in connection with some specific problems in Hilbert space operator theory.

4.2. Assume that A is a fixed C^*-algebra. Given $n \geq 2$ an integer let $\mathcal{E} = \mathcal{E}_n(A)$ (resp. $\mathcal{P} = \mathcal{P}_n(A)$) be the space of all extended n-flags (resp. n-flags) of A. Then $\mathcal{E}$ is a complex manifold and $\mathcal{P}$ is an almost complex manifold. For $\mathcal{E}$ the complex structure is exactly that one induced by the complex structure of A^n, and for $\mathcal{P}$ we consider the canonical almost complex structure defined in Section 2.

We shall need the following standard result (see for example [14, Lemma 2.15]).

Lemma. *Let e be an idempotent of A. Then there exists a unique hermitian idempotent p of A, such that*

$$ep = p, \quad pe = e. \tag{4.1}$$

We denote the unique hermitian idempotent p associated to e as in (4.1) by $p = \pi(e)$. In [14] there are four alternative descriptions of $\pi(e)$. A simple one which suffices for our purposes is

$$\pi(e) = e(1 + e - e^*)^{-1}. \tag{4.2}$$

4.3. We shall next prove that the map $e \mapsto \pi(e)$ preserves the natural order relation on the set of idempotents of A. Recall that given two idempotents e and f of A, we write $e \leq f$ if and only if $ef = fe = e$. In such a case $f - e$ is also an idempotent of A and $f - e \leq f$, too.

Lemma. *If e and f are idempotents of A and $e \leq f$, then $\pi(e) \leq \pi(f)$.*

PROOF. From $fe = e$ and $e\pi(e) = \pi(e)$ one obtains $f\pi(e) = \pi(e)$. It follows that $\pi(f)f\pi(e) = \pi(f)\pi(e)$. But $\pi(f)f = f$, hence $\pi(f)\pi(e) = \pi(f)f\pi(e) = f\pi(e) = \pi(e)$. By taking adjoints one finds also that $\pi(e) = \pi(e)\pi(f)$. Therefore $\pi(e) \leq \pi(f)$. □

4.4. We next define a map $\Pi = (\Pi_1, \ldots, \Pi_n)$ from $\mathcal{E} = \mathcal{E}_n(A)$ onto $\mathcal{P} = \mathcal{P}_n(A)$ as follows. For each $E = (e_1, \ldots, e_n)$ in $\mathcal{E}$ we set

$$\Pi_1(E) = \pi(e_1) \tag{4.3}$$

$$\Pi_j(E) = \pi(e_1 + \cdots + e_j) - \pi(e_1 + \cdots + e_{j-1}) \quad (2 \leq j \leq n). \tag{4.4}$$

By Lemma 4.3 clearly $\Pi : \mathcal{E} \to \mathcal{P}$ is well defined. Moreover, formula (4.2) gives an explicit description of Π. The map Π defined by (4.3) and (4.4) above will be referred to as *the Gram-Schmidt map* from $\mathcal{E}$ onto $\mathcal{P}$. A basic property of Π is the following.

Theorem. *The Gram-Schmidt map is a holomorphic map.*

PROOF. According to Corollary 3.4 we have to prove that Π satisfies condition (3.17). Let

$$\tilde{\Pi}_j = \Pi_1 + \Pi_2 + \cdots + \Pi_j \quad (1 \leq j \leq n).$$

Since $\Pi_1 + \Pi_2 + \cdots + \Pi_n = 1$ (the constant map), condition (3.17) takes the form

$$\left(1 - \tilde{\Pi}_j(E)\right) \bar{\partial} \tilde{\Pi}_j(E) = 0 \tag{4.5}$$

for each $1 \leq j \leq n$ and any $E = (e_1, \ldots, e_n) \in \mathcal{E}$. From (4.3), (4.4) and (4.2) we have

$$\tilde{\Pi}_j(E) = \pi(e_1 + \cdots + e_j) = (e_1 + \cdots + e_j)(1 + e_1 + \cdots + e_j - - e_1^* - \cdots - e_j^*)^{-1}. \tag{4.6}$$

Now let $T = (t_1, \ldots, t_n)$ be an arbitrary tangent vector to $\mathcal{E}$ at a fixed point $E = (e_1, \ldots, e_n)$. From (4.6) it follows that the $\bar{\partial}$ derivative of $\tilde{\Pi}_j$ acts according to the next formula

$$\bar{\partial} \tilde{\Pi}_j(E)(T) = \tilde{\Pi}_j(E) \left(t_1^* + \cdots + t_j^*\right) \left(1 + e_1 + \cdots + e_j - e_1^* - \cdots - e_j^*\right)^{-1}. \tag{4.7}$$

But clearly $\left(1 - \tilde{\Pi}_j(E)\right) \tilde{\Pi}_j(E) = 0$, hence (4.5) follows from (4.7). □

4.5 Remark. The Gram-Schmidt map $\Pi : \mathcal{E} \to \mathcal{P}$ has another interesting property, namely, it enables us to find a new characterization of holomorphic maps from a finite dimensional complex manifold Ω into $\mathcal{P}$. More precisely, one obtains that a smooth map $P : \Omega \to \mathcal{P}$ is holomorphic, if and only if, for each point $\omega_0 \in \Omega$, there exist an open neighborhood Ω_0 of ω_0 in Ω and a complex analytic map $E : \Omega_0 \to \mathcal{E}$ such that $\Pi \circ E = P \mid \Omega_0$. The proof uses essentially the same techniques developed in [14] for the case of the Grassmann manifolds. Complete details of the proof, as well as other results concerning holomorphic maps into flag manifolds, are presented in a subsequent paper [12], with a special emphasis on the Cowen-Douglas theory (see [5], [6], and [7]).

We now prove the fact that the canonical almost complex structure J on $\mathcal{P}$ is uniquely determined by the Gram-Schmidt map Π.

Theorem. *The almost complex structure J and the Gram-Schmidt map Π are related by the equation*

$$J_P(T) = \mathrm{d}\Pi(P)(\mathrm{i}T) \tag{4.8}$$

for any $P \in \mathcal{P}$ and $T \in T_P\mathcal{P}$.

PROOF. Since $\Pi : \mathcal{E} \to \mathcal{P}$ is holomorphic we know that

$$\mathrm{d}\Pi(E)(\mathrm{i}T) = J_{\Pi(E)} \circ \mathrm{d}\Pi(E)(T) \tag{4.9}$$

for all $E \in \mathcal{E}$ and $T \in T_E\mathcal{E}$. Assume that $E = P \in \mathcal{P}$ and $T \in T_P\mathcal{P} \subset T_P\mathcal{E}$. Since $\Pi \mid \mathcal{P} = \mathrm{id}_{\mathcal{P}}$ we have $\Pi(P) = P$, and $\mathrm{d}\Pi(P)(T) = T$, therefore (4.8) follows from (4.9). □

5. The Integrability of the Canonical Almost Complex Structure

5.1. As a matter of fact, J is a complex structure on $\mathcal{P}$. By general arguments (see, for instance, [17]), this will follow if we prove that, given any point $P \in \mathcal{P}$, there exist an open neighborhood $\mathcal{Q}$ of P in $\mathcal{P}$, a complex Banach space B, and a real analytic map $\Phi : \mathcal{Q} \to B$, which satisfy the next conditions:

(i) Φ is a diffeomorphism from $\mathcal{Q}$ onto an open subset of B;

(ii) for each $Q \in \mathcal{Q}$ the derivative $(\mathrm{d}\Phi)(Q) : T_Q\mathcal{P} \to B$ of Φ at Q satisfies

$$(\mathrm{d}\Phi)(Q)(J_Q(T)) = \mathrm{i}(\mathrm{d}\Phi)(Q)(T) \quad (T \in T_Q\mathcal{P}). \tag{5.1}$$

The proof below was improved by a remark on our original argument due to D. R. Wilkins. Assume that $P \in \mathcal{P}$ is fixed and let $B = N_P^-$ (see Section 2). Since N_P^- is a closed (complex) subspace of A, the space B inherits a structure of a complex Banach space from A. Next we introduce a map $\Theta : B \to T_P\mathcal{E}$ by $\Theta = \partial_P \mid B$, where $\partial_P : A \to A^n$ is defined according to formula (1.8). Let $E_P : T_P\mathcal{E} \to \mathcal{E}$ be the map that asigns to each $T \in T_P\mathcal{E}$ an extended n-flag $E_P(T) = E = (e_1, \dots, e_n)$ defined by

$$e_j = (\exp \varepsilon_P(T))p_j(\exp \varepsilon_P(-T)) \quad (1 \le j \le n). \tag{5.2}$$

The composition map $E_P \circ \Theta : B \to \mathcal{E}$ associates to any $z \in B$ the extended n-flag $E_P \circ \Theta(z) = E = (e_1, \dots, e_n)$ where

$$e_j = (\exp(z))p_j(\exp(-z)) \quad (1 \le j \le n). \tag{5.3}$$

Clearly $E_P \circ \Theta$ is a complex analytic map from B into A^n. Therefore, for any $z \in B$, the derivative $\mathrm{d}(E_P \circ \Theta)(z)$ is a complex linear map from $T_zB = B$ into $T_E\mathcal{E}$, where $E = E_p \circ \Theta(z)$, that is,

$$\mathrm{d}(E_p \circ \Theta)(z)(\mathrm{i}w) = \mathrm{i}\ \mathrm{d}(E_p \circ \Theta)(z)(w), \tag{5.4}$$

for all $w \in B$.

Now let us consider the Gram-Schmidt map $\Pi : \mathcal{E} \to \mathcal{P}$. According to Collorary 3.4 and Theorem 4.4 above Π has the next basic property:

$$(\mathrm{d}\Pi)(E)(\mathrm{i}T) = J_{\Pi(E)} \circ \mathrm{d}\Pi(E)(T), \tag{5.5}$$

for all $E \in \mathcal{E}$ and $T \in T_E\mathcal{E}$.

We define a smooth map $\Psi : B \to \mathcal{P}$ by $\Psi = \Pi \circ E_P \circ \Theta$. From (5.4) and (5.5) above it follows that the derivative $(\mathrm{d}\Psi)(z) : B \to T_{\Psi(z)}\mathcal{P}$ satisfies the condition

$$(\mathrm{d}\Psi)(z)(\mathrm{i}w) = J_{\Psi(z)} \circ (\mathrm{d}\Psi)(z)(w), \tag{5.6}$$

for all $z \in B$ and $w \in B$.

Clearly $\Psi(0) = P$. We claim that if $\mathcal{N}$ is a sufficiently small neighborhood of 0 in B, then Ψ is a diffeomorphism from $\mathcal{N}$ onto an open neighborhood $\mathcal{Q}$ of P in

$\mathcal{P}$. This claim, combined with (5.6),implies that the map $\Phi = \Psi^{-1} : \mathcal{Q} \to \mathcal{N}$ is a diffeomorphism which satisfies condition (5.1), and, consequently, ends the proof of the fact that J is a complex structure on $\mathcal{P}$. In order to prove our claim it is enough to show that $(\mathrm{d}\Psi)(0) : B \to T_P\mathcal{P}$ is an isomorphism. Actually we shall prove the equality

$$(\mathrm{d}\Psi)(0)(w) = \partial_P(w - w^*) \quad (w \in B), \tag{5.7}$$

and we already know that this map is an isomorphism from $B = N_P^-$ onto $T_P\mathcal{P}$, as a composition of the isomorphism $w \mapsto w - w^*$ from B onto $(A_P)_{\mathrm{sh}}$, and the isomorphism $x \mapsto \partial_P(x)$ from $(A_P)_{\mathrm{sh}}$ onto $T_P\mathcal{P}$. To this end note first that (5.3) and the definition of Ψ imply $(\mathrm{d}\Psi)(0)(w) = (\mathrm{d}\Pi)(P)(T)$, where $T = (t_1, \ldots, t_n) \in T_P(\mathcal{E})$ is related to w by $t_j = [w, p_j]$ $(1 \le j \le n)$. Next, from the explicit formulas (4.3) and (4.4) which define the Gram-Schmidt map Π it follows that the derivatives of the components Π_j $(1 \le j \le n)$ of Π at P, are given, respectively, by

$$(\mathrm{d}\Pi_1)(P)(T) = t_1 - p_1(t_1 - t_1^*) \tag{5.8}$$

and, for $j \ge 2$, by

$$(\mathrm{d}\Pi_j)(P)(T) = t_j - p_j(t_1 + \cdots + t_j - t_1^* - \cdots - t_j^*) - (p_1 + \cdots + p_{j-1})(t_j - t_j^*). \tag{5.9}$$

If $t_k = [w, p_k]$ $(1 \le k \le n)$, where $w \in N_P^-$, then

$$\begin{aligned} t_1 - p_1(t_1 - t_1^*) &= [w, p_1] - p_1[w + w^*, p_1] \\ &= [w, p_1] - p_1(w + w^*)p_1 + p_1(w + w^*) = [w, p_1] + p_1 w^* \\ &= [w - w^*, p_1], \end{aligned}$$

since $p_1 w = w^* p_1 = 0$. If $j \ge 2$, we have

$$\begin{aligned} p_j(t_1 + \cdots + t_j - t_1^* - \cdots - t_j^*) &= p_j[w + w^*, p_1 + \cdots + p_j] \\ &= p_j(w + w^*)(p_1 + \cdots + p_j) - p_j(w + w^*) \\ &= -p_j w^*, \end{aligned}$$

where the last equality is a consequence of the fact that $w \in N_P^-$ and $w^* \in N_P^+$. Similarly we obtain that

$$\begin{aligned} (p_1 + \cdots + p_{j-1})(t_j - t_j^*) &= (p_1 + \cdots + p_{j-1})[w + w^*, p_j] \\ &= (p_1 + \cdots + p_{j-1})(w + w^*)p_j = w^* p_j. \end{aligned}$$

From (5.9) we find

$$(\mathrm{d}\Pi_j)(P)(T) = [w, p_j] + p_j w^* - w^* p_j = [w - w^*, p_j].$$

On the other hand

$$\partial_P(w - w^*) = ([w - w^*, p_1], \ldots, [w - w^*, p_n]),$$

hence (5.7) is completely proved.

REFERENCES

[1] E. ANDRUCHOW AND D. STOJANOFF, *Nilpotent operators and systems of projections*, J. Operator Theory **20** (1988), 359–374.

[2] A. BOREL AND F. HIRZEBRUCH, *Characteristic classes and homogeneous spaces. I*, Amer. J. Math. **80**(1958), 458–538.

[3] G. CORACH, H. PORTA, AND L. RECHT, *Differential geometry of systems of projections in Banach algebras*, Pacific J. Math. **143** (1990), 209–228.

[4] G. CORACH, H. PORTA, AND L. RECHT, *The geometry of spaces of projections in C^*-algebras*, Advances in Math. (to appear).

[5] M. J. COWEN AND R. G. DOUGLAS, *Complex geometry and operator theory*, Acta Math. **141** (1978), 187–261.

[6] M. J. COWEN AND R. G. DOUGLAS, *Operators possessing an open set of eigenvalues*, in: Colloquia Math. **35**, North Holland, 1980, pp. 323–341.

[7] R. CURTO AND N. SALINAS, *Generalized Bergman kernels and the Cowen-Douglas theory*, Amer. J. Math. **106** (1984), 447–488.

[8] M. MARTIN, *Almost product structures and derivations*, Bull. Math. Soc. Sci. Math. Roumanie **23** (1979), 171–176.

[9] M. MARTIN, *An operator theoretic approach to analytic functions into the Grassmann manifold*, Math. Balkanica, **1** (1987), 45–58.

[10] M. MARTIN, *Projective representations of compact groups in C^*-algebras*, in: Operator Theory: Advances and Applications, Vol. 43, Birkhäuser Verlag, Basel, 1990, pp. 237–253.

[11] M. MARTIN AND N. SALINAS, *Differential geometry of generalized Grassmann manifolds in C^*-algebras*, in: Operator Theory: Advances and Applications, Vol. 80, Birkhäuser Verlag, Basel, 1995, pp. 206–243.

[12] M. MARTIN AND N. SALINAS, *Flag manifolds and the Cowen-Douglas theory* (Preprint).

[13] H. PORTA AND L. RECHT, *Minimality of geodesics in Grassmann manifolds*, Proc. Amer. Math. Soc. **100** (1987), 464–466.

[14] N. SALINAS, *The Grassmann manifold of a C^*-algebra and hermitian holomorphic bundles*, in: Operator Theory: Advances and Applications, Vol. 28, Birkhäuser Verlag, Basel, 1988, pp. 267–289.

[15] N. R. WALLACH, *Harmonic Analysis on Homogeneous Spaces*, Marcel Dekker, Inc., New York, 1973.

[16] H. C. WANG, *Closed manifolds with homogeneous complex structure*, Amer. J. Math. **76** (1954), 1–32.

[17] D. R. WILKINS, *The Grassmann manifold of a C^*-algebra*, Proc. of the Royal Irish Acad. **90A**(1990), 99–116.

[18] K. YANG, *Almost Complex Homogeneous Spaces and Their Submanifolds*, World Scientific Publishing Co., Singapore, 1987.

Department of Mathematics
University of Kansas
Lawrence, KS 66045

Received: August 23rd, 1995.

Operator Theory:
Advances and Applications, Vol. 104

An Algebraic Characterization of Boundary Representations

PAUL S. MUHLY* AND BARUCH SOLEL†

Dedicated to Carl Pearcy

ABSTRACT. We show that boundary representations of an operator algebra may be characterized as those (irreducible) completely contractive representations that determine Hilbert modules that are simultaneously orthogonally projective and orthogonally injective. As a corollary, we conclude that if an operator algebra is an admissable subalgebra of its C^*–envelope, in the sense of Arveson, then it has a completely isometric representation such that the associated Hilbert module is simultaneously orthogonally projective and orthogonally injective.

1. INTRODUCTION

Thanks to the fundamental discovery of Blecher, Ruan, and Sinclair [3], one may define an *operator algebra* to be a (unital or approximately unital[1]) Banach algebra endowed with an L^∞–matrix norm structure with respect to which multiplication is a completely contractive bilinear map. Such an algebra A, say, may be completely isometrically isomorphically embedded in the algebra of operators on some Hilbert space and, therefore, comes equipped with an essentially unique C^*–envelope, denoted $C^*(A)$.[2] One is ineluctably led, then, to study the completely contractive representations of operator algebras on Hilbert space and to analyze them in terms of the C^*–representations of their C^*–envelopes. This is the setting of modern dilation theory.

*Supported in part by grants from the National Science Foundation and the U.S.-Israel Binational Science Foundation.

†Supported in part by the U.S.-Israel Binational Science Foundation and by the Fund for the Promotion of Research at the Technion.

[1]We always assume that the identity has norm one in the unital case, and in the approximately unital case we assume that there is an approximate identity with norm at most one.

[2]The existence of such an envelope was conjectured by Arveson in [2] and finally proved by Hamana in [5]. Both Arveson and Hamana restrict their attention to *unital* operator algebras, *unital* representations, and *unital* completely positive maps. However, the theory works in the nonuntial case, provided one restricts attention, as we shall, to those representations that are nondegenerate and to those completely positive maps on nonunital C^*–algebras (mapping into $L(\mathcal{H})$) that are contractive and approximately unital in the sense that they carry (contractive) approximate identities to contractions converging strongly to $I_{\mathcal{H}}$.

It is often revealing to identify a representation with a Hilbert module. Specifically, if $\rho : A \mapsto L(\mathcal{H})$ is a representation of the operator algebra A on the Hilbert space $\mathcal{H}$, then we view $\mathcal{H}$ as a (left) module over A via the formula $a\xi := \rho(a)\xi$. When it is helpful to emphasize the algebra A when considering Hilbert module $\mathcal{H}$, we shall write ${}_A\mathcal{H}$. Likewise, when it is helpful to tie a Hilbert module to a representation, we shall write $\mathcal{H}_\rho$ or $\rho_{\mathcal{H}}$. While there are a variety of types of representations and associated Hilbert modules that one might want to study, we will restrict our attention in this note to those that are completely contractive. Thus, *all representations and Hilbert modules considered here are assumed to be completely contractive.* (And again, for emphasis, they will always be nondegenerate or, equivalently, essential.) Module maps are simply (bounded linear) intertwining operators for the representations and Hilbert module isomorphisms are simply unitary module maps.

As was promulgated initially by Douglas and Paulsen in [4], dilation theory suggests that one should try to *resolve* general Hilbert modules in terms of "simpler" Hilbert modules. Just what the "simpler" Hilbert modules ought to be has so far been left undecided. The perspective adopted in [4] and by us in [6] suggests that the "simpler" Hilbert modules should be found *among* the so-called Shilov modules. A *Shilov module* over the operator algebra A is a Hilbert module $\mathcal{S}$ with property that there is a C^*–representation $\pi : C^*(A) \mapsto L(\mathcal{K})$ and a subspace $\mathcal{M} \subseteq \mathcal{K}$ such that $\rho_{\mathcal{S}}$ is unitarily equivalent to the representation $\rho_{\mathcal{M}}$ defined by the formula

$$\rho_{\mathcal{M}}(a) = \pi(a)|\mathcal{M},$$

$a \in A$. To say the same thing differently, $\mathcal{S}$ is a Shilov module over A iff there is a Hilbert module over $C^*(A)$, ${}_{C^*(A)}\mathcal{K}$, such that $\mathcal{S}$ is isomorphic to a submodule of ${}_A\mathcal{K}$. As we shall see in a moment, Arveson's interpretation of Sarason's notion of semi-invariant subspaces [7] asserts that every Hilbert module is a quotient of two Shilov modules and this suggests that Shilov modules are "the resolvers of choice". However, as was discovered in [6], it appears that in general one needs to restrict attention to a subclass of the Shilov modules when seeking resolutions. This is the subclass of orthoprojective Hilbert modules.

To define this notion[3], it is helpful to reflect a little on the concepts of "submodule", "quotient module" and "short exact sequence". The reason for this is that the category of Hilbert modules is rather far from being an abelian category and it is helpful to take stock of the categorical properties that are at our disposal. If ${}_A\mathcal{M}$ is a *submodule* of ${}_A\mathcal{N}$, then certainly one wants $\mathcal{M}$ to be a subspace of $\mathcal{N}$ in the Hilbert space sense. Once this is decided, then the quotient module ${}_A\mathcal{N}/{}_A\mathcal{M}$ is realized at the Hilbert space level as $\mathcal{N} \ominus \mathcal{M}$ and the module structure is given by the *compressed* action. Admittedly, this is so elementary that operator algebraists hardly give it a thought, but the fact that subspaces of Hilbert spaces have preferred complements is one of the key facts that distinguishes operator algebra from general ring theory.

[3]Details and amplification of the discussion to follow may be found in [6].

In pure algebra, one frequently expresses the fact that a module $\mathcal{M}$ contains a submodule $\mathcal{K}$ with quotient $\mathcal{N}$ by asserting that there is a short exact sequence with module maps Ψ and Φ,

$$0 \to \mathcal{K} \xrightarrow{\Psi} \mathcal{M} \xrightarrow{\Phi} \mathcal{N} \to 0,$$

meaning that the map Ψ has zero kernel, the range of Ψ is the kernel of Φ, and the range of Φ is all of $\mathcal{N}$. To do this in our theory, we must in addition require $\mathcal{K}$, $\mathcal{M}$, and $\mathcal{N}$ to be Hilbert modules over an operator algebra A, and require Ψ and Φ to be *partially isometric* module maps. This, of course, makes Ψ an isometry and Φ a co-isometry. To emphasize these constraints, we say that the short exact sequence is *isometric.*

There is really nothing fancy going on here. To say that the short exact sequence is isometric is to say that as a Hilbert space, $\mathcal{M}$ is the *orthogonal* direct sum $\mathcal{K} \oplus \mathcal{N}$, and that matricially we may write $\rho_{\mathcal{M}}$ as

$$\begin{bmatrix} \rho_{\mathcal{K}} & D \\ 0 & \rho_{\mathcal{N}} \end{bmatrix},$$

where the map D carries A into the bounded operators mapping $\mathcal{N}$ into $\mathcal{K}$ and satisfies the equation $D(ab) = D(a)\rho_{\mathcal{N}}(b) + \rho_{\mathcal{K}}(a)D(b)$. That is, D is a derivation. It should be remarked that severe constraints are placed on D by forcing all the representations to be completely contractive. We believe that this is a fundamental fact that deserves a lot of attention.

In pure algebra, the short exact sequence is said to *split* if there is a module map $\Phi' : \mathcal{N} \to \mathcal{M}$ with the property that $\Phi \circ \Phi'$ is the identity on $\mathcal{N}$. In this event, $\mathcal{M}$ is isomorphic to the algebraic direct sum $\mathcal{K} \oplus \mathcal{N}$. Since in our theory we want direct sums to be *orthogonal* direct sums at the Hilbert space level, we are led to the definition: A short exact isometric sequence is *orthogonally* split if there is a *contractive* module map Φ' such that $\Phi \circ \Phi'$ is the identity on $\mathcal{N}$. It is easy to see that this happens if and only if Φ^* is a module map, if and only if the initial space of Φ is a submodule of $\mathcal{M}$. Alternatively, this happens if and only if D is the zero map. Note that in pure algebra one may express exact sequences in terms of derivations also and, then, the condition that a short exact sequence splits is tantamount to the condition that the derivation is *inner* in the sense that there is a map $X : \mathcal{N} \to \mathcal{K}$ such that $D(a) = \rho_{\mathcal{K}}(a)X - X\rho_{\mathcal{N}}(a)$, $a \in A$. The point to keep in mind is that inner derivations need not lead to completely contractive representations.

Definition 1.1. *A Hilbert module ${}_A\mathcal{P}$ over an operator algebra A is called* orthogonally projective (or orthoprojective) *in case every short exact isometric sequence*

$$0 \to_A \mathcal{K} \to_A \mathcal{M} \to_A \mathcal{P} \to 0$$

is orthogonally split. Likewise, a Hilbert module ${}_A\mathcal{I}$ is called orthogonally injective (or orthoinjective) *in case every short exact isometric sequence*

$$0 \to_A \mathcal{I} \to_A \mathcal{M} \to_A \mathcal{N} \to 0$$

is orthogonally split.

Evidently, these are the "orthogonal" versions of one formulation of the algebraic notions of projective and injective modules. There are others and, in contrast with what happens in algebra, there are distinctions among them. We will not enter into these here except to say that a stronger notion of "orthogonal projectivity" plays a fundamental role in the analysis of commutant lifting problems.

If one specializes the operator algebra to be the disc algebra, then Hilbert modules are specified by specifying contraction operators. That is, a Hilbert module $\mathcal{H}$ over the disc algebra is completely determined by specifying the contraction $T_{\mathcal{H}} = \rho_{\mathcal{H}}(z)$ on $\mathcal{H}$. It is then quite easy to see that $\mathcal{H}$ is orthogonally projective (resp. orthogonally injective) if and only if $T_{\mathcal{H}}$ is an isometry (resp. co-isometry) [6, Example 2.7]. For this reason and for the theorem that we will prove momentarily, we regard orthogonally projective Hilbert modules as a natural *algebraic* generalization of isometry operators. Similarly, orthogonally injective Hilbert modules generalize co-isometries.

It is worth noting that just as isometries and co-isometries are adjoints of one another, the same, essentially, is true of orthogonally projective Hilbert modules and orthogonally injective Hilbert modules. Indeed, if $\mathcal{H}$ is a Hilbert module over an operator algebra A with associated representation $\rho_{\mathcal{H}}$, then defining σ by the formula

$$\sigma(a) = (\rho_{\mathcal{H}}(a^*))^*, \tag{1}$$

$a \in A^*$, where the adjoint on elements of A is calculated in $C^*(A)$, yields a representation and Hilbert module over A^*. It is then easy to see that $\mathcal{H}$ is orthogonally projective iff the Hilbert module associated with σ is orthogonally injective. It follows that for many parts of the elementary theory, at least, one may focus attention on one class or the other. However, the theories are not entirely parallel, even when the operator algebra A is isomorphic to A^*. One need only reflect on how differently isometries and co-isometries are treated, to see this.

In [6, Proposition 3.2], we showed that every orthogonally projective Hilbert module is a Shilov module. However, the converse is not true. The quotient algebra $\mathcal{T}_3/\mathcal{J}$, where $\mathcal{T}_3$ is the algebra of upper triangular 3×3 matrices and $\mathcal{J}$ is the ideal consisting of those matrices whose only possible non-zero entry is in the $3,3-$position, has Shilov modules that are not orthogonally projective [6, Example 3.10]. In fact, in [6] we left unsettled the question of whether non-zero orthogonally projective Hilbert modules always exist! However, we did show there in Theorem 4.3, that if A is an incidence algebra or one of the infinite dimensional generalizations contained in the C^*-algebra of an $r-$discrete principal groupoid, then every Shilov module is orthogonally projective. Although the existence question remains unsettled, we show here how to improve considerably upon this result.

To this end, recall that Arveson calls an *irreducible* representation π of $C^*(A)$ a *boundary representation* for A if the only completely positive map of $C^*(A)$ that agrees with π on A is π itself [2, Definition 2.1.1]. It is convenient for us to drop the assumption of irreducibility for boundary representations and say that an *arbitrary* C^*–representation of $C^*(A)$ is a boundary representation for A if it coincides on A with no other completely positive map on $C^*(A)$. We then have

Theorem 1.2. *Let $\mathcal{H}$ be a Hilbert module over an operator algebra A and let ρ be the associated representation. Then ρ is the restriction to A of a boundary representation of $C^*(A)$ for A if and only if $\mathcal{H}$ is both orthogonally projective and orthogonally injective.*

PROOF. Suppose $\mathcal{H}$ is both orthogonally projective and orthogonally injective. We first show that ρ is a C^*- representation restricted to A. Since $\mathcal{H}$ is orthogonally projective and completely contractive, $\mathcal{H}$ is a Shilov module, by [6, Proposition 3.2]. Thus, there is a C^*–representation π of $C^*(A)$ on a Hilbert space $\mathcal{K}$ and an isometric module map $\Psi : \mathcal{H} \mapsto \mathcal{K}$. By hypothesis, $\mathcal{H}$ is orthogonally injective and so the range of Ψ reduces π. Thus Ψ implements a unitary equivalence between ρ and the restriction of π to the range of Ψ. Hence ρ is a C^*–representation restricted to A.

To show that ρ is a boundary representation, suppose σ is a completely positive extension of $\rho|A$ to all of $C^*(A)$, and let $\sigma(\cdot) = \Phi^*\pi(\cdot)\Phi$ be the minimal Stinespring dilation of σ. (This π is not necessarily the same π of the preceding paragraph.) This means that π is a C^*- representation of $C^*(A)$ on a Hilbert space $\mathcal{K}$, that $\Phi : \mathcal{H} \mapsto \mathcal{K}$ is a Hilbert space isometry such that $\sigma(a) = \Phi^*\pi(a)\Phi$ for all $a \in C^*(A)$, and that the smallest reducing subspace for $\pi(C^*(A))$ containing $\Phi\mathcal{H}$ is all of $\mathcal{K}$. In particular, this means that for $a \in A$, $\rho(a) = \sigma(a) = \Phi^*\pi(a)\Phi$. Since ρ is a representation of A, the range of Φ is a semi-invariant subspace for $\pi(A)$. We need to work through the "Sarason representation" of this semi-invariant subspace carefully. (See [2, Page 156 and Appendix A.1] and [7].) Let P be the projection of $\mathcal{K}$ onto $\Phi\mathcal{H}$. Thus $P = \Phi\Phi^*$. Let $\mathcal{K}_1$ be the smallest invariant subspace for $\pi(A)$ containing $\Phi\mathcal{H}$. Thus, since π is assumed to be nondegenerate, $\mathcal{K}_1 = (\pi(A)\Phi\mathcal{H})^{cl}$. Let P_1 be the orthogonal projection of $\mathcal{K}$ onto $\mathcal{K}_1$. We write π_1 for the representation of A obtained by restricting $\pi(A)$ to $\mathcal{K}_1$, so that $\pi_1(a) = \pi(a) \mid \mathcal{K}_1$ for all a in A. Alternatively, we may think of $\pi_1(a) = \pi(a)P_1$, for $a \in A$. Also, we set $\Phi_1 = P_1\Phi$. This is the same thing as viewing Φ as a map from $\mathcal{H}$ to $\mathcal{K}_1$ and gives $\Phi_1^* = \Phi^*P_1$. Finally, let $\mathcal{K}_2 = \mathcal{K}_1 \ominus (\Phi\mathcal{H})$ and let P_2 be the orthogonal projection of $\mathcal{K}$ onto $\mathcal{K}_2$. Sarason's insight is that $\mathcal{K}_2$ is invariant for $\pi_1(A)$ (and for $\pi(A)$). Of course $P_1 = P + P_2$. We show that $\Phi_1^* : \mathcal{K}_1 \mapsto \mathcal{H}$ is a module map, i.e., $\rho(a)\Phi_1^* = \Phi_1^*\pi_1(a)$ for all $a \in A$. Indeed, for $a \in A$, $\rho(a)\Phi_1^* = \rho(a)\Phi^*P_1 = \Phi^*\pi(a)\Phi\Phi^*P_1 = \Phi^*\pi(a)PP_1$. Now, $\mathcal{K}_2$ is invariant for every $\pi(a)$ and so $\pi(a)P_2 = P_2\pi(a)P_2$ for all $a \in A$. Furthermore, $\Phi^*P_2 = 0$, since the initial projection of Φ^* is P which is orthogonal to P_2. Thus we find that $\Phi^*\pi(a)PP_1 = \Phi^*\pi(a)(P + P_2)P_1 = \Phi^*\pi(a)P_1 = \Phi^*P_1\pi(a)P_1$, where the last equation is justified by the fact that $\mathcal{K}_1$ is invariant for $\pi(A)$. However, the last term of these equations is none other than $\Phi_1^*\pi_1(a)$. This proves our assertion. But Φ_1^* is a co-isometric map and since $\mathcal{H}$ is orthogonally projective by hypothesis, we conclude that Φ_1 is a module map, too. That is, $\Phi_1\rho(a) = \pi_1(a)\Phi_1$ for all $a \in A$; or, what is the same, $P_1\Phi\rho(a) = \pi(a)P_1\Phi$. Since the range of Φ is contained in $\mathcal{K}_1$, we can drop P_1 altogether and write $\Phi\rho(a) = \pi(a)\Phi$, for all $a \in A$. This shows that $\Phi : \mathcal{H} \mapsto \mathcal{K}$ is an isometric module map. Since, however, ρ is orthogonally injective, we conclude that the range of Φ reduces $\pi(A)$. But the range of Φ reduces $\pi(A)$ iff it reduces $\pi(C^*(A))$. By our minimality assumption, this means that the range of Φ is all of $\mathcal{K}$ and, therefore, that Φ is a *unitary* operator such that $\rho(a) = \Phi^*\pi(a)\Phi$ for all $a \in A$. Since ρ and π are C^*–representations, this equation must persist for all $a \in C^*(A)$. That is, $\rho = \sigma$. Thus, ρ is a boundary representation.

For the reverse implication, suppose that ρ is the restriction to A of a boundary representation of $C^*(A)$ for A. We shall view ρ either as a map on A or on $C^*(A)$. We show first that $\mathcal{H}$ is orthogonally projective. Suppose

$$0 \mapsto \mathcal{N} \mapsto \mathcal{M} \overset{\Phi}{\mapsto} \mathcal{H} \mapsto 0$$

is a short exact isometric sequence determined by Hilbert modules $\mathcal{N}$ and $\mathcal{M}$. We want to show that the initial projection P_1 of Φ commutes with $\rho_{\mathcal{M}}$. Since $\mathcal{M}$ is completely contractive, there is a C^*–representation π of $C^*(A)$ on a Hilbert space $\mathcal{K}$ and a co-isometry $\Psi : \mathcal{K} \mapsto \mathcal{M}$ such that $\rho_{\mathcal{M}}(a) = \Psi\pi(a)\Psi^*$ for all a in A. But then, since $\Phi\rho_{\mathcal{M}}(a) = \rho(a)\Phi$, $a \in A$, by assumption, we find that $\Phi\rho_{\mathcal{M}}(a)\Phi^* = \rho(a)$ for all $a \in A$, and that $\Phi\Psi$ is a co-isometry such that $\Phi\Psi\pi(a)(\Phi\Psi)^* = \rho(a)$ for all $a \in A$. On $C^*(A)$, $\Phi\Psi\pi(\cdot)(\Phi\Psi)^*$ is a completely positive map that agrees with ρ on A. Since ρ is a boundary representation, we conclude that $\Phi\Psi\pi(a)(\Phi\Psi)^* = \rho(a)$ for all $a \in C^*(A)$. Thus the initial space of $\Phi\Psi$ reduces π and $\Phi\Psi$ implements an equivalence between ρ and π restricted to this initial space. Let P_2 be the initial projection of $\Phi\Psi$. Then for $a \in A$, we have

$$\begin{aligned}
\rho_{\mathcal{M}}(a)P_1 &= (\Psi\pi(a)\Psi^*)(\Phi^*\Phi) \\
&= (\Psi\pi(a)\Psi^*)(\Phi^*\Phi)(\Psi\Psi^*) \\
&= \Psi\pi(a)(\Psi^*\Phi^*\Phi\Psi)\Psi^* \\
&= \Psi\pi(a)P_2\Psi^* \\
&= \Psi P_2\pi(a)\Psi^* \\
&= \Psi(\Psi^*\Phi^*\Phi\Psi)\pi(a)\Psi^* \\
&= (\Psi\Psi^*)(\Phi^*\Phi)(\Psi\pi(a)\Psi^*) \\
&= P_1\rho_{\mathcal{M}}(a)
\end{aligned}$$

where the second and seventh equations are justified by the fact that Ψ is a co-isometry with range $\mathcal{M}$. This shows that $\mathcal{H}$ is orthogonally projective, as promised.

To see that $\mathcal{H}$ is orthogonally injective, simply note that ρ is also a boundary representation for A^*. Thus $\mathcal{H}$ is orthogonally projective for A^* by what we just showed. This, of course, implies that $\mathcal{H}$ is orthogonally injective for A and completes the proof. □

Definition 1.3. *An operator algebra A is called* admissible *in case there is a faithful family of irreducible boundary representations of $C^*(A)$ for A.*

This terminology was introduced by Arveson [2, Definition 2.2.1], but his definition was somewhat different. However, thanks to his Theorem 2.2.3 and Hamana's Theorem 4.4 in [5], the two definitions agree. We believe that the following problem is open.

Problem. Is every operator algebra admissible?

Arveson raised this question in [2] because of its relation to the existence problem for the so-called Shilov boundary ideal of a C^*-algebra that is generated by an operator algebra. (Showing the existance of the Shilov boundary ideal is the same as showing the existance of the C^*-envelope.) Hamana [5, Theorem 4.4] solved that problem without reference to boundary representations at all. The relation between this problem and our work is clarified in

Corollary 1.4. *If A is an admissible operator algebra, then A has Hilbert modules that are simultaneously orthogonally projective and orthogonally injective, and have the additional property that the corresponding representations are completely isometric.*

PROOF. By definition, $C^*(A)$ has a faithful family of boundary representations for A, and each of these corresponds to a Hilbert module that is simultaneously orthoprojective and orthoinjective by Theorem 1.2. The result, then, follows from Proposition 3.5 of [6], which asserts that the orthogonal direct sum of orthoprojective Hilbert modules is orthoprojective. (The same is true for orthoinjective Hilbert modules by considering adjoints and the representations described in Equation (1).) □

In ring theory, one calls an algebra *QF-3* if it has a faithful module that is both projective and injective [1, Theorem 31.6]. Thus, our results suggest that admissable operator algebras have a "QF-3" character to them. Whether anything can be made of this, and in particular, whether this may be used to determine if all operator algebras are admissable, we do not know. Orthogonally projective and orthogonally injective modules bear certain obvious relationships with their unalloyed, algebraic progenitors, but the fact that for incidence algebras the collections of orthogonally projective Hilbert modules and Shilov modules coincide [6, Theorem 4.3] shows that there are striking differences. We regard one of the principal challenges of the theory to be the development of characterizations of orthogonally projective Hilbert modules. We would like something along the lines of the algebraic assertion that a projective module is a direct summand of a free module. However the result just cited and the evidence adduced in [6] to the effect that there are no well defined free objects in our category indicate that characterizations, if they exist, will require more sophisticated analysis.

Acknowledgment. We would like to record our thanks to David Blecher for helpful comments on this note.

REFERENCES

[1] F. ANDERSON AND K. FULLER, *Rings and Categories of Modules* (2nd edition), Springer-Verlag, New York-Berlin-Heidelberg, 1992.

[2] WM. B. ARVESON, *Subalgebras of C^*-algebras*, Acta Mathematica **123** (1969), 141–224.

[3] D. BLECHER, Z-J. RUAN, AND A. SINCLAIR, *A characterization of operator algebras*, J. Functional Anal. **89** (1990), 188–201.

[4] R. G. DOUGLAS AND V. PAULSEN, *Hilbert Modules over Function Algebras*, Pitman Research Notes in Mathematics Series #217, Longman Scientific & Technical, Essex, 1989.

[5] M. HAMANA, *Injective envelopes of operator systems*, Publ. R.I.M.S. **15** (1979), 773–785.

[6] P. MUHLY AND B. SOLEL, *Hilbert modules over operator algebras*, Memoirs of the Amer. Math. Soc. **117** #559 (1995).

[7] D. SARASON, *On spectral sets having connected complement*, Acta Sci. Math. (Szeged) **26** (1965), 289–299.

PAUL S. MUHLY
Department of Mathematics
The University of Iowa
Iowa City, IA 52242

BARUCH SOLEL
Department of Mathematics
Technion
32000 Haifa, Israel

Received: August 23rd, 1995.

Operator Theory:
Advances and Applications, Vol. 104

Joint Spectrum and Nonisometric Functional Calculus

ALFREDO OCTAVIO & SRDJAN PETROVIC

Dedicated to Professor Carl Pearcy on the occasion of his sixtieth birthday, with love and admiration

ABSTRACT. Let T be a completely nonunitary contraction on a Hilbert space $\mathcal{H}$ and assume that the spectrum of T contains the unit circle. We show that, in this situation, there exists a contraction S that commutes with T, and such that the joint (Taylor) spectrum of the pair (S,T) contains the two-dimensional torus, but the functional calculus generated by the pair has a nontrivial kernel. This improves a result in [12].

Let $\mathcal{H}$ be an infinite dimensional, complex, separable Hilbert space and let $\mathcal{L}(\mathcal{H})$ denote the algebra of all bounded linear operators acting on $\mathcal{H}$. If S and T belong to $\mathcal{L}(\mathcal{H})$ and commute with each other, we say that a pair (S,T) is *nonsingular* if the sequence

$$0 \to \mathcal{H} \xrightarrow{\delta_1} \mathcal{H} \oplus \mathcal{H} \xrightarrow{\delta_2} \mathcal{H} \to 0$$

is exact, where $\delta_1(x) = (Sx, Tx)$, $x \in \mathcal{H}$ and $\delta_2(x,y) = Sy - Tx$. A pair of complex numbers (λ_1, λ_2) belongs to the *Taylor joint spectrum* of (S,T) (notation: $(\lambda_1,\lambda_2) \in \sigma(S,T)$), if $(S-\lambda_1, T-\lambda_2)$ is not nonsingular (cf. [8]).

Our notation is consistent with [12], [11], and [9]. As usual, $\mathbb{C}$ is the complex plane, $\mathbb{D}$ is the open unit disk, and $\mathbb{T} = \partial\mathbb{D}$. Also, $\mathbb{D}^2$ is the bidisk (the Cartesian product of two copies of $\mathbb{D}$) and $\mathbb{T}^2$ is the torus (the Cartesian product of two copies of $\mathbb{T}$). Finally, $H^\infty(\mathbb{D})$ [resp. $H^\infty(\mathbb{D}^2)$], is the Banach algebra of bounded analytic functions on $\mathbb{D}$ [resp. $\mathbb{D}^2$].

It is a famous result of Sz.-Nagy and Foias (cf. [13], Theorem III.2.1) that if T is a completely nonunitary contraction in $\mathcal{L}(\mathcal{H})$, then there exists a contractive algebra homomorphism Φ_T from $H^\infty(\mathbb{D})$ into $\mathcal{L}(\mathcal{H})$(see also [4], Theorem 4.1). This homomorphism, usually referred to as the Nagy-Foias functional calculus, has played an important role in the development of the structure theory of linear operators on Hilbert space as expounded in the classical book [13].

As a generalization of this result for a pair of commuting contractions (S,T), several authors have provided conditions under which there exists an algebra homomorphism $\Phi_{S,T}$ from $H^\infty(\mathbb{D}^2)$ into $\mathcal{L}(\mathcal{H})$, satisfying, at least, the following properties:

- $\Phi_{S,T}(p) = p(S,T)$, for any complex polynomial p in two variables.

- $\|\Phi_{S,T}(h)\| \leq \|h\|_\infty$, for any $h \in H^\infty(\mathbb{D}^2)$.
- $\Phi_{S,T}$ is weak* continuous.

The last property means that $\Phi_{S,T}$ is continuous when $H^\infty(\mathbb{D}^2)$ and $\mathcal{L}(\mathcal{H})$ are provided with their respective weak* topologies, which arise from the fact that $\mathcal{L}(\mathcal{H})$ can be identified with the dual of the trace-class, while a predual of $H^\infty(\mathbb{D}^2)$ is described in [11].

Some sufficient conditions for the functional calculus to exist are:

(A) Both contractions are completely nonunitary (cf. [5]).
(B) The spectral measure associated with the pair is absolutely continuous with respect to some representing and measure (cf. [9] and [11]).
(C) Each member of the pair satisfies

$$\sup_{\|p\|\leq 1} |\langle p(S,T)A^n x, y\rangle| \to 0, \quad \text{as } n \to \infty \text{ for all } x,\, y \in \mathcal{H},$$

when substituted for A in the expresion above (see [2] for details).

One knows (cf. [9]) that if S is completely nonunitary and T is absolutely continuous (i.e., the spectral measure of its minimal unitary dilation is absolutely continuous with respect to Lebesgue measure on $\mathbb{T}$), then the pair (S,T) satisfies condition (B). Therefore, (B) is more general than (A). Recently, the first author and M. Kosiek have established the equivalence of (B) and (C). From now on we shall assume that the pair (S,T) satisfies one of these conditions and we shall write $h(S,T)$ instead of $\Phi_{S,T}(h)$.

It is an old result of Apostol (cf. [1]) that if T is an absolutely continuous contraction in $\mathcal{L}(\mathcal{H})$ such that $\sigma(T) \supset \mathbb{T}$ and the Nagy-Foias functional calculus is not an isometry, then T has a nontrivial hyperinvariant subspace (i.e., a closed subspace $\mathcal{M}$ of $\mathcal{H}$, such that $S\mathcal{M} \subset \mathcal{M}$ for any S in the commutant of T). There is no known version of this result for a pair of commuting contractions (S,T) such that $\sigma(S,T) \supset \mathbb{T}^2$. Actually, even the particular case when the functional calculus has a nontrivial kernel is well understood only in the one variable case. Recall that a completely nonunitary contraction $T \in \mathcal{L}(\mathcal{H})$ belongs to the class C_0 and has *minimal (inner) function* $m_T \in H^\infty(\mathbb{D})$ if $m_T(T) = 0$ and every nonzero function $f \in H^\infty(\mathbb{D})$ such that $f(T) = 0$ has m_T as a divisor. For a full account of this theory we recomend [3]. Sz.-Nagy and Foias ([13]) established the existence of a contraction $T \in C_0$ such that $\sigma(T) \supset \mathbb{T}$. In [12] the first author has proved the following extension of this result.

Theorem 1. *Let $T \in \mathcal{L}(\mathcal{H})$ be a completely nonunitary contraction with $\sigma(T) \supseteq \mathbb{D}$. Then there is a completely nonunitary contraction S that commutes with T and a function $h \in H^\infty(\mathbb{D}^2)$ such that $\sigma(S,T) \supseteq \mathbb{T}^2$ and $h(S,T) = 0$.*

It is the main result of this paper that the assertion above is true for a larger class of operators.

Theorem 2. *Let $T \in \mathcal{L}(\mathcal{H})$ be a completely nonunitary contraction with $\sigma(T) \supseteq \mathbb{T}$. Then there is a completely nonunitary contraction S that commutes with T and a function $h \in H^\infty(\mathbb{D}^2)$ such that $\sigma(S,T) \supseteq \mathbb{T}^2$ and $h(S,T) = 0$.*

Before proceeding with the proof of Theorem 2 we shall need some definitions and preliminary results. If J is a subset of a metric space (X, d) and ε is a positive number, we denote by J_ε the set of points $x \in X$ such that $d(x, J) < \varepsilon$. The Hausdorff metric is defined on the family of compact subsets of X as

$$d_H(J,K) = \inf\{\varepsilon > 0 \;:\; K_\varepsilon \supset J,\, J_\varepsilon \supset K\}.$$

Let $\{S_n\}$ be a sequence of mutually commuting operators acting on $\mathcal{H}$. One knows (cf. [10]) that if $S_n \to S$ in the norm topology, then $\sigma(S_n) \to \sigma(S)$ in the Hausdorff metric. The following is a generalization of this result in two variables.

Lemma 3. *Let $\mathcal{F} = \{\mathcal{S}_\backslash\} \cup \{\mathcal{T}_\backslash\}$ be a commuting family of operators acting on $\mathcal{H}$. Furthermore, assume that there exist S and T in $\mathcal{L}(\mathcal{H})$ such that $\|T_n - T\| \to 0$ and $\|S_n - S\| \to 0$. Then $\sigma(S_n, T_n) \to \sigma(S,T)$ in the Hausdorff metric.*

PROOF. Let $\mathcal{B}$ be the commutative Banach subalgebra of $\mathcal{L}(\mathcal{H})$ generated by $\mathcal{F}$, S, and T and let M be its maximal ideal space. By Theorem 3.3 of [8] we can associate with $\mathcal{B}$ a nonempty compact set $M_\sigma \subset M$ in such a way that for any $A, B \in \mathcal{B}$

$$\sigma(A,B) = \{(\hat{A}(\phi), \hat{B}(\phi)) \;:\; \phi \in M_\sigma\},$$

where, as usual, $\hat{A}$ and $\hat{B}$ are the Gelfand transforms of A and B, respectively.

Since S_n converges to S in norm, $\hat{S_n} \to \hat{S}$ uniformly in the space of continuous functions on M_σ, denoted $C(M_\sigma)$, and the same holds for $\hat{T_n}$ and $\hat{T}$. Thus, the sequence of sets

$$\sigma(S_n, T_n) = \{(\hat{S_n}(\phi), \hat{T_n}(\phi)) \;:\; \phi \in M_\sigma\}$$

converges in the Hausdorff metric to

$$\sigma(S,T) = \{(\hat{S}(\phi), \hat{T}(\phi)) \;:\; \phi \in M_\sigma\}.$$

This completes the proof. □

Remark 4. *We notice that Theorem 3.3 of [8] can be used to prove an N variable version of Lemma 3, for every $N \in \mathbb{N}$. Of course, when $N = 1$, the set M_σ is exactly the maximal ideal space M.*

We briefly recall that the operator S in Theorem 1 is obtained as $g(T)$, where $g \in H^\infty(\mathbb{D})$ is any function with the property that

$$\{(g(\lambda), \lambda) : \lambda \in \mathbb{D}\}^- \supseteq \mathbb{T}^2.$$

Clearly, such a function is universal (i.e., does not depend on T). Several such functions have been constructed in the literature of Cluster Set Theory. One such construction can be found in [12].

Next we notice that the proof of Theorem 1 holds if the hypothesis $\sigma(T) \supset \mathbb{D}$ is replaced by the weaker hypothesis that $\sigma(T)$ contains an annulus $\Omega = \{z \in \mathbb{D} : 0 < r < |z| < 1\}$.

PROOF OF THEOREM 2

Consider the sequence $\{\phi_n\}_{n=1}^{\infty} \subset L^{\infty}[0,1]$, defined as

$$\phi_n(x) = \begin{cases} x + 1 - 1/n, & 0 < x \leq 1/n, \\ 1, & 1/n < x < 1, \end{cases}$$

for every $n \in \mathbb{N}$. As usual, M_{ϕ_n} denotes the operator of multiplication by ϕ_n acting on $L^2[0,1]$. Finally, for every $n \in \mathbb{N}$, we define $T_n = M_{\phi_n} \otimes T$, a bounded linear operator acting on the Hilbert space $L^2[0,1] \otimes \mathcal{H}$. One knows (cf. [6]) that $\sigma(T_n) = \sigma(M_{\phi_n})\sigma(T)$ and thus, since $\sigma(M_{\phi_n}) = [1-1/n, 1]$, the set $\sigma(T_n)$ contains the annulus $\Delta_n = \{z \in \mathbb{C} : 1/n < |z| < 1\}$. Using Theorem 1, the remark above, and recalling the universal property of the function g, we obtain a pair of completely nonunitary contractions $(g(T_n), T_n)$ whose spectrum contains $\mathbb{T}^2$. Clearly, $T_n \to I \otimes T$ and $g(T_n) \to g(I \otimes T) = I \otimes g(T)$, both in the norm of $L^2[0,1] \otimes \mathcal{H}$. Now, Lemma 3 implies that $\sigma(g(T_n), T_n) \to \sigma(I \otimes g(T), I \otimes T)$ in the Hausdorff metric and by Lemma 2.1 of [7], $\sigma(I \otimes g(T), I \otimes T) = \sigma(g(T), T)$. Therefore, $\sigma(g(T), T) \supset \mathbb{T}^2$, and the theorem is proved. □

The authors wish to express their gratitude to the Instituto Venezolano de Investigaciones Cientificas (IVIC) and Indiana University for partial support during the period this paper was being written, as well as to Prof. Raul Curto for his guidance and help.

REFERENCES

[1] C. APOSTOL, *Ultraweakly closed operator algebras*, J. Operator Theory **2** (1979), 49–61.

[2] C. APOSTOL, *Functional calculus and invariant subspaces*, J. Operator Theory **4** (1980), 159–190.

[3] H. BERCOVICI, *Operator theory and arithmetic in H^{∞}*, Mathematical Surveys, vol. 26, Amer. Math. Soc., Providence, R.I., 1988.

[4] H. BERCOVICI, C. FOIAS, AND C. PEARCY, *Dual algebras with applications to invariant subspaces and dilation theory*, CBMS Regional Conf. Ser. in Math., vol. 56, Amer. Math. Soc., Providence, R.I., 1985.

[5] E. BRIEM, A. M. DAVIE, AND B. K. ØKSENDAL, *Functional calculus for commuting contractions*, J. London Math. Soc. **7** (1973), 709–718.

[6] A. Brown and C. Pearcy, *Spectra of tensor products of operators*, Proc. Amer. Math. Soc. **17** (1966), 162–166.
[7] Z. Ceausescu and F. H. Vasilescu, *Tensor products and the joint spectrum in Hilbert spaces*, Proc. Amer. Math. Soc. **72** (1978), 505–508.
[8] R. Curto, Applications of several complex variables to multiparameter spectral theory, Surveys of recent results in operator theory, vol. II, ch. 2, pp. 25–90, Longman, London, 1988.
[9] M. Kosiek, A. Octavio, and M. Ptak, *On the reflexivity of pairs of contractions*, Proc. Amer. Math. Soc. **123** (1995), 1229–1236.
[10] J. D. Newburgh, *The variation of spectra*, Duke Math. J. **18** (1951), 165–176.
[11] A. Octavio, *Coisometric extension and functional calculus for pairs of contractions*, J. Operator Theory **31** (1994), 67–82.
[12] A. Octavio, *On the joint spectrum and H^∞-functional calculus for pairs of commuting contractions*, Proc. Amer. Math. Soc. **114** (1992), 497–503.
[13] B. Sz.-Nagy and C. Foias, *Harmonic analysis of operators on Hilbert space*, North-Holland, Amsterdam, 1970.

Alfredo Octavio
Departmento de Matemáticas
IVIC
Caracas, Venezuela
E-mail: aoctavio@ivic.ivic.ve

Srdjan Petrovic
Department of Mathematics
Indiana University
Bloomington, IN 47405
E-mail: petrovic@iu-math.math.indiana.edu

Received: August 23rd, 1995.

OPERATOR THEORY: ADVANCES AND APPLICATIONS

BIRKHÄUSER VERLAG

Edited by

I. Gohberg,

School of Mathematical Sciences, Tel-Aviv University, Ramat Aviv, Israel

'his series is devoted to the publication of current research in operator theory, with particular empha-is on applications to classical analysis and the theory of integral equations, as well as to numerical nalysis, mathematical physics and mathematical methods in electrical engineering.

---

0 I. Gohberg / H. Langer (Eds)
Operator Theory and Boundary Eigenvalue Problems. International Workshop in Vienna, July 27–30, 1993.
1995. ISBN 3-7643-5275-2

1 H. Upmeier
Toeplitz Operators and Index Theory in Several Complex Variables
1996. ISBN 3-7643-5282-5

2 T. Constantinescu
Schur Parameters, Factorization and Dilation Problems
1996. ISBN 3-7643-5285-X

3 A.B. Antonevich
Linear Functional Equations. Operator Approach.
1995. ISBN 3-7643-2931-9

4 L.A. Sakhnovich
Integral Equations with Difference Kernels on Finite Intervals
1996. ISBN 3-7643-5267-1

5 Y.M. Berezansky / G.F. Us / Z.G. Sheftel
Functional Analysis. Volume I.
1996. ISBN 3-7643-5344-9

6 Y.M. Berezansky / G.F. Us / Z.G. Sheftel
Functional Analysis. Volume II.
1996. 3-7643-5345-7

7 I. Gohberg / P. Lancaster / P.N. Shivakumar (Eds)
Recent Developments in Operator Theory and Its Applications. International Conference in Winnipeg, October 2–6, 1994.
1996. ISBN 3-7643-5414-5

8 J. van Neerven (Ed.)
The Asymptotic Behaviour of Semigroups of Linear Operators
1996. ISBN 3-7643-5455-0

9 Y. Egorov / V. Kondratiev
On Spectral Theory of Elliptic Operators
1996. ISBN 3-7643-5390-2

90 A. Böttcher / I. Gohberg (Eds)
Singular Integral Operators and Related Topics.
Joint German-Israeli Workshop, Tel Aviv, March 1–10, 1995.
1996. ISBN 3-7643-5466-6

91 A.L. Skubachevskii
Elliptic Functional Differential Equations and Applications.
1997, ISBN 3-7643-5404-6

92 A.Ya. Shklyar
Complete Second Order Linear Differential Equations in Hilbert Spaces
1997. ISBN 3-7643-5377-5

93 Y. Egorov / B.-W. Schulze
Pseudo-Differential Operators, Singularities, Applications
1997. ISBN 3-7643-5484-4

94 M.I. Kadets / V.M. Kadets
Series in Banach Spaces. Conditional and Unconditional Convergence.
1997. ISBN 3-7643-5401-1

95 H. Dym / V. Katsnelson / B. Fritzsche / B. Kirstein (Eds)
Topics in Interpolation Theory
1997. ISBN 3-7643-5723-1

96 D. Alpay / A. Dijksma / J. Rovnyak / H. de Snoo
Schur Functions, Operator Colligations, and Reproducing Kernel Pontryagin Spaces
1997. ISBN 3-7643-5763-0

97 M.L. Gorbachuk / V.I. Gorbachuk
M.G. Krein's Lectures on Entire Operators
1997. ISBN 3-7643-5704-5

98 I. Gohberg / Yu. Lyubich (Eds)
New Results in Operator Theory and Its Applications
The Israel M. Glazman Memorial Volume
1997. ISBN 3-7643-5775-4

99 T. Ayerbe Toledano / T. Dominguez Benavides / G. López Acedo
Measures of Noncompactness in Metric Fixed Point Theory
1997. ISBN 3-7643-5794-0

100 C. Foias / A.E. Frazho / I. Gohberg / M.A. Kaashoek
Metric Constrained Interpolation, Commutant Lifting and System
1998. ISBN 3-7643-5889-0

101 S.D. Eidelman / N.V. Zhitarashu
Parabolic Boundary Value Problems
1998. ISBN 3-7643-2972-6

102 I. Gohberg / R. Mennicken / C.Tretter (Eds)
Differential and Integral Operators. International Workshop on Operator Theory and Applications, IWOTA 95, in Regensburg, July 31–August 4, 1995.
1998. ISBN 3-7643-5890-4

103. I. Gohberg / R. Mennicken / C. Tretter (Eds)
Recent Progress in Operator Theory. International Workshop on Operator Theory and Applications, IWOTA 95, in Regensburg, July 31–August 4, 1995.
1998. ISBN 3-7643-5891-2